Agricultural Materials as Renewable Resources

ACS SYMPOSIUM SERIES 647

Agricultural Materials as Renewable Resources

Nonfood and Industrial Applications

Glenn Fuller, EDITOR
Agricultural Research Service
U.S. Department of Agriculture

Thomas A. McKeon, EDITOR
Agricultural Research Service
U.S. Department of Agriculture

Donald D. Bills, EDITOR
Agricultural Research Service
U.S. Department of Agriculture

Developed from a symposium sponsored
by the Division of Agricultural and Food Chemistry

American Chemical Society, Washington, DC

Library of Congress Cataloging-in-Publication Data

Agricultural materials as renewable resources: nonfood and industrial applications / Glenn Fuller, editor, Thomas A. McKeon, editor, Donald D. Bills, editor.

p. cm.—(ACS symposium series, ISSN 0097–6156; 647)

"Developed from a symposium sponsored by the Division of Agricultural and Food Chemistry at the 209th National Meeting of the American Chemical Society, Anaheim, California, April 2–7, 1995."

Includes bibliographical references and indexes.

ISBN 0–8412–3455–8

1. Agricultural processing—Congresses. 2. Plant products—Congresses. 3. Animal products—Congresses.

I. Fuller, Glenn, 1929– . II. McKeon, Thomas A., 1949– III. Bills, Donald D., 1932– . IV. American Chemical Society. Division of Agricultural and Food Chemistry. V. American Chemical Society. Meeting (209th: 1995: Anaheim, Calif.) VI. Series.

S698.A36 1996
660.6′—dc20

96–31694
CIP

This book is printed on acid-free, recycled paper.

PRINTED IN THE UNITED STATES OF AMERICA

Advisory Board

ACS Symposium Series

Foreword

THE ACS SYMPOSIUM SERIES was first published in 1974 to provide a mechanism for publishing symposia quickly in book form. The purpose of this series is to publish comprehensive books developed from symposia, which are usually "snapshots in time" of the current research being done on a topic, plus some review material on the topic. For this reason, it is necessary that the papers be published as quickly as possible.

Before a symposium-based book is put under contract, the proposed table of contents is reviewed for appropriateness to the topic and for comprehensiveness of the collection. Some papers are excluded at this point, and others are added to round out the scope of the volume. In addition, a draft of each paper is peer-reviewed prior to final acceptance or rejection. This anonymous review process is supervised by the organizer(s) of the symposium, who become the editor(s) of the book. The authors then revise their papers according to the recommendations of both the reviewers and the editors, prepare camera-ready copy, and submit the final papers to the editors, who check that all necessary revisions have been made.

As a rule, only original research papers and original review papers are included in the volumes. Verbatim reproductions of previously published papers are not accepted.

ACS BOOKS DEPARTMENT

Contents

Preface

AS THE CONSTRAINTS OF LIVING ON A PLANET where resources are finite become increasingly evident, it is becoming necessary to look again toward renewable materials. Plants obtain energy from the sun to recycle carbon and hydrogen into organic matter. The majority of products from agriculture will probably always be used for food and feed, but many new and attractive possibilities exist for application of agricultural by-products and new or modified species of plants and animals to satisfy human needs. The day will come when we will use the products of agriculture much more fully than we do now.

At the 209th national meeting of the American Chemical Society in Anaheim, California, in April 1995, we organized a symposium entitled "Agriculture as a Renewable Source of Raw Materials". The impetus for this program came from our desire to bring together researchers from a wide range of commodity interests and scientific disciplines to discuss current research on both old and new uses of agricultural products. Most of the topics were directed toward nonfood utilization of crops, but a few speakers also touched on applications of biopolymers to food products, including edible and biodegradable packaging. From the 21 presentations in the symposium came the 17 chapters in this volume, which with such limited space is not intended to be comprehensive, but which we hope will be sufficient to inform the reader. The chapters, which are combinations of review and original research, introduce and describe the state of the art in each subject and provide references for follow-up by the interested researcher.

The chapters in this book discuss a wide range of topics including longstanding applications of materials such as jute, cotton fibers, and animal hides; new applications of starch, sugar, and vegetable oils; and use of genetic engineering and biotechnology in the creation of new drugs, industrial chemicals, and human proteins. Subjects also include use of waste biomass and production of biomass for fuel. The material should be useful to students and entrepreneurs as well as researchers in the field of agricultural science.

GLENN FULLER
THOMAS A. MCKEON
Agricultural Research Service
U.S. Department of Agriculture
Albany, CA 94710

DONALD D. BILLS
Agricultural Research Service
U.S. Department of Agriculture
Beltsville, MD 20705–2350

OVERVIEW

Chapter 1

Nonfood Products from Agricultural Sources

Glenn Fuller[1], Thomas A. McKeon[1], and Donald D. Bills[2]

[1]Western Regional Research Center, Agricultural Research Service, U.S. Department of Agriculture, 800 Buchanan Street, Albany, CA 94710
[2]Beltsville Agricultural Research Center, Agricultural Research Service, U.S. Department of Agriculture, 10300 Baltimore Avenue, Beltsville, MD 20705–2350

In the 12,000 years since the beginning of the Neolithic Revolution, agriculture has been one of the most important of human activities. Since the United States became independent, the country's demographics has changed from a nation primarily of farmers to one in which two percent of the population produce the farm commodities needed by the rest of us. While food comprises the bulk of agricultural products, there have always been non-food farm products significant to the economy. Concern for the environment, the trade balance, and reliable sources of supply have caused increased research to utilize agricultural commodities. This chapter presents an overview of some current and potential applications of farm products for fibers, fuels, feedstocks, pharmaceuticals, fine chemicals and polymers. Molecular genetics has created the capability to develop plants and animals, able to grow in the temperate climate of the United States, which can be utilized for production of valuable chemical products.

The United States, with less than seven percent of the world's land, and less than five percent of the world's population, produced 12.6 percent of the world's agricultural commodities in 1988 (1). At the beginning of the current decade there were some 2.1 million farmers in the country, growing crops on 1.6 million farms. Although only a little over two million people are actually engaged in production of agricultural crops, agricultural and related industries provide jobs for more than 21 million people involved in transporting, processing, manufacturing and sales of agricultural goods. American farmers are very efficient. Each farm worker in 1988 supplied enough food and fiber for more than 100 people in the United States and other countries. Output per acre increased forty percent between 1967 and 1987. The farm share of income from food and fiber, however, has been declining. Most of any increased prices for food during the last two decades have gone to transportation, manufacturing and marketing. Farms have excess capacity and there is a strong incentive to grow new crops which will provide higher income. A number of circumstances have produced new interest in directing agricultural production toward non-food end products.

Many incentives to use agricultural products for manufacture come from concern for the environment and the desire to conserve non-renewable resources such as petroleum. While the petroleum supply is finite, agriculture will provide crops and livestock every year, as long as there is land, water and sunlight. Two major categories of products for manufacture are of current and future interest: (a) by-products from food production and (b) materials made from crops grown specifically for non-food use. An example of the former comes from the meat processing industry, in which cattle hides, horns, glandular tissue, bones, blood, tallow, and even gallstones are used for various non-food purposes (2). One of the long-used examples of the second category is cotton, which is grown for fiber, and which in modern times provides food and feed as by-products from the seed in the form of vegetable oil and cottonseed meal. A specific crop may also be grown for direction into either food or non-food use, as is corn (maize), which is a food, an animal feed, and a source of starch and fermentation alcohol. Economics continues to drive crop production and use. The ideal crop for non-food use would be needed in relatively high volume for manufacturing processes providing high added value. Many profitable crops do not completely meet these criteria, and there is incentive for research to provide improved uses and new crops. Plants grown for production of medicinal drugs often have a high mark-up in price but are usually needed in relatively low volume. If fermentation alcohol is to be used in motor fuel (3), the selling price must be very low, though the amount needed is extremely large. One very desirable goal is to find profitable uses for cellulosic wastes of variable composition that could justify their transportation and processing. Economical processes for converting lignified cellulosic materials to fermentable sugar are a desired research goal (Penner, *et. al.*) . This book presents an overview of some of the processes in current use, and of some research to develop new uses for plants and animals.

Historical Aspects

Although there are arguments concerning the identity of the oldest profession, the growing of crops and the herding of animals are certainly among the oldest human activities leading to development of human society as we know it. Until about 10,000 B.C. humans lived by hunting, fishing and gathering their food and materials for clothing and shelter. Then, after the last ice age came the Neolithic Revolution. The dog was domesticated to become not only a source of food but, more importantly, a servant to humans. The warming climate in both the eastern and western hemispheres produced conditions that favored fast-growing plants. Much of mankind gave up the nomadic life, settling in areas in which crop plants could grow. By 9,000 B.C. there were settled villages in northern Israel (4). Over the next 2,000 years, planting of crops and selection of seed for the best plant varieties developed. Soon after, sheep and goats were domesticated. Finally by about 6,000 B.C., cattle were brought under control. Milk, as well as meat, became a diet staple. By about the fifth millennium B.C., crop irrigation began, and large centers of population developed in areas where annual flooding renewed the soil. Some of the first manufacturing processes were developed to preserve and prepare food. By trial and error the use of microorganisms began -- for making beer, wine and cheese as well as leavened bread.

Although the impetus for agriculture came from the need for food for an increasing population, people found many uses for agricultural by-products, even in prehistoric times. Leather was tanned for clothing and fibers from sheep and plants were spun and woven into cloth. Vegetable oils were burned in lamps to provide light and heat, and soaps were made from fats and wood ashes. Herbal medicine has been practiced for thousands of years; effects of certain plant components were known to have physiological effects which, although not well understood at the time, have proven to be useful in modern medicine (see Kinghorn and Seo). Though there were many advances in technology from 1,000 B.C. to 1500 A.D., the practice of agriculture changed very little. Food preparation evolved and spices from the East were increasingly used for preservation and to provide variety. In fact, potential profits from the spice trade were a large factor in the explorations that led to discovery of the New World. That discovery brought about introduction and widespread use of new crops such as potatoes, corn (maize), potatoes, tomatoes and tobacco, although the addition of these crops did not change agricultural practices or, with the exception of tobacco and some crops for medicinal use, contribute to non-food uses of crops. The beginning of the nineteenth century saw continuing use of beeswax for candles, whale oil and vegetable oils for lamps, wood for household fuel, and animal fats for lubrication of machinery. However, the industrial revolution had begun, and with it came profound changes in science and technology, which affected the practice of agriculture. Early in the century Nicholas Appert had invented the canning process to preserve food, and though the scientific explanation for why it prevented spoilage was wrong, the invention contributed greatly to the variety and safety of the food supply.

Soon after the middle of the nineteenth century, change accelerated. In 1866 Mendel published the results of his experiments with pea plants, which initiated the science of genetics, and supplied the basis for plant and animal breeders to improve agricultural production in a systematic way. At about the same time Pasteur, in his research on microorganisms explained food spoilage and many important aspects of food processing, *e.g.* pickling, brewing and bread making. When E. L. Drake brought in the first oil well in Titusville, Pennsylvania, in 1859 (5), it signalled the end of many agricultural fuels for heating and lighting. Petroleum-derived oil and natural gas soon became plentiful and cheap. The availability of suitable petroleum fractions led to widespread use of the internal combustion engine for transportation and other jobs. The horse was replaced as a primary means of travel and transport, and use of many agricultural products for industrial purposes was discontinued. Still, early in the twentieth century many of our best innovators believed that crops could be used as raw materials for a variety of industrial products. In 1934, Henry Ford's new V-8 cars contained half a bushel of soybeans in the form of paint, plastic gear shift knobs and horn buttons. Ford's researchers produced textile fibers from soybeans (6) and in 1941 Ford exhibited a prototype automobile with a plastic body composed largely of soybean- derived products (7). Thomas Edison searched for a domestic source of rubber and found that goldenrod might be such a source. Many advocates were found for the practice of "chemurgy", defined as the development of new industrial chemical products from organic raw materials, especially those of agricultural origin.

In 1935 Wheeler McMillen, longtime editor of the *Farm Journal*, formed and directed a Farm Chemurgic Council, supported by many outstanding leaders such as Henry Ford, Irenee duPont and Dr. Carl T. Compton. The efforts of this council, together with the need to use surplus crops during the depression years caused many profound changes in agricultural policy of the United States, embodied in the Agricultural Adjustment Act of 1938. A small section of this act created four laboratories in the United States Department of Agriculture to find new chemical and technical uses for farm commodities, particularly those with "regular or seasonal surpluses". These four laboratories opened in 1940 in Wyndmoor, Pennsylvania; Albany, California; Peoria, Illinois; and New Orleans, Louisiana, and are still in operation. Much of their most successful research consisted of improvements in food applications and food preservation, but they have also accomplished important industrial research. During World War II, investigations at the Peoria Laboratory developed a fermentation process for penicillin production using a medium consisting largely of corn steep liquor, a by-product of wet milling. When supplies of natural rubber, more desirable than the synthetic Buna S and Buna N for many uses, were cut off, sources such as Russian dandelion, goldenrod and guayule were investigated. Three million pounds of guayule rubber were produced, but work on guayule was abandoned when *Hevea* rubber again became available. Now, guayule may find a new niche in non-allergenic latex products (Cornish and Siler). The USDA utilization laboratories have continued to cooperate with farmers and industry to develop new crops and enhance the uses of old ones. Recent developments in molecular genetics now create the potential for a multitude of new crops and new uses.

Biological Potential of Plants and Animals

Availability of Plant and Animal Species. E. O. Wilson points out (8) that only three percent of the more than 220,000 species of flowering plants in the world have been examined to determine if they have useful alkaloids. Yet a quarter of all prescriptions dispensed by pharmacies in the United States are substances extracted from plants. About 90 percent of all plants have not yet been given a name. Many of these plants are likely to be extinct before they are ever investigated and a somewhat similar situation is true of animals. Principal reasons for extinction of species are: destruction of physical habitat, displacement by introduced species, alteration of habitat by chemical pollutants, hybridization with other species and subspecies, and overharvesting. Not only is it important from an environmental standpoint to try to prevent the loss of species, but we do not know what we are losing in the manner of germplasm potentially important to our agriculture. The science of molecular genetics has opened the possibility not only to incorporate desired characteristics into new cultivars by classical breeding, but also to bring new components into plants or animals from totally different species.

Genetic Manipulation in Agriculture. Molecular genetics provides the capability to modify the characteristics of plants and animals in ways that are beneficial to humankind. This potential is now being realized with the introduction of genetically engineered crops and production of recombinant pharmaceuticals. Attaining these successes could only come about by overcoming many technological hurdles (not to mention regulatory matters). Some desirable traits introduced into engineered crops include disease resistance and pest resistance in various plants, short chain fatty acids in canola oil (9), and delayed ripening in tomatoes (10). The pharmaceutical market for "human" enzymes, peptide hormones and other proteins, produced by animals or bacteria is already significant. Three chapters of this book (May *et. al.*, Paleyanda *et. al., and* McKeon *et. al.*) discuss genetic modification of plants and animals. There also remains tremendous potential to make proteins, oils and carbohydrate polymers using genetically modified bacteria. But as each hurdle is overcome it becomes an additional steppingstone for the genetic engineer. As the path to development of "designed crops" and transgenic animals becomes more routine, research can then shift its focus to applications of the technology.

Utilization of Agricultural Products

The intent of this volume is to cover some of the many facets of use and potential use of agricultural products, particularly for non-food uses. While the term "food and fiber" is often used to sum up farm production, the contents of the book show that feeds, fuels, feedstocks, fine chemicals and pharmaceuticals should be added. Only a few of the many possible topics are covered here, but these subjects illustrate the renewed interest and newly discovered potential for agriculture as a source of raw materials.

Fibers. In the sense of providing protection from the elements, "fiber" provides one of the most basic human needs. The first product to meet this need was animal hide, at first from wild animals which were hunted, and later from domesticated livestock. A tanning process that better preserved hides was developed long before the common era, and has not fundamentally changed. As late as the 1830's the hides of cattle were often a more important cash producer than the meat. Richard Henry Dana in *Two Years Before the Mast* describes a salting process to preserve hides for shipping that is almost the same as the current process in the chapter by Marmer. There are environmental problems associated with the salting and tanning processes. Using molecular modeling of the protein that forms the basis for hide structure, the leather group at the Eastern Regional Research Center has developed processes that greatly reduce the volume of heavy metal waste and examined solvent-free techniques for finishing the hide. Substitution of potassium chloride for salt also promises to solve some of the brine disposal problems.

Coarse fibers such as hemp, sisal and jute, have been important in many countries with developing economies. Jute has been largely displaced by plastic fibers, but it is still produced in India. Although generally considered a source of crude fiber, suitable for linoleum backing, this reputation has arisen as a result of the difficulty met in retting the fiber. B. Wood *et. al.* describe an enzymatic retting process which will yield jute fibers with better quality and strength.

Cotton is the major plant source of fine textiles. It is grown in many areas and is a major agricultural commodity in California and the Southern States. Though cotton textiles as produced have fine qualities -- dyability and ease of spinning and weaving, research has added many improvements. Increased comfort, shrink resistance, retention of shape, flame retardancy, and resistance to microbial rot and soiling have all added value to cotton fabrics (Andrews and Reinhardt). Other sources of plant fiber are also available, including flax of which producers still grow varieties used to produce fine linens. Silk, a proteinaceous fiber of animal origin, is still in demand as a high-quality easily processable fiber.

The short fibers of wood pulp have been widely used in paper manufacture. Tree farming for wood pulp has to some extent supplanted cutting of old forests, but tree plantations are a long-term investment. Kenaf is a fast growing relative of cotton and okra, which can be grown as an annual crop in the South and Southwest to produce high yields of pulp suitable for newsprint. Currently we import about $4 billion worth of newsprint annually (11)and kenaf grown in the United States could provide a new domestic source of pulp. Pilot programs to produce kenaf paper are in operation in Texas, Mississippi and other states.

Fuel. The concept of fuel was once covered by what kept the home fires burning, including wood for heat and oil for light. The definition has expanded to include whatever keeps the power plants generating and engines firing. The need for massive quantities of fuel means the requirement for organized production. The fuels most easy to deliver for use in internal combustion engines are liquids or gases. Petroleum fractions are thus most widely used for transportation and natural gas has been the fuel of choice for stationary generators. Now, however, we are taking another look at solid fuels. Wood can be made to burn cleaner than most coals. To use biomass as a major source of fuel for power generation, huge tree plantations are necessary. The chapter by B. McCown *et. al.* deals with the susceptibility of such plantations to pest infestations. They have addressed this problem using poplar genetically engineered to produce an insect toxin and have applied principles of population genetics to reduce the risk of insect resistance to the toxin. Almost all crop plants produce large amounts of biomass. Unfortunately, with a few exceptions such as sugar cane bagasse, which can be burned at the refinery, costs of transport do not justify using this biomass for fuel.

Vegetable oils have a long history of use for food and fuel. Burning vegetable oil has provided a source of light over the centuries, but has been displaced by petroleum oil. The chapter by Bagby illustrates the potential to burn vegetable oils or their derivatives in diesel fuel. Research in this field has brought to light both economic and technological problems to be solved in order for this use to be practical.

One other fuel receiving political support is fermentation alcohol. The alcohol, derived from grain, is now under consideration as a component of clean-burning gasoline (3). It has the added advantage of being a good anti-knock additive. Sugar production represents an interesting combination of energy from biomass and production of fuel. For sugarcane, the biomass after processing (bagasse) is burned for cogeneration and local power (Clarke and Edye). The sugar produced from cane or beet can be fermented into ethanol, which is a potential gasoline component and is being considered for direct use in engines. Corn starch is another major source for conversion to ethanol in the United States.

While sugarcane and sugarbeets are grown for sugar, much of the carbohydrate produced by photosynthesis is incorporated into lignified cellulose, which gives rigidity to the plant. This material makes up a majority of debris remaining after harvest and after processing; in some cases this is burned for energy, but most often it is burned on site to eliminate it as a source of plant pathogens. Methods being developed to depolymerize agricultural waste into fermentable sugars (Penner *et. al.*) could be especially useful since they convert useless waste into a commodity.

Chemicals. In addition to meeting basic needs, plants have great potential as chemical factories. They are able to convert carbon dioxide to solid matter via photosynthesis. Through eons of evolution and centuries of human selection, the synthetic capacity of certain plants has been directed to the production of materials that have specific industrial uses in quantities that make it preferable to obtain the product from a plant source. The advent of genetic engineering serves to accelerate human selection and expand the number of different substances produced.

There are important chemical uses for sucrose, in addition to sugar for table use as fermentation to ethanol. Sugar is often fermented to food chemicals such as citrate, and some microbes are used to convert sugars to biodegradable polymers, *e.g.* dextran (Fett *et. al.*). The hydroxyl groups on sucrose provide sites for esterification. It serves as a monomer for synthesis of polymers and a source of fatty acyl esters that are detergents and non-caloric dietary fats.

Just as vegetable oils may be used to replace petroleum-based diesel fuel, they also provide a source of industrial chemicals for use in manufacturing lubricants, coatings and plastics. An example is jojoba oil, actually a wax composed largely of long-chain (C_{20-22}) fatty acids esterified to long -chain alcohols. Jojoba is a replacement for sperm whale oil which used to be a source of specialty lubricant additives. The crop, grown in arid regions is now used in a variety of cosmetics. Numerous substituents that are incorporated in certain fatty acids present the opportunity to develop them as sources of specialty chemicals (McKeon *et. al.*). Some common oil crops have been genetically engineered to produce oils containing large quantities of specific fatty acids, such as laurate for detergents. Researchers are now attempting to engineer plants to produce oils that meet the chemical needs of different industries.

Polymers and Plastics. Natural rubber is a *cis*-1,4-polyisoprene polymer that has numerous commercial uses, and is a strategic defense material. Despite many efforts to develop a commercial source, able to grow in temperate climates, we are still dependent on developing countries for natural rubber from *Hevea brasiliensis*. There are new approaches to creating alternative sources of natural rubber from other plants by genetic engineering to enhance yield. Meanwhile, guayule is finding a new niche in latex products which are hypoallergenic (Cornish and Siler).

Accumulation of plastics poses an increasing problem for municipal landfills. The development of a greater variety of biodegradable polymers has potential for reducing landfill problems and using agricultural products as replacements for petroleum-based products in packaging. There are a number of polymers based on biologically produced materials, including cellulose, starch and protein from agricultural products, as well as microbial-produced polymers. While there remain problems associated with moisture and oxygen diffusion in the use of these materials, their mechanical properties can be matched with those of synthetic polymers (Krochta and DeMulder). The future of these polymers will rely on finding suitable uses and finding ways to ameliorate their undesirable properties.

There is another potential use for some biodegradable polymers. In preparation of fruits of fruits and vegetables the skin is removed, exposing the surface to enzyme-catalyzed oxidation as well as non-enzymatic browning which affect appearance and flavor of "lightly-processed" produce. The proper combination of polymeric films to coat cut produce will prolong the desirable appearance and protect it from wieght loss and fungal infection(Pavlath *et. al.*).

Starch is the major storage carbohydrate of plants and its use as a basic unit for polymer design is well-established. Corn milling produces over 4.5 million pounds of starch annually, mostly for non-food purposes. There are also large amounts of wheat starch available. Starch can be converted into many forms, including microcellular foams, which have low densities, large pore volume and surface area, and high compressive strength (Glenn *et. al.*). These properties impart varied potential uses for this novel structural material. The National Center for Agricultural Utilization Research in Peoria has developed an interesting cross-linked starch product called super slurper, which can absorb many times its weight of water. This product is used in applications, such as disposable diapers, where its absorbent properties are useful.

Pharmaceuticals. Natural products obtained from plants or drugs obtained from such compounds provide much of the pharmaceutical market. Plants have always been important sources of medicinals. Aspirin, the most widely-used pharmaceutical, was synthesized from salicycilic acid found in meadowsweet. The list of drugs from plants ranges from atropine, through codeine, morphine and reserpine, to vinblastine (8). The chapter by Kinghorn and Seo points out the problem of maintaining stable and reliable supplies of the plants, often necessitating cultivation. Using ginseng and *Ginkgo* as examples, they describe the medicinal uses and maintenance of cultivated stock.

One chapter in this book (May *et.al.*) explores the possibilities of using genetic technology to provide transgenic plants that produce pharmaceuticals. Pathways that produce secondary metabolites can be altered to allow production of other desired compounds that serve as drugs or intermediates in drug synthesis. An exciting prospect lies in the production of foreign proteins in plants. Incorporation of proteins that bear antigens to human or animal pathogens can provide low cost, easy delivery of immunization worldwide, without the need for maintaining sterile and stable conditions as for vaccines.

Farm animals have served important biomedical uses, with cows as sources of the first smallpox vaccine in the nineteenth century, and pigs as a source of insulin for diabetics in the 1920's. More recently, other drugs have been extracted from these animals, including heparin, steroids, and medicine to treat parathyroid deficiencies (2). Now, investigators are using transgenic animals for medical purposes. With increasing concern over contamination from the human blood supply, animals can provide ideal bioreactors that produce physiologically active human proteins. The authors demonstrate the utility of transgenic animals in producing and secreting the anti-coagulant protein, human Protein C, in milk of mice and pigs. With the availability of increasing numbers of sequences for proteins with human therapeutic value underlies the importance of this technique for development of protein pharmaceuticals.

Conclusions

Agriculture is an increasingly valuable source of raw materials. It has the advantage of being flexible, particularly with annual crops, as well as renewable and versatile. Plant and animal breeding, as well as genetic engineering allow the possibility of many new products from agriculture. Increased use of agricultural raw materials must overcome some obstacles, especially economic ones. Petroleum resources are finite, but oil is a reliable source and is often cheaper than agricultural products. Prices for food are variable, and are sometimes quite high. For both political reasons as well as economic ones, crops for food use will generally be the priority. Crops also depend on weather; hence their supply may be more variable than manufacturers would like. Despite this, farmers are looking for more cash crops, especially since many commodities are likely to lose their subsidies in the future. There are likely to be many more products from farms in the decades to come.

Literature Cited

1. U. S. Department of Agriculture, *1990 Fact Book of Agriculture;* USDA Miscellaneous Publication No. 1063, Washington, DC, 1991.
2. Zane, J. P.; *New York Times*, May 5, 1996: Section 3, p. 5
3. Peaff, G. *Ethanol Mandate;* In *Chem. & Eng. News*, July 11, 1994; p.4.
4.Tannehill, R., *Food in History*; Crown Trade Paperbacks: New York, NY, 1989; p.21.
5. Beaton, K., *Enterprise in Oil*; Appleton-Century-Crofts; New York, NY, 1989; pp. 10-17.
6. Ferrell, J.; In *Ag Industrial Material Products*; New Uses Council, St. Louis, MO; March 1994, p. 13.
7.Boyer, R. A., *Ind. Eng. Chem.*, **1940**, 1549-51.
8. Wilson, E. O., *The Diversity of Life*; W. W. Norton, New York, NY, 1992; Chapter 13.
9.Voelker, T. A., Worrell, A. C., Anderson, L., Bleibaum, J., Fan, C., Hawkins, D. J., Radke, S. E., and Davies, H. M.; *Science,* **1992**, 257: 72-74.
10. Sheehy, R. E., Kramer, M., and Hiatt, W. R., *Proc. Natl. Acad. Sci. USA;* **1988**, 85: 8805-8809.
11. Kelley, H. W., *Always Something New*; USDA Miscellaneous Publication No. 1507; Washington, DC, 1993; p. 96.

Applications in Materials Science

Chapter 2

Acid-Catalyzed Hydrolysis of Lignocellulosic Materials

Michael H. Penner[1], Andrew G. Hashimoto[2], Alireza Esteghlalian[2], and John J. Fenske[1]

[1]Department of Food Science and Technology and [2]Department of Bioresource Engineering, Oregon State University, Corvallis, OR 97331–6602

Lignocellulosic materials include those agricultural and forestry related products, harvest residues and processing wastes whose composition is dominated by lignified plant cell walls. These materials represent a major, renewable, source of organic carbon. A large percentage of this carbon is incorporated in the material's polysaccharide fractions. Many of the potential industrial uses for lignocellulosic materials are dependent on identifying economical processes for the depolymerization and dissolution (saccharification) of these polysaccharides. One such process, which involves the saccharification of component polysaccharides via acid-catalyzed hydrolysis, continues to receive considerable attention. The chemical mechanisms for the actual hydrolytic events that depolymerize the polysaccharides during this process appear to be the same as those defined for model glycopyranosides and glycofuranosides. However, the rates of hydrolysis or dissolution of lignocellulosic polysaccharides during dilute acid-catalyzed hydrolysis are generally quite different from those of soluble model glycosides. The factors governing the rates of dissolution of these polysaccharides are thus related to the macromolecular interactions in and the molecular architecture of lignocellulosic materials.

The Nature of Lignocellulosic Materials

Lignocellulosic biomass includes those biomaterials whose composition is dominated by lignified cell walls of vegetative plants. As such, the three principle components of this material are hemicellulose, cellulose and lignin. Much of the lignocellulosic material available is considered waste or a by-product. Annually, 300 million tons of agricultural residues-half of it is corn stover-are generated in the USA alone (1). Also, 42 million tons of recoverable woody material is discarded every year in municipal solid waste, construction and demolition waste, and timber processing waste(2). Other sources of

0097–6156/96/0647–0012$15.00/0

lignocellulosic material include wheat and rice straw, as well as fast growing herbaceous plants such as switchgrass which are well-established in North America.

The polysaccharide component of lignocellulosic biomass is a major source of fermentable carbohydrate. However, for fermentation to occur, the polysaccharides in these materials must be broken down into readily available mono, di or oligosaccharides. A major product resulting from the fermentation of lignocellulose derived carbohydrate is ethanol, which is of interest as an alternative liquid transportation fuel. The carbohydrate fraction of biomass, particularly the sugars resulting from the saccharification of biomass, is also a potential starting material for the chemical/biochemical production of a host of other value-added products, including ethylene, furfural and functional biopolymers. The economical utilization of the polysaccharide fraction of lignocellulosic materials is largely dependent on identifying relatively inexpensive processes for the solubilization and depolymerization of these polysaccharides. These saccharification processes may be chemically and/or biologically based. One saccharification process which continues to receive considerable attention involves the hydrolysis of the hemicellulose fraction of the biomass in dilute acid followed by enzyme-catalyzed hydrolysis of the cellulose (3).

In this paper we present information which is fundamental to understanding the chemical nature of the acid-catalyzed hydrolysis of lignocellulosics. Our intent is to bring together and introduce relatively basic information that is pertinent to analyzing and interpreting the wide spectrum of data currently being generated from studies on the dilute acid-treatment of lignocellulosics.

The macrocomponent composition of three lignocellulosic materials is given in Table I. The table contains values for a representative hardwood (cottonwood), softwood (pine), and herbaceous (wheat straw) feedstock. The glucan, xylan, mannan, etc. values in the table are based on quantification of the monosaccharides resulting from the complete hydrolysis of cell wall polysaccharides. Hence, although the values are reported as homoglycans, the sugars generated during the hydrolysis were derived from a mixture of homoglycans and heteroglycans. With this in mind, it is still reasonable to use the glucan value as an approximation of the cellulose content and the xylan value as an approximation of xylan content. It is apparent that the major components of dried lignocellulosic feedstocks are cellulose, hemicellulose (approximated by summing all glycans other than glucans) and lignin.

Lignified cell walls are often compared with reinforced concrete (*4*), with the cellulose microfibrils, hemicellulose and lignin being analogous to metal reinforcing rods, concrete and a component to improve bonding, respectively. This gross analogy may be appropriate for our purposes since the actual architectural arrangement of polymers within the lignified cell wall is not yet clear.

Cellulose. The reinforced concrete analogy emphasizes the point that cellulose, the predominant polysaccharide in a cell wall, provides the wall's structural framework. The cellulose molecule is a linear homopolysaccharide of β-D-glucopyranose units linked 1,4. Molecular weight measurements indicate that wood cellulose has a degree of polymerization corresponding to approximately 10,000 glucose units per molecule (*5*). Individual cellulose molecules are not the structural elements found in nature, rather the individual cellulose molecules associate to form paracrystalline cellulose microfibrils. It is these cellulose microfibrils which are considered to be the basic structural elements of native cellulose (*6*). Microfibril dimensions vary depending on their source; those in wood have square cross sections with lateral widths of 3-4 nm (*6*). In lignified cell walls the cellulose microfibrils are found embedded in and intimately associated with the wall matrix, whose primary macromolecular components are the hemicellulose polysaccharides and lignin (*7*).

Table I. Polysaccharide and lignin content of three lignocellulosic materials [a,b]

Component	Lignocellulosic Material		
	Straw	Cottonwood	Pine
glucan	31.93	42.38	42.37
xylan	18.95	13.03	5.94
arabinan	2.08	0.25	1.28
mannan	0.15	2.05	11.02
galactan	0.56	0.33	2.29
lignin[c]	22.79	22.70	27.10
sum of above	76.46	80.74	90.00

[a] values are percentages on a dry weight basis
[b] data taken from Puls and Schuseil (1992)
[c] measured as Klason lignin

Hemicellulose. The term hemicellulose does not refer to a specific molecule, but rather to a group of polysaccharides with similar functional properties. The term is commonly used in reference to the cellulose-associated polysaccharides of the plant cell wall which are soluble in aqueous alkali (*8*). Thus, the term encompasses a host of polysaccharides with a range of molecular structures. In general, hemicelluloses are branched heteropolysaccharides. Quantitatively, the most important of the hemicelluloses in hardwoods and herbaceous feedstocks are the xylans. The term is applied to those polysaccharides for which the main chain (backbone) of the polymer is composed of 1,4 glycosidically linked β-D-xylopyranosyl units. Both hardwood and herbaceous xylans are branched. The majority of the branches are single monosaccharide units attached via glycosidic linkages at the number 2 or 3 carbon of the main chain xylopyranosyl units. In hardwoods, the branches are predominantly 4-O-methyl-α-D-glucuronopyranosyl moieties attached at position 2 of the main chain xylose units (*9*). It is estimated that ten percent of the xylopyranosyl units are so substituted. The xylans of grasses also have these acidic side chains, although they are fewer in number. Overall, grass xylans are more heavily branched than their hardwood counterparts. This is because grass xylans also possess L-arabinofuranosyl side chains; these units are glycosidically linked to the main chain sugars at either the number 2 or 3 position (*10*). The data of Table I suggest an average of one arabinofuranosyl unit per nine xylopyranosyl units in straw. The arabinofuranosyl side chains may exist as such, or they may be further coupled via ester linkages to low molecular weight phenolic acids, primarily ferulic and p-coumaric acids. Approximately 7% of the arabinosyl side chains in barley straw xylan were shown to be linked to phenolic acids via the hydroxyl at position 5 (*11*). Xylans of both hardwoods and grasses are known to contain O-bound acetyl groups. The main chain xylopyranosyl units may be acetylated at either the 2 or 3 position. Considering mature plants, about one-half of the main-chain sugar units are thought to be acetylated in both hardwoods (*12, 13*) and grasses (*14*). Partial structures intended to illustrate linkages prevalent in hardwood and grass derived xylans are presented in Figures 1 and 2, respectively.

Lignin. Lignin is the major aromatic polymer found in the cell wall and middle lamella of lignified plant tissues. It is made up of covalently linked phenyl propane-like units (C_6C_6). Lignin biosynthesis is believed to involve the coupling of its monomer precursors via a free radical mechanism (*5*). This mechanism allows for the polymer to couple at random, linking the structural units through a variety of covalent carbon-carbon and carbon-oxygen linkages. Quantitatively, the most important linkages in lignin appear to be alkyl-aryl ethers as well as alkyl-aryl and aryl-aryl carbon-carbon bonds (*15*). Lignin is always found in intimate contact with other cell wall constituents. Direct covalent linkages between lignin and hemicellulosic sugars have been reported (*16*). Other covalent lignin-hemicellulose linkages are believed to arise via phenolic acid bridges (*17, 18*). In both cases, lignin-polysaccharide crosslinking appears to involve primarily ether and ester bonds. Lignification, and its concomitant crosslinking, would appear to reduce the mobility and flexibility of wall polysaccharides and simultaneously increase the relative hydrophobicity of the wall matrix.

Mechanisms of Acid-Catalyzed Hydrolysis of Glycosides.

Hydrolysis of glycopyranosides. The general mechanism for the acid-catalyzed hydrolysis of glycopyranosides is presented in Figure 3. The glycoside is activated by the rapid and reversible protonation of the glycosidic oxygen. The activated glycoside next undergoes a relatively slow, overall rate-limiting, unimolecular heterolysis to yield the glycopyranosyl cation (a cyclic oxo-carbenium ion in this case) and the aglycone. A water molecule then attacks at the C1 position to form the protonated hemiacetal, which subsequently loses a proton. The result is the production of a neutral reducing sugar. The hydrolysis is thus a specific acid-catalyzed SN1-like reaction. Acid catalysis is necessary to prevent the formation of an unfavorable ion-pair during the heterolytic fission of the C1-glycosidic oxygen bond.

It has been known for well over a century that glycosides of different structure undergo hydrolysis at different rates. The relative rate of hydrolysis of any given glycoside is determined by both its glycose and aglycone components. The importance of the glycose component was demonstrated in a striking way by Masserchalt and co-workers (*19*), who showed that under similar reaction conditions, 2-methoxy-tetrahydropyran, the simplest model for a glycopyranoside, is hydrolyzed approximately 106-fold faster than methyl-α-D-glucopyranoside (Table II).

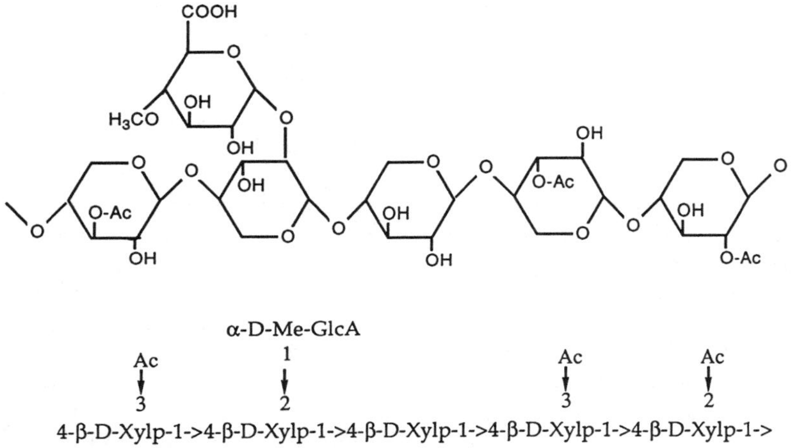

Figure 1. Partial chemical structure of the O-acetyl-4-O-methylglucuronoxylans characteristic of hardwoods.

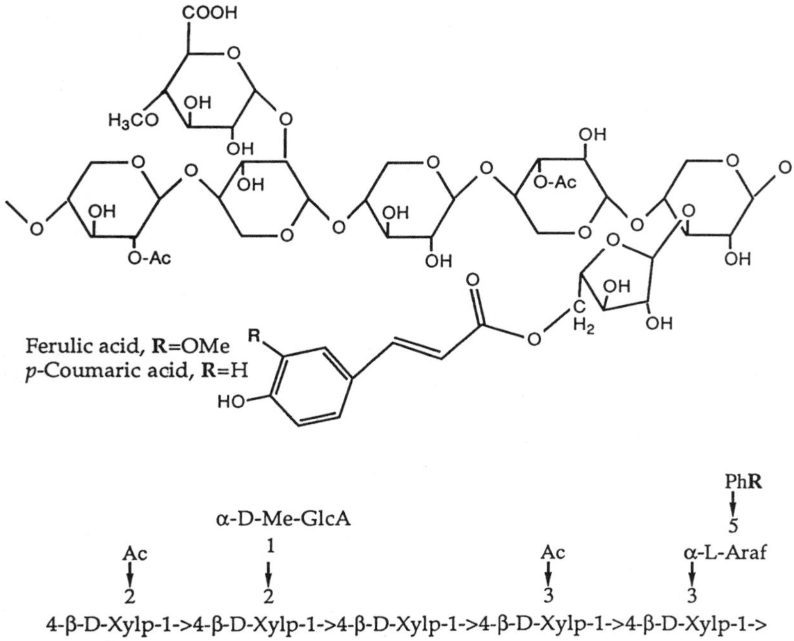

Figure 2. Partial chemical structure of the arabinoxylans characteristic of grasses.

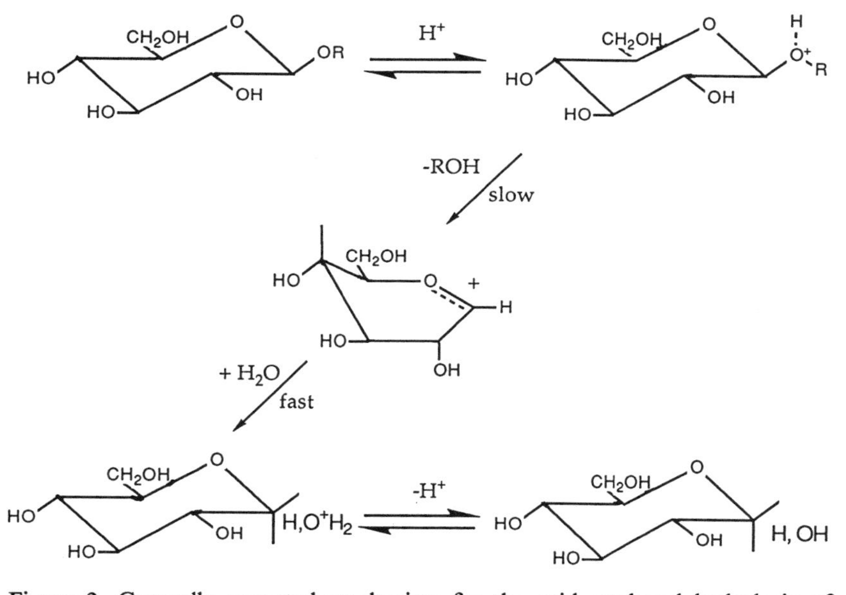

Figure 3. Generally accepted mechanism for the acid-catalyzed hydrolysis of glycopyranosides, as proposed by Edward(1955). The hydrolysis of a generic β-D-xylopyranoside is illustrated.

Table II. Effect of pyranose ring hydroxyl groups on the susceptibility to hydrolysis of methylglycopyranosides[a,b]

compound	position of -OH groups in ring[c,d]	10^6 k (sec^{-1})[e]	relative rate of hydrolysis[f] (k/ko)
tetrahydro-2-methoxypyran	none	94,200	31.4 x 10^6
tetrahydro-3-hydroxy-2-methoxypyran	2	87	29,000
methyl-2-deoxy-alpha-D-glucopyranoside	3,4,5	33	11,000
methyl-alpha-D-xylopyranoside	2,3,4	0.019	.33
methyl-3-deoxy-alpha-D-glucopyranoside	2,4,5	0.010	3.33
methyl-alpha-D-glucopyranoside	2,3,4,5	0.0030	1.0

[a] reaction conditions were $H_o = +0.40$ (where H_o is the Hammett acidity function - an H_o of 0.4 is approximately that of a 0.4 *N* HCl solution) and 30°C
[b] data taken from Dyer *et al.* (1962) and references therein
[c] for purposes of comparison, ring numbering in this column is based on that of a glycopyranose (i.e. the ring oxygen bridges the anomeric C1 and C5 carbons)
[d] position 5 indicates the presence of a $-CH_2OH$ group, all others are -OH groups
[e] k_o equals the rate constant for the hydrolysis of methyl-alpha-D-glucopyranoside
[f] k is the first order rate constant for hydrolysis under the conditions specified above

Obviously, 2-methoxy-tetrahydropyran represents an extreme case, where the C2, C3, and C4 hydroxyls and the C5 hydroxymethyl group of the parent sugar have been replaced with hydrogen. However, other glycoside analogues having intermediate structures with respect to the number and position of hydroxyl groups were also shown to vary widely in their susceptibility to acid hydrolysis (Table II). Much less variability is observed when comparing hydrolysis rates of glycosides which differ only with respect to configuration and/or substituent at the C5 position (Table III). Methyl α- and β-D-glucopyranosiduronic acids were shown to be only slightly more stable than their corresponding neutral glycosides. Glycosides varying only in anomeric configuration have rates of hydrolysis which differ by approximately 2-fold, with the anomer having the aglycone in the equatorial position being hydrolyzed faster (Table III). Changes in the structural chemistry of a glycoside are likely to modify its susceptibility to hydrolysis through either inductive effects or steric constraints imposed on ring structures (*20*). For simple glycosides, it appears that the largest changes in hydrolysis rates are associated with changes in the inductive electron-withdrawing effects of ring substituents, particularly those at C2. These inductive effects are of importance with respect to the stabilization/destabilization of the oxo-carbenium ion transition state implicated in Figure 3. Steric constraints are also important with respect to the formation of the reaction's oxo-carbenium ion intermediate. The formation of this intermediate presumably requires a conformational change as depicted in Figure 3. Steric constraints which increase the resistance to rotation about the C2-C3 and C4-C5 bonds, rotation about which is necessary to go from the chair to the half-chair conformation, are associated with decreased rates of hydrolysis(*21*).

Table III. First order rate constants (k) and activation energies (E_a) for the hydrolysis of selected methylglycopyranosides in 0.5 *M* sulfuric acid [a]

methyl glycopyranoside of	10^6 k $(sec)^{-1}$ [b]	E_a (kcal mol^{-1})
β-D-glucose	6.25	32.5
β-D-xylose	28.6	33.9
β-D-glactose	23.3	32.0
β-D-glucuronic acid	4.14	29.3
α-D-glucose	2.85	35.1
α-D-xylose	13.9	33.4
α-D-galactose	13.6	34.0
α-D-glucuronic acid	1.93	30.2

[a] data taken from Timell *et al.* (1965)
[b] first order rate constant at 70°C

Hydrolysis of glycofuranosides. Two mechanisms have been observed for the acid catalyzed hydrolysis of aldofuranosides (*22*, *23*). In both mechanisms there is an initial pre-equilibrium protonation which activates the glycoside; this activated glycoside then undergoes an overall rate-limiting heterolytic cleavage of the glycosyl-oxygen bond. The two pathways differ with respect to the site of protonation (activation) and the nature of the intermediate oxo-carbenium ion. One pathway is analogous to the mechanism described above for glycopyranosides (Figure 4, path a). In this case, the glycofuranoside is activated by protonation at the glycosidic oxygen, followed by fission of the glycosyl-oxygen bond, resulting in a cyclic oxo-carbenium ion intermediate and the aglycone. The glycofuranosyl cation then reacts with water, eventually yielding a neutral reducing sugar. In the second mechanism (Figure 4, path b) the glycoside is activated through protonation of the ring oxygen. The activated glycoside then undergoes ring-opening via heterolysis of the ring oxygen-C1 bond, producing an acyclic oxo-carbenium ion intermediate. This intermediate is rapidly hydrolyzed, eventually yielding the free aglycone and the neutral reducing sugar. The pathway that predominates for the hydrolysis of a glycofuranoside

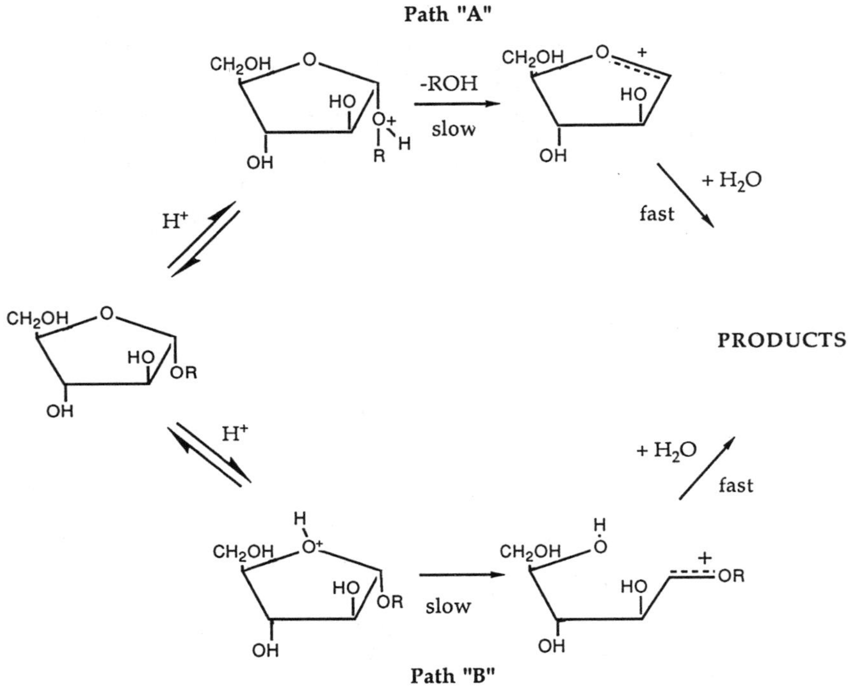

Figure 4. Alternative pathways for the hydrolysis of glycofuranosides, as discussed by Lonnberg *et al.*(1977). The hydrolysis of a generic α-arabinofuranoside is illustrated.

has been shown to be dependent on the configuration of the glycone and the electronegativity of the aglycone (*24, 23*). β-xylofuranosides with electropositive or weakly electronegative aglycones were shown to be hydrolyzed primarily via the acyclic mechanism, while those with highly electronegative aglycones were hydrolyzed via the "cyclic" mechanism. The situation is somewhat different for the α-arabinofuranosides, in which case the cyclic mechanism is favored for alkyl aglycones which are only weakly electronegative (*i.e.* aglycones that would favor the acyclic mechanism when conjugated to β-xylofuranosides).

Acid-catalyzed hydrolysis of disaccharides, oligosaccharides and polysaccharides. The glycosides discussed in this section have in common the feature that their aglycone is a carbohydrate itself, either a monosaccharide or a combination of monosaccharides. In this situation, it is reasonable to assume that the disaccharide or oligosaccharide in question will behave similarly to the corresponding aliphatic glycoside. However, pairs of disaccharides and oligosaccharides and their corresponding aliphatic glycosides do not always exhibit the same relative reactivities. For example, at 70°C in 0.5 *M* sulfuric acid, the rate of hydrolysis of maltose is approximately 2.5-fold greater than that for cellobiose (Table IV), yet under these conditions the relative rates associated with the analogous methyl glycosides are reversed; the rate of hydrolysis of methyl-β-glucopyranoside is approximately 2.2-fold greater than that of methyl-α-glucopyranoside (Table III).

The rate constants associated with the hydrolysis of selected 1,4-linked disaccharides are listed in Table IV. The differences in the rate constants are relatively small, with the extreme values differing by less than 10-fold. Xylobiose was found to be the most readily hydrolyzed of these neutral disaccharides, while cellobiose was shown to be relatively resistant to acid hydrolysis. Thus, the rate constants for these two disaccharides, both having obvious relevance to lignocellulosic substrates, were found to differ by approximately 7-fold.

Table IV. First order rate constants (k) and activation energies (E_a) for the hydrolysis of selected disaccharides in 0.5 *M* sulfuric acid [a]

Disaccharide	10^6 k (sec^{-1})[b]	E_a (kcal mol^{-1})
4-O-β-D-xylopyranosyl-D-xylose (xylobiose)	69.8	32.7
4-O-β-D-glucopyranosyl-D-glucose (cellobiose)	9.63	31.5
4-O-α-D-glucopyranosyl-D-glucose (maltose)	23.6	32.7
4-O-β-D-mannopyranosyl-D-mannose (mannobiose)	16.8	32.7
4-O-β-D-glucopyranosyl-D-mannose	9.4	32.0
4-O-β-D-galactopyranosyl-D-glucose (lactose)	22.5	33.0

[a] data taken from Timell (1964)
[b] first order rate constant at 70°C

Wolfram *et al.* (*25*) measured the hydrolysis rates of reducing disaccharides of D-glucopyranose which differ only with respect to position of linkage (Table V). They demonstrated that linkage position alone may account for a 4-fold difference in hydrolysis rate. In their work, the 1,6 linkages (both anomeric configurations) were most resistant to acid hydrolysis. Their data also reveals that when the aglycone is attached via a secondary hydroxymethyl group (e.g. positions 2, 3, and 4) then the disaccharide with the α-D-configuration is more susceptible to acid hydrolysis than the corresponding disaccharide having the β-D-configuration.

Table V. First order rate constants (k) and activation energies (E_a) for the hydrolysis of D-glucopyranosyl-D-glucopyranosides in 0.1 *N* hydrochloric acid [a]

Disaccharide	anomer (Linkage)	10^6 k (sec^{-1})[b]	E_a (kcal mol^{-1})
kojibiose	α(1→2)	14.6	33.0
sophorose	β(1→2)	11.7	28.9
nigerose	α(1→3)	17.8	27.2
laminaribiose	β(1→3)	9.9	30.0
maltose	α(1→4)	15.5	31.5
cellobiose	β(1→4)	6.6	30.8
isomaltose	α(1→6)	4.0	33.8
gentiobiose	β(1→6)	5.8	33.8

[a] data taken from Wolfram *et al.*(1963)

[b] first order rate constant at 80°C

Hydrolysis rate studies using model compounds. Studies with cellooligosaccharides (*26*), xylooligosaccharides (*27*) and maltooligosaccahrides (*28*) have demonstrated that the glycosidic linkages making up homooligosaccharides are not all equivalent with respect to their susceptibility to acid hydrolysis. The glycosidic linkage at the non-reducing terminus of cellooligosaccharides and maltooligosaccharides were shown to undergo hydrolysis at a rate comparable to that of the corresponding disaccharide, which is from 1.5 (cellooligosaccharide) to 1.8 (maltooligosaccharide) -fold faster than that of the other glycosidic linkages within these molecules (Table VI). Studies with xylooligosaccharides indicate that one of the terminal glycosidic bonds of the molecule is hydrolyzed at approximately 1.8 times the rate of the molecule's other glycosidic linkages. It is likely that the terminal bond which undergoes the faster rate of hydrolysis in xylooligosaccharides is the nonreducing one, thus being analogous to the cellooligosaccharides and maltooligosaccharides. A kinetic model based only on the measured rate of hydrolysis of xylobiose, using the xylobiose rate constant for one of the terminal glycosidic linkages and a rate constant 1.8-fold less for all other glycosidic linkages, has been shown to accurately predict xylose yields during acid hydrolysis of xylooligosaccharides (*27*). The reason for the faster rate of hydrolysis of the nonreducing terminus appears to be related to the relative ease with which an unsubstituted D-glycopyranose residue may approach the half-chair conformation of the reaction's transition state compared with the more substituted glycones that make up the other glycosidic linkages (*21*).

Table VI. Relative rates of hydrolysis of internal and terminal linkages in homo-oligosaccharides

compound	reaction conditions	k_{nrt}[a] (min^{-1})	k_{int}[b] (min^{-1})	ratio (k_{nrt}/k_{int})	reference
cellotriose	0.5 *N* H_2SO_4 120°C	0.076	0.050	1.5	Feather and Harris, 1967
malto-oligosaccharides	0.01 *N* H_2SO_4 95°C	3.05×10^{-4}	1.66×10^{-4}	1.8	Weintraub and French, 1970
xylo-oligosaccharides	0.5 *N* H_2SO_4 100°C	0.120	0.065	1.85	Kamiyama and Yoshio, 1979

[a] First order rate constant for glycosidic linkage at non-reducing end of oligimer
[b] First order rate constant for all internal glycosidic linkages and the linkage at the reducing terminus

Factors influencing hydrolysis rates. The rate of hydrolysis of a polysaccharide is expected to be influenced by all of the factors discussed above. These include substituent configuration, ring size, conformation, anomeric configuration, position of linkage and the presence of functional groups. It is tempting to consider that the rate of hydrolysis of a homopolysaccharide may be predicted based on the measured rate of hydrolysis of the corresponding disaccharide--as shown for the xylooligosaccharides. This concept is based on the rationale that all glycosidic linkages in a polysaccharide, with the exception of the one at the nonreducing terminus, will have essentially the same rate constant for acid hydrolysis, and that this rate constant is a simple function of the rate constant for the corresponding disaccharide. However, this rationale may not be applicable in many cases because it ignores long-range inter- and intra-molecular interactions that may be of relevance in polysaccharide systems. These types of interactions are expected to significantly influence the flexibility of a polysaccharide chain. The data from studies of aliphatic and aryl glycosides and those with disaccharides suggest that inter and intra-molecular interactions which reduce the flexibility of the glycone are likely to alter its rate of hydrolysis (*21*). Analogous to this, it has been cautiously suggested that anything which restricts the flexibility of a polysaccharide chain is likely to decrease its rate of hydrolysis (*20*). Hence, even in cases of the homogeneous hydrolysis of a homopolysaccharide, it seems reasonable that one may observe rate constants for polysaccharide hydrolysis which deviate substantially from those of corresponding di-, tri- and tetra- oligosaccharides.

The effect of intra- and inter-molecular interactions on the rates of polysaccharide hydrolysis have been considered with respect to heterogeneous reaction systems, *i.e.* where the polysaccharide is insoluble in the aqueous solvent. Studies using heterogeneous reaction systems suggest that rates of polysaccharide hydrolysis decrease as the extent of intermolecular interactions increase (*20*). It has been observed that the rate of xylan hydrolysis in a homogeneous reaction system is approximately 4-fold that of cellulose, while the rate of xylan hydrolysis in a heterogeneous reaction system was observed to be up to 70-times as fast as that of cellulose (*29*). The large difference in the relative reactivity of these two polysaccharides in homogeneous versus heterogeneous reaction systems has been attributed to differences in their extents of intermolecular interactions. The degree to which an insoluble polysaccharide is involved in inter- and intra-molecular interactions is likely to affect its inherent flexibility and its solvent accessibility. Decreases in polysaccharide flexibility and lower solvent accessibility are both expected to decrease the polysaccharide's rate of hydrolysis. The polymer's flexibility is important due to the necessity for conformational changes to occur during the progress of the hydrolytic reaction. Solvent accessibility influences the number of exposed glycosidic bonds that can be proton-activated and subsequently hydrolyzed.

Dilute Acid-catalyzed Hydrolysis of Lignocellulosic Materials.

Overview of dilute acid hydrolysis. The dilute acid-catalyzed hydrolysis of lignocellulosic materials typically occurs in heterogeneous reaction mixtures composed of an insoluble lignocellulosic feedstock in an acidic aqueous phase. Hydrolysis under these conditions results in a solid phase comprised primarily of cellulose and lignin as well as a liquid phase containing a complex mixture of products, including neutral sugars, acidic sugars, oligosaccharides, acetic acid, sugar degradation products, and other acid-soluble compounds. The products of greatest relevance to our present discussion are xylose and glucose, since these two neutral sugars are the principle products resulting from extensive hydrolysis of biomass polysaccharides. The majority of the xylose equivalents in biomass are contained in the relatively acid-labile xylan fraction of the hemicellulose. The vast majority of the glucose equivalents are in cellulose, a relatively acid-recalcitrant polysaccharide.

Kinetic models of dilute acid hydrolysis. Kinetic models for the acid-catalyzed hydrolysis of biomass polysaccharides are often assumed to follow first order kinetics. In its simplest form, this behavior may be depicted schematically as in 1.

$$\text{Polysaccharide(P)} \xrightarrow{k_p} \text{Monosaccharide(M)} \qquad 1$$

The corresponding rate equation for this scheme is given in 2.

$$dP_{total} / dt = -k_p (P_{total}) \qquad 2$$

where:

P_{total} equals the total amount of polysaccharide P in the lignocellulosic residue

t is time

k_p is an apparent first order rate constant for polysaccharide dissolution

Equation 2 indicates that the rate of hydrolysis of a biomass polysaccharide is directly proportional to the amount of the polysaccharide present. In a pure sense, P of equation 2 would correspond to the number of glycosidic linkages in molecules of type P, and M would correspond to the number of reducing ends generated by the hydrolysis of P. However, these numbers are not readily measured when working with complex lignocellulosic substrates. A more accessible measure for estimating P is made by determining the number of anhydroglycose equivalents associated with the lignocellulosic substrate. This measure of P is typically accomplished by isolating the insoluble lignocellulosic residue, subjecting it to complete hydrolysis and then quantitatively measuring the resulting monosaccharides. Thus, the rate constant of equation 2 is most typically calculated from measures of the rate of dissolution of the anhydroglycose units that make up the polysaccharide in question, and not from the rate of hydrolysis of the glycosidic linkages specific to that polysaccharide. For example, if xylan hydrolysis is being evaluated, then rate constants are based on the time-dependent loss of anhydroxylose units from the insoluble phase of the reaction mixture. Rate constants calculated in this way are often used with reference to the "rate of hydrolysis" of a polysaccharide. This usage is based on the assumption that the rate of dissolution of P will be a function of the rate of hydrolysis of P. There are cases where this assumption is not warranted (*30*).

A more accurate description of the hydrolysis (and dissolution) of polysaccharides in lignocellulosic materials is depicted in scheme 3. This scheme emphasizes that, in most cases, the polysaccharide will be hydrolyzed to soluble monosaccharides at least partially via soluble oligosaccharides.

$$\text{Polysaccharide}_{\text{insol.}} \xrightarrow{k_p} \text{oligosaccharide}_{\text{sol.}} \xrightarrow{k_o} \text{monosaccharide}_{\text{sol.}} \qquad 3$$

The most elementary case for scheme 3 would be if $k_p = k_o$. Then a single rate constant would apply to the splitting of all glycosidic linkages, a scenario which is more likely to be applicable to homogeneous reaction systems rather than to the heterogeneous reaction systems employed in the dilute acid-catalyzed hydrolysis of lignocellulosics. If the kinetic parameters for polysaccharide hydrolysis are based on measurements of the amount of polysaccharide remaining in the insoluble phase, as discussed above, then the extent of polysaccharide dissolution via oligosaccharides versus monosaccharides is not detected. In cases where hydrolysis occurs randomly along a polysaccharide chain, and assuming dissolution is a function of the chain's degree of polymerization, it is likely that a given number of hydrolytic events in the initial phase of the reaction will generate relatively less polymer dissolution than the same number of hydrolytic events occurring in a later phase of the reaction. In this regard, scheme 3 does not account for those hydrolytic events which do not generate a soluble product. Hydrolytic events of this nature will occur, with their result being the production of chains of shorter degrees of polymerization along with new reducing ends that remain associated with the insoluble feedstock.

The use of reaction scheme 1, if interpreted with respect to hydrolysis specifically, implies that the hydrolysis of all of the glycosidic linkages within the polysaccharide can be adequately described by a single rate constant. This is of interest since the glycosidic linkages within any given polysaccharide fraction are likely to show some heterogeneity with respect to their chemical nature (glycone and aglycone chemistry) and/or their physical environment (amorphous versus crystalline regions). These properties, as discussed in previous sections, are known to significantly effect rates of glycoside hydrolysis. It thus seems prudent to assume that the rate constants used to describe the dissolution of particular lignocellulosic polysaccharide fractions are average values that are influenced by a set of kinetically distinct hydrolytic reactions; each of these reactions having its own unique rate constant for hydrolysis. Rate constants calculated from dissolution data may, therefore, be of only limited use when considering the fundamental mechanisms involved in the hydrolysis of lignocellulosic polysaccharides. This is not to say that these rate constants are of no mechanistic value, since evidence has been presented to suggest that a direct proportionality exists between observed dissolution rate constants and the rate constants associated with rate-controlling chemical reactions (*31*).

Reaction scheme 1 has obvious limitations with respect to its application in mechanistic studies. However it has been applied repeatedly and successfully to describe the kinetics of the acid-catalyzed dissolution of xylan (3) and cellulose (32). In order to optimize rates of polysaccharide dissolution it is essential to know the acid-dependency and the temperature-dependency of pertinent rate constants. The effect of temperature on reaction kinetics is adequately accounted for by assuming that the first order rate constant of equation 2 obeys the general law of Arrhenius, which correlates the rate of a process with the temperature of the system, as shown in 4. Equation 4 is an appropriate expression for the incorporation of temperature-dependency on rates of hydrolysis under conditions of constant acid concentration.

$$k_p = Ae^{(-E_a/RT)} \qquad 4$$

where: k_p is as in equation 2
A is the preexponential factor (a constant)
E_a is the energy of activation
R is the ideal gas constant
T is absolute temperature

If acid concentrations are varied, then the effect of acid concentration on reaction kinetics must also be considered. Acid dependency is generally handled by incorporating an acid concentration term in the general equation used to describe the appropriate rate constant, such as in 5. Equation 5 is an appropriate expression for the incorporation of acid-dependency on rates of acid hydrolysis under conditions of constant temperature.

$$k_p = WC^n \qquad 5$$

where: k_p is as in 2
W and n are constants
C is an acid concentration term

In most cases one is interested in optimizing both acid concentration and temperature, so the parameters introduced in equations 4 and 5 are both employed to account for the effect of temperature and acid on reaction rate, as shown in 6.

$$k_p = AC^n e^{(-E_a/RT)} \qquad 6$$

where: k_p is as in equation 2
A, E_a, R, and T are as in 4
C and n are as in 5

This form of rate equation was described by Seaman in 1945 with respect to the acid hydrolysis of wood cellulose (*33*) and has been successfully used to account for the acid- and temperature-dependency of polysaccharide dissolution from a variety of lignocellulosic sources in more recent years (3).

In many cases a single equation of form 2 is not adequate to describe the full time course of polysaccharide dissolution from lignocellulosic materials. In these cases the rate of dissolution, be it for xylan or for cellulose, is better described by considering that the polysaccharide exists in two distinct phases - an easily hydrolyzed phase and a recalcitrant phase, as depicted in scheme 7.

$$P_{easy} \xrightarrow{k_{easy}} \text{Monosaccharide} \xleftarrow{k_{difficult}} P_{difficult} \qquad 7$$

where: P_{total} is as in equation 2
P_{easy} is the amount of easily degraded polysaccharide P
$P_{difficult}$ is the amount of diffucult-to-degrade polysaccharide P

The total amount of a given polysaccharide in the feedstock is equal to the sum of the easy-to-hydrolyze and the difficult-to-hydrolyze portions of that polysaccharide as shown in equation 8.

$$P_{total} = P_{easy} + P_{difficult} \qquad 8$$

where: P_{total} is as in equation 2
P_{easy} is the amount of easily degraded polysaccharide P
$P_{difficult}$ is the amount of difficult-to-hydrolyze polysaccharide P

The dissolution of any such heterogeneous polysaccharide (in total) is expected to be biphasic. It is generally modelled by assuming two simultaneous, parallel, first order reactions, as in 9.

$$dP_{total}/dt = (-k_{easy})(P_{easy}) + (-k_{diffficult})(P_{difficult}) \qquad 9$$

where: P_{total} and t are as in equation 2
P_{easy} and $P_{difficult}$ are as in equation 8
k_{easy} and $k_{difficult}$ are apparent first order rate constants for dissolution of the easy-to-hydrolyze and difficult-to-hydrolyze fractions of polysaccharide P

Rationales for kinetic models. The use of parallel first order reactions to model xylan dissolution appears to have been introduced by Kobayashi *et al.* (*34*). The chemical and/or physical basis for the observed biphasic nature of xylan dissolution is not agreed upon. Maloney *et al.* (*35*) have presented the following possibilities: (a) there may be mass transport limitations within the reaction mass that cause rate variations during the course of hydrolysis, (b) portions of the xylan polymers may possess different intrinsic reactivities which could change during the course of reaction and (c) the reaction may not be homogeneous, but dependent on a xylan-water interface that could change during the course of the reaction.

The biphasic nature of cellulose dissolution has been attributed to the amorphous (analogous to $\text{Polysaccharide}_{easy}$) and crystalline (analogous to $\text{Polysaccharide}_{difficult}$) regions in cellulose microfibrils (*36*). It is interesting to note that under selected conditions the kinetics of hemicellulose deacetylation (*37*) and lignin dissolution (*38*) have also been described as biphasic. As indicated above, the dissolution of lignocellulosic polysaccharides does not correspond to biphasic kinetics under all conditions. Grohman *et al.* (*39*) found that xylan dissolution from wheat straw could be described using a single, first order rate constant at temperatures above 140°C, yet parallel first order equations were necessary to describe the biphasic dissolution observed at temperatures equal to or below 140°C.

Equations of the form discussed above have been used to model the time course of hydrolysis of both cellulose (*40*) and xylan (*3*). These equations are empirical in the sense that the true mechanisms underlying the hydrolysis of lignocellulosic polysaccharides are not yet known. Empirical models of polysaccharide dissolution do not necessarily have to take the form of those presented above, as several alternative models have appeared in the literature (*40*).

Relative rates of hydrolysis of cellulose and hemicellulose. Even though the mechanisms governing the rates of hydrolysis and dissolution of lignocellulosic polysaccharides are not clearly understood, one can make some reasonably safe generalizations with regard to the relative reactivity of the xylan and cellulose fractions of biomass. The relatively high susceptibility to acid-catalyzed hydrolysis of hemicellulose compared to cellulose is widely accepted. For example, Simmonds *et al.* (*41*) showed that less than 20% of the xylan fraction, compared with greater than 98% of the α-cellulose fraction, remained associated with a hardwood feedstock following the incubation of the feedstock as a water slurry at 170°C for 150 min. This process of hydrolyzing the feedstock in heated water is sometimes termed "autohydrolysis" since it does not involve the addition of mineral acid. However, autohydrolysis is still acid-catalyzed due to the generation of acetic acid via deacetylation of the feedstock at elevated temperatures. Converse *et al.* (*42*) have reported kinetic parameters for the hydrolysis of the xylan and cellulose fractions of a representative hardwood under conditions where the dissolution kinetics of each polymer was consistent with a simple first order reaction, as depicted in scheme 1 (Table VII). The calculated rate constant of xylan dissolution (4.105 min-1) was approximately 77-fold greater than that of cellulose dissolution (0.053 min-1) under conditions of 1% (w/w) added sulfuric acid and 180°C. The half-life of the xylan component under these conditions was approximately 10 seconds, compared to a half-life of approximately 13 minutes for cellulose. Kim and Lee's study (*43*) of the acid-catalyzed hydrolysis of a hardwood sawdust treated at temperatures ranging from 120 to 140°C provides a representative measure of the kinetic parameters associated with xylan hydrolysis under conditions where xylan dissolution shows biphasic kinetics (Table VIII). Rate constants for the hydrolysis of the easily degraded and difficult-to-degrade xylan fractions were 0.025 min-1 and 0.0027 min-1 for reaction conditions of 1% acid and 130°C, respectively, based on the kinetic parameters of Table VIII. Under these conditions, the dissolution rate constants for the two xylan fractions differed by slightly more than 9-fold. The half-lives corresponding to the two rate constants are approximately 28 min. (easily degraded fraction) and 258 min. (difficult-to-degrade fraction). When judging the relative rates of hydrolysis and dissolution of the different lignocellulosic components, it is important to remember that, due to differences in the acid-dependency and temperature-dependency of the corresponding rate constants, these relative values will change as the severity of the reaction system is changed.

Table VII. Kinetic parameters for the dilute acid-catalyzed dissolution of xylan and cellulose from a hardwood; based on equation 6 of text [a,b,c]

hardwood fraction	preexponential factor, "A" (min^{-1})	acid dependency constant, "n"	activation energy "E_a" ($kcal\ mol^{-1}$)
xylan	9.11×10^{12}	0.73	25.5
cellulose	2.50×10^{16}	1.15	36.5

[a] data taken from Converse *et al.* (1989)

[b] equation (6) of text: $k = AC^n \exp(-Ea/RT)$, where k is the first order rate constant for dissolution, C is acid concentration in % (w/w), other terms as indicated above

[c] parameters calculated from experimental data covering 160-265°C and 0.2-2.4 wt% H_2SO_4

Table VIII. Kinetic parameters for the dilute acid-catalyzed dissolution of xylan fractions from a hardwood; parameters based on equation (6) of text [a,b,c]

xylan fraction	preexponential factor, "A" (min^{-1})	acid dependency constant, "n"	activation energy "E_a" ($kcal\ mol^{-1}$)
easy to degrade	1.04×10^{14}	1.54	28.7
difficult to degrade	0.06×10^{14}	1.19	28.2

[a] data taken from Kim and Lee (1987)

[b] scheme (6) of text: $k = AC^n \exp(-Ea/RT)$, where k is the first order rate constant for dissolution, C is acid concentration in % (w/w), other terms as indicated above

[c] parameters calculated from experimental data covering 120-140°C and 0.53-4.5 wt% H_2SO_4.

The majority of studies now appearing on the acid-catalyzed hydrolysis of lignocellulosics are directed toward the optimization of specific process schemes. These experiments invariably make use of empirical models, such as that described above. It is probable that these studies will identify an array of useful reaction conditions, each being potentially applicable to industrial processes. The actual conditions (time, temperature, acid concentrations, acid type, etc.) considered optimum for industrial purposes will obviously be effected by the overall economics of the process. It may be some time before relatively detailed mechanistic models are verified for the hydrolysis of lignocellulosics. It seems unlikely that a more detailed understanding of the chemical mechanisms underlying the hydrolytic and dissolution events will lead to major improvements in the ability to model xylan or cellulose dissolution. However, it is reasonable to assume that a better understanding of the fundamental chemistry involved in the hydrolysis of lignocellulosics will lead to novel approaches to the utilization of lignocellulosic residues.

Literature cited

1. Schell, D; Walter, P.; Johnson, D. *App Biol. Biot.* **1992**, 34-35, pp 659-663.
2. McKeerer, D. B. In *Proceedings of the Second Biomass Conference ofthe Americas.* **1995**, Portland, OR.
3. McMillan, J.D. Processes for Pretreating Lignocellulosic Biomass: A Review, National Renewable Energy Laboratory - U.S. Department of Energy, **1992**, NREL/TP-421-4978.
4. Fujita, M.; Harada, H. In *Wood and Cellulosic Chemistry*; Hon, D.N.S.; Shiraishi, N., Eds.; Marcel Dekker, Inc. New York, New York, **1991**, pp 3-57.
5. Sjostrom, E. *Wood Chemistry, Fundamentals and Applications*, Academic Press: San Diego, California, **1981**, 49-67.
6. Chanzy, H. In *Cellulose Sources and Exploitation:Industrial Utilization, Biotechnology and Physico-Chemical Properties*; Kennedy, J.F., Phillips, G.O., Williams, P.A., Eds.; Ellis Horwood: New York, New York, **1990**, pp 3-12.
7. Atalla, R.H. In *Enzyme Structure, Biochemistry Genetics and Application;* Suominen, P. and Reinikainen, Eds.; Foundation for Biotechnical and Industrial Research: City, state, **1993**, Vol. 8, pp 25-39.
8. Wilkie, K.C.B. *Adv. Carb. Chem. Biochem.* **1979**, *36*, 215-264.
9. Puls, J.; Schuseil, J. In *Hemicellulose and Hemicellulases* Coughlan, M.P.; Hazelwood, G.P; Eds.; Portland Press: London, England, UK, **1992**, pp 1-26.
10. Aspinall, G.O. *Adv. Carb. Chem.* **1959,** *14*, 429-468.
11. Mueller-Harvey, I.; Hartley, R.D. *Carbohydr. Res.* **1986**, *148*, 71-85.
12. Timell, T.E. *Wood Sci. Technol.* **1967**, *1*, 45-70.
13. Shimizu, K. In *Wood and Cellulose Chemistry*; Hon, D.N.S.; Shiraishi, N., Eds.; Marcel Dekker, Inc.: New York, New York, **1991**, pp 177-214.
14. Bacon, J.S.D.; Gordon, A.H.; Morris, E.J. *Biochem. J.* **1975**, *149*, 485-487.
15. Sakakibara, A. In *Wood and Cellulose Chemistry*; Hon, D.N.S.; Shiraishi, N., Eds.; Marcel Dekker, Inc.: New York, New York, **1991**, pp 113-176.
16. Monties, B. *Meth. Plant Biochem.* **1989**, *1*, 113-157.
17. Lam, T.B.T.;Iiyama, K.; Stone, B.A.*Phytochemistry* **1992**, *31*, 1179-1183

18. Iiyama, K.; Lam, T.B.T.; Stone, B.A. *Phytochemistry* **1990**, *29*, 733-737.
19. Dyer, E.; Glaudemans, C.P.J.; Koch, M.J.; Marchessault, R.H. *J. Chem. Soc.* **1962,** 3361-3364.
20. Bemiller, J.N. *Adv. Carb. Chem.* **1967**, *22*, 25-108.
21. Feather, M.S.; Harris, J.F. *J. Org. Chem.* **1965**, *30*, 153-157.
22. Benner, R.; Hodson, R.E. *Appl. Envir. Microb.* **1985**, *50*:4, 971-976.
23. Lonnberg, H.; Kulonpaa, A. *Acta. Chem. Scan. A* **1977**, *31*, 306-312.
24. Lonnberg, H.; Kankaanpera, A.; Haapakka, K. *Carb. Res.* **1977**, *56*, 277-287.
25. Wolfrom, M.L.; Thompson, A.; Timberlake, C.E. *Cereal Chem.* **1963**, *40*, 82-86.
26. Feather, M.S.; Harris, J.F. *J. Am. Chem. Soc.* **1967**, *89*, 5661-5664.
27. Kamiyama, Y.; Sakai, Y. *Carb. Res.* **1979**, *73*, 151-158.
28. Weintraub, M.S.; French, D. *Carb. Res.* **1970**, *15*, 251-262.
29. Konkin, A.A.; Kaplan, N.I.; Rogovin, Z.A. *Zhur. Priklad. Khim.* **1955**, *28*, 729-34.
30. Stewart, C.M.; Williams, E.J. *Chem. Ind.* **1955**, 1350-1351.
31. Springer, E.L. *TAPPI* **1966,** *49*, 102-106.
32. Grethlein, H.E. *J. Appl. Chem. Biotechnol.* **1978**, *28*, 296-308.
33. Seaman, J.F. *Ind. Eng. Chem.* **1945**, *37*, 43-52.
34. Kobayashi, T.; Sakai, Y. *Bull. Agric. Chem. Soc. Japan* **1956**, *20*, 1-7.
35. Maloney, M.T.; Chapman, T.W.; Baker, A.J. *Biotech. Bioeng.* **1985**, *27*, 355-361.
36. Sharples, A. *Trans. Faraday Soc.* **1957**, *53*, 1003-1013.
37. Erins, P.; Cinite, V.; Jakobsons, M.; Gravitis, J. *Appl. Polymer Symp. No. 28*, **1976**, 1117-1138.
38. Springer, E.L.; Harris, J.F.; Neill, W.K. *TAPPI* **1963**, *46*, 551-555.
39. Grohmann, K.; Torget, R.; Himmel, M. *Biotech. Bioeng. Symp. No. 15*, **1985**, 59-80.
40. Fan, L.T.; Gharpuray, M.M.; Lee, Y.H. *Cellulose Hydrolysis*, Springer-Verlag: City, state, **1987**, pp 121-149.
41. Simmonds, F.A.; Kingsbury, R.M.; Martin, J.S. *TAPPI* **1955,** *38*, 178-186.
42. Converse, A.O.; Kwarteng, I.K.; Grethlein, H.E.; Ooshima, H. *Appl. Biochem. Biotech.* **1989**, *20/21*, 63-78.
43. Kim, S.B.; Lee, Y.Y. *Biotech. Bioeng. Symp. No. 17*, **1987**, 71-84.

Chapter 3

Enhanced Cotton Textiles from Utilization Research

B. A. Kottes Andrews and Robert M. Reinhardt

Southern Regional Research Center, Mid-South Area, Agricultural Research Service, U.S. Department of Agriculture, New Orleans, LA 70179

Cotton, the most important vegetable fiber used in spinning, is a member of the genus *Gossipium* and is widely grown in warm climates the world over. Cotton, a renewable resource, has many good qualities, such as comfort, breathability and dyeability. Utilization research has provided other value added properties. Elucidation of mechanisms for the reactions of cellulose has paved the way for product development. Significant contributions in the areas of mercerization, rot resistance, antibacterial treatments, flame retardancy, durable press, antisoiling treatments, dyeing, water absorption, and comfort have been provided by research conducted at the Southern Regional Research Center.

In 1938 the U. S. Congress, in the Agricultural Adjustment Act, provided for the establishment of four regional research laboratories (*1*). These four laboratories were directed "to conduct research into and to develop new scientific, chemical, and technical uses and new and extended markets and outlets for farm commodities and products and byproducts." One of these laboratories is now named the Southern Regional Research Center (SRRC), New Orleans, LA, and is part of the Agricultural Research Service of the U. S. Department of Agriculture. The farm commodities originally selected for research in this laboratory included cotton, peanuts, and sweet potatoes.

Because cotton is one of the most important plant crops in the U. S., the major research effort of the new laboratory was directed toward increasing the use of the fiber. The cotton fiber has a short taper at the tip and is twisted frequently along its entire length, with occasional reversals in the direction of twist. Cotton is essentially 95% cellulose which is the reactive species that can be modified to impart improved physical properties to the fiber and fabric. The noncellulosic material is composed of waxes, proteins, pectinacious substances and sugars.

The gross morphology of cotton refers to the relatively large structural elements, the primary wall, secondary wall and lumen, as well as a cuticle and winding layer. The microfibrillate structure includes pores, channels and cavities that play an important role in the chemical modification of cotton. Microfibrils of the secondary wall are 10-49 nm in width, and these in turn are composed of elementary fibrils, or crystallites, 3-6 nm wide.

The initial cotton program, carried out by three research groups, addressed establishment of the relationship of the chemical and physical structure of the cotton fiber to technically significant properties, development of new or improved cotton products to meet specific use requirements, and improvement of cotton fiber properties by chemical and physical treatments without loss of fiber integrity (*2*). The results of this early research met the needs of the time; many of the products and their successors are still in use today.

The research, always relevant to the textile industry, has changed throughout the years with evolving commodity and industry requirements. Tracing the development of improved products and textiles from modification of the cotton fiber discloses a pattern of basic and applied research with knowledge acquired from early research used as building blocks for later developments. Highlighted below are some of the achievements from chemical research more widely adopted by industry.

Slack Mercerization

Although mercerization of cotton under tension had long been known to produce lustrous fabrics and enhance dyeability, research on slack mercerization at SRRC led to the development of unique products. As early as 1945, a product was developed that revolutionized the surgical bandage concept and is still in use almost 50 years later. Goldthwait and Kettering invented the cotton stretch bandage, of surgical gauze, that contained crimps and kinks which produced not only increased shrinkage but an easily stretchable gauze that clings rather than loosens in place (*3, 4*).

Much later, concepts of comfort and action stretch became important to wearers of the tight fitting garments of the 1960's. Slack mercerization provided all-cotton fabric with a response to the challenge of fabrics containing synthetic stretchable yarns in the filling direction (*5, 6*). Further refinement extended the process to novel cotton fabric constructions such as lace (*7*).

Rot Resistance Treatments

Some of the earliest uses of cotton outside the apparel and household applications were in tirecord, tentage, sandbags and press covers. Much early research concentrated on extending the usable life of these products. Partial acetylation successfully produced cotton with resistance to the effects of high temperatures or continued heat, and high resistance to mildew and microbiological rotting; a drawback was the added cost per yard (*8-10*).

Other rot resistance and antifungal treatments developed over the years included treatment with salts of copper, zirconium, tributyltin, tributylplumbum, and phenylmercurics (*11-15*). The "Zirchrome Process" received much industry interest (*16*). Non-salt based treatments for resistance to microbiological deterioration included finishes from cyanoethylation (*17*) and from treatment with the formic acid colloid of methylolmelamine (*18*). The latter was particularly suitable for use in tobacco shade cloth.

Antibacterial Treatments

The success of zirconium and other salts against mildew and microbiological rotting prompted research with these compounds in antibacterial treatments for cotton (*19*). Zirconyl acetate was found to effect peroxide binding to cotton. These initial studies have led to finishes with improved durability to laundering by substituting zinc acetate as the bonding agent for hydrogen peroxide in the "Permox" treatment (*20*). A slow release of active oxygen is believed responsible for the protection against odor- or infection-producing bacteria.

Water Repellency

A type of impermeable fabric, called self-sealing fabric, was the subject of early work. The uses of these fabrics were in tentage, water bags and fire hoses. Research areas included preventing the passage of water through cloth by application of hydroxyethylcellulose to the fibers to give a permanent finish which swelled when wet (*21*). Textile engineering researchers devised strictly mechanical means to produce impermeable fabrics through use of fine, immature cottons to reduce the size of the interstices; and by use of a special loom attachment to increase the number of picks per inch (*22*). Later research focused on esterification and etherification reactions to chemically bind water-repellent groups to cotton cellulose. Weaver, Schuyten, Frick, and Reid studied the reactions of various stearamidomethyl derivatives with alcohols, including cellulose (*23-26*). Formation of a stearamidomethyl ether from stearamidomethylpyridinium chloride proceeds:

$$[C_{17}H_{35}CONHCH_2NC_5H_5]^+Cl^- + ROH \rightarrow C_{17}H_{35}CONHCH_2OR + HCl + C_5H_5N$$

Fluorochemical textile treatments were also investigated at SRRC, including chromium coordination complexes of perfluorocarboxylic acids (*27*), polyethyleneimines having perfluoroacyl side chains (*28-30*), and epoxides having a perfluorinated side chain (31). Bullock and Welch investigated crosslinked silicone films as possible wash-wear, water-repellent finishes for cotton (*32*).

Flame Resistance Treatments

Achievements in flame retardant technology by scientists at SRRC have made profound contributions to the safety of our civilian and military population. Flame retardant cotton garments prepared by SRRC processes have been used by our astronauts. Early research established the foundation for the chemical theory of cotton flame retardancy (*33*). Criteria for potential flame retardants were established. The greatest advances have been made with chemicals containing phosphorus and the most important developments are with agents based on tetrakis(hydroxymethyl)phosphonium chloride (THPC):

$$(HOCH_2)_4P^+Cl^-$$

It is prepared by the reaction of phosphine (PH_3) with formaldehyde and HCl. Reeves and Guthrie initially found that THPC reacted with aminized cotton and conferred flame resistance (*34*). Further refinements eliminated the costly aminization step by application of THPC to cotton with relatively inexpensive materials such as methylolmelamine and urea. A host of other materials containing active hydrogen atoms was shown to form insoluble polymers with THPC in the cotton fiber thereby imparting flame retardancy (*35*).

Improvements in the THPC-based process were the result of many successful research efforts at SRRC. Chemical fixation of a water-soluble precondensate of THPC and amides with ammonia and/or ammonium hydroxide improved the degree of flame retardancy by the additional nitrogen content in the finish (*36, 37*). The THPC precondensate treatment using ammonia gas to polymerize the precondensate in the fiber represents the largest commercial use of flame retardants in the U. S. today. This is a direct development from SRRC.

The mechanism of the phosphorus-based flame retardant finishes on cotton is very complex and has been summarized in a series of papers (*38-41*). Durability of the THPOH-based finishes to laundering was found to depend on the oxidation state of the phosphorus (*42*).

Non-durable flame retardants based on boric acid and its derivatives found application in treatments of cotton batting for bedding and upholstery (*43, 44*). Cooperation with Cotton Incorporated led to a large scale reactor for the process (*45*). Treatments with boric acid and methyl borate derivatives, developed by Knoepfler and coworkers, achieved commercial success.

Although smolder resistance was found to follow a different mechanism than flame resistance, boric acid-based treatments also were quite effective for imparting smolder resistance (*46*).

Durable Press Treatments

Researchers at SRRC have been in the forefront of the development of wrinkle resistant cotton fabrics. A substantial segment of the studies to establish cellulose crosslinking as the mechanism for wrinkle resistance and the reactivity of cotton cellulose in general was carried out at SRRC. Cotton is enjoying a resurgence in the fiber market today largely because of the renewed popularity of 100% cotton wrinkle resistant garments. This resurgence has allowed cotton to recapture usages and markets lost to products of synthetic fibers and cotton/synthetic blend fabrics, and again become the fiber of greatest consumption in the U. S.

Frick, Kottes Andrews, and Reid used analyses for total nitrogen and formaldehyde contents of fabrics treated with formaldehyde and N-methylolamide cellulose reactants to estimate the size of crosslinks. Formaldehyde crosslinks were found to be monomeric; those from methylolamides, oligomeric (*47*). Increases in resiliency and corresponding losses in strength/toughness properties were related to the extent of crosslinking. Crosslinking was established as the primary reaction between dimethylolethyleneurea and cotton; little, if any homopolymer was formed

(*48*). Rau, Roberts, and Rowland isolated and identified formaldehyde crosslinked components from ball-milled cellulose modified with formaldehyde (*49*).

A definitive body of research on the mechanisms of etherification crosslinking of cellulose was conducted by researchers at the SRRC and by Petersen in Germany. Both reactivity of crosslinking agents in etherification of cellulose and resistance of these cellulose crosslinks to hydrolysis were found by Reeves, *et al.* to depend on the electron density around the amidomethyl ether group, suggesting a carbocation mechanism for reaction under acid conditions (*50*). A different mechanism was proposed by Andrews, Arceneaux and Frick for those amidomethylol agents that also can crosslink cellulose under alkaline conditions (*51*). Vail proposed the reversible reactivity-hydrolysis relationship in chemical finishing of cotton with N-methylol reactants under acidic conditions (*52*). Petersen contributed greatly to the understanding of the relationship between crosslinkers, catalysts, liberation of formaldehyde and properties of the finished fabrics (*53-56*).

Although in most cases cellulose crosslinking is responsible for improvements in cotton resiliency, some exceptions were discovered. Improvements in resiliency have been imparted by esterification with monofunctional long chain fatty acid chlorides (*57, 58*), and by deposition of crosslinkable polysiloxanes (*59*), and other elastic polymers.

Of new crosslinking agents for cotton developed at SRRC, the most successful, and still in use today, is the methylolcarbamate class of agents:

$$ROC(O)N(CH_2OH)_2$$

developed by Frick, Arceneaux, and Reid (*60, 61*). An advantage of carbamate agents is a lack of yellowing, important for white goods. Etherification of the methylol groups in these agents reduces the formaldehyde release in the agents, in textile processing operations, and in finished fabric to levels in compliance with federal regulations (*62, 63*).

Until the early eighties, etherification of methylolamides with methanol was the technique used to reduce formaldehyde release in agents and subsequent finishes. A joint publication by SRRC and WestPoint Pepperell proposed pad bath addition (blending) of low molecular weight glycols to modify the methylolamides *in situ* at the time of cure and lower formaldehyde release (*64*). This modification is less expensive than the process in which methanol is reacted with the methylolamides as an etherification step in agent synthesis. Blending is used extensively by the textile industry, for example in finishing with glycolated dimethyloldihydroxyethyleneurea (DMDHEU).

Finishing with formaldehyde itself played a large role in the SRRC research program. The vapor phase formaldehyde process was developed and refined (*65, 66*). The process has been commercialized by the American Textile Processing Company. Other, wet, processes for formaldehyde finishing were developed, namely the Form D and Form W processes, that improved cotton's wet wrinkle recovery properties (*67*).

Systems were devised with less harsh curing conditions to produce smooth drying fabrics comparable to those from European "feucht-vernetzung" (*68*). Catalysis played an important role in process modification (*69, 70*). Frick and Pierce developed a magnesium chloride/citric acid catalyst that is the catalyst of choice throughout the world in DMDHEU-based finishing of cotton for durable press (*71*).

Additives to change fabric handle were the subjects of SRRC research also. Replacement of low glass transition temperature polyacrylates with emulsified polyethylene, the subject of a patent granted to Reinhardt, Mazzeno, and Reid, produced a fabric with a soft hand, but without tendency to pick up oily soil (*72*). Recovering the abrasion resistance lost by finishing of cellulosics for durable press has been an industry problem addressed by SRRC research. Harper and coworkers have not only optimized additive type and amount but also have explored variations in fabric construction for improved abrasion resistance (*73-76*). Processes such as "wet fix" and other polymerization/crosslinking treatments were developed toward this end (*77, 78*). Grafting of polymers onto cotton, either alone, or in combination with crosslinking was also researched at the SRRC. Graft copolymerization reactions of unsaturated monomers with cotton at free-radical sites formed on the cellulose molecule were studied extensively (*79*). A pilot-scale process was devised for continuous photoinitiated copolymerization with N-methylolacrylamide for preparation of durable press cotton (*80*).

In today's regulatory climate, replacement of formaldehyde-containing agents with those that are environmentally friendly is of major interest to the U. S. textile industry. Non-formaldehyde agents, however, have been considered throughout the history of textile finishing. In the early days, the reason was to eliminate sites in the finish that could attract positive chlorine from hypochlorite bleaching during laundering and thus cause discoloration or scorch damage on subsequent heating. Later, elimination of noxious formaldehyde was the goal.

McKelvey and Benerito pioneered crosslinking of cotton with epoxide agents (*81*). Welch developed silicone treatments for smooth drying cottons (*82*). The most widely used non-formaldehyde agent for durable press today is 4,5-dihydroxy 1,3-dimethylimidazolidinone (DHDMI or DMUG), the reaction product of symmetrical dimethylurea and glyoxal. A patent covering that compound is held by Vail and Murphy (*83*). DHDMI is shown in Figure 1.

In the 1960's a patent was granted to Rowland, Welch, and Brannan for improvements in dimensional stability and crease resistance of cellulosics through reactions with polycarboxylic acids (*84*). However, the catalysts used at that time did not bring these fabric properties up to durable press levels. Recent research by Welch and Andrews showed that high levels of resiliency could be obtained with esterification crosslinking if alkali metal salts of inorganic phosphorus-containing acids were used as catalysts (*85-88*). With these catalysts, the cellulose esterification is believed to proceed through an anhydride mechanism. It involves two carboxylic acid groups in the agent for the esterification reaction and regeneration of one carboxylic acid group. This mechanism limits successful crosslinking to those acids containing three or more non-sterically hindered carboxylic acid groups. The two polycarboxylic acids of commercial interest for the process are cis-1,2,3,4-butanetetracarboxylic acid (BTCA) and citric acid (CA), shown in Figure 2.

BTCA, because of its tetrafunctionality, is one of the most efficient cellulose crosslinkers; however, a major drawback to its use is cost. Citric acid, on the other hand, although not as reactive toward cellulose, is not only relatively inexpensive,

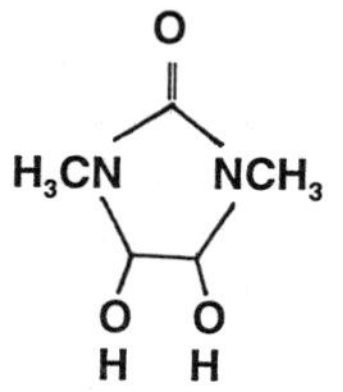

Figure 1. Reaction product of dimethylurea and glyoxal, 4,5-dihydroxy-1,3-dimethyl-2-imidazolidinone.

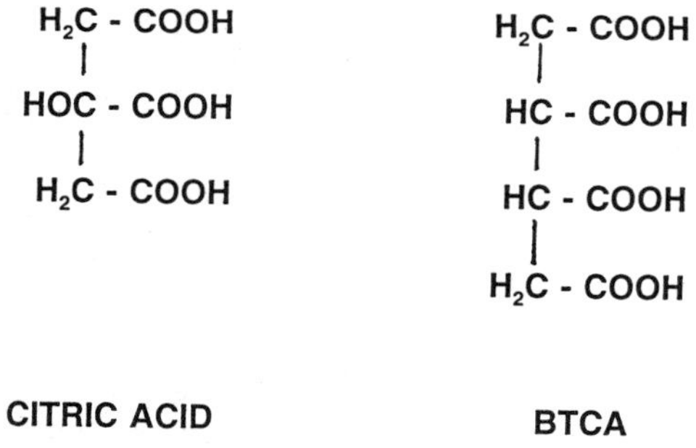

Figure 2. Polycarboxylic acids of commercial interest in crosslink finishing of cotton: left, citric acid; right, cis-1,2,3,4-butanetetracarboxylic acid.

but also has been approved for food use. Treatment with a combination of these two acids produces a non-formaldehyde wrinkle-resistant fabric that is both economically and technically competitive with the formaldehyde-containing agents currently in use (*89, 90*).

As an outgrowth of SRRC research, use of polycarboxylic acids also has been extended to non-textile applications. Agents with acid functionality can provide an alternative to formaldehyde based crosslinkers used in wet strength treatments for paper (*91, 92*). Absorbent cellulosic materials for use in diaper and personal care products are prepared from cellulose fibers crosslinked with polycarboxylic acids (*93, 94*).

Although ammonia mercerization of cotton fabric does not involve crosslinking, this treatment can impart satisfactory smoothness to certain fabrics. Development of a chainless mercerization process and basic research on the changes to the cellulose structure were carried out at SRRC (*95*).

Antisoiling Treatments

Excellent research on the location of soil in cotton fabrics was performed in the 1950's by the microscopy group st SRRC. Influence of chemical modification and of surface polymers on deposition of dry soil was established (*96*). Wet soiling studies by Mazzeno and coworkers isolated soiling problems in fabrics containing smooth drying finishes that contained soft tacky polymers as hand modifiers as the cause of irreversible soiling (*97*). Mentioned earlier was the research by Reinhardt that used emulsified polyethylene as a softener to overcome this problem. Other research by Reeves, Drake, and their coworkers demonstrated that carboxymethylcellulose, durably bound to cotton fabric from methylolamide crosslinking formulations, improved antisoiling characteristics of resin treated cottons (*98, 99*). Oil repellant finishes based on fluorochemicals were also developed (*100, 101*).

Dyeing Studies

A substantial amount of important research that contributed significantly to the development of dyeing of cotton has been conducted at SRRC. This information has not always been thought of *per se* as dyeing research, but it has been instrumental in advancing the chemistry and technology of the dyeing of cotton and other fibers.

Two studies that contributed greatly to the chemistry of reactive dyes were conducted at SRRC. In one, the reactions of cellulose with cyanuric chloride were elucidated by Warren and coworkers (*102*). The reactivities of the three chlorine groups of the triazine ring were explored and stabilities of residual groups for further modification were shown. These reactions are the basis of mono- and dichloro-triazine dyes which command much of the market for reactive dyes. In the other study, the chemistry of reactions of cellulose with sulfatoethyl derivatives was established by Guthrie (*103*). This chemistry provided the basis for reactive dyes of the sulfatoethylsulfone class.

Crossdyeing and multicolor dyeing effects were produced by treatments combining finishing with crosslinkers and polymers or lattices (*104, 105*) and by finishing/partial hydrolysis dyeing schemes (*106*). Surface coatings were developed to allow heat transfer printing of cotton with disperse dyestuffs (*107, 108*). These treatments enabled cotton to meet the challenge of transfer printed thermoplastic materials such as polyester.

Similarly, patents were issued on other processes in which coreactants or additives in crosslinking produced fabrics that could be dyed with basic dyes (*109*) and with acid dyes (*110, 111*). Hydroxyalkyl sulfonic or phosphinic acid grafts introduced along with crosslinking gave fabrics that could be dyed with basic dyes (*112*). Dyeability with disperse dyestuffs was even demonstrated from treatment with crosslinking agent and a suitable polymer (*113*).

Recent efforts in research on dyeing and finishing have concentrated on the preparation of durable press cottons that are dyeable after the finishing treatment and are thus suitable for dyeing in the garment form (*114*). Garment dyeing has become more popular and the textile market segment devoted to garment processing has increased in importance.

Most successful of the approaches to providing a dyeable, durable press cotton fabric has been that of including a suitable additive in otherwise conventional pad-dry-cure crosslinking treatments. Two additives have been identified as most promising for this purpose, choline chloride and triethanolamine:

$HOCH_2CH_2N^+(CH_3)_3\ Cl^-$ and $N(CH_2CH_2OH)_3$

choline chloride — triethanolamine

Both are nitrogenous compounds that react with the crosslinking agent during the finishing treatment and become a part of the fabric/finish matrix. The bound agent provides a cationic site to which anionic dyes can attach.

Harper and coworkers have described the basic concept and many variations in processes employing choline chloride and similar quaternary derivatives (*115-117*). Finishing with triethanolamine and tetrakis(2-hydroxyethyl)ammonium chloride as additives to methylolamide and polycarboxylic acid crosslinking has been described by Blanchard and coworkers (*118-122*).

Absorption and Swellability Studies

The successful development of swellable fibers was accomplished at SRRC by partial carboxymethylation of cotton linters, yarns, and fabric. Initially, the goal was decreased water permeability by the wetted fabric (*123, 124*). Utility of carboxymethylated cotton in other areas, such as improved physical properties, resin acceptability, and varied dyeing characteristics, was explored later (*125, 126*). Alkali solubility of partially etherified cottons has found a use in lace manufacture (*127*).

Thermally Adaptable Fabrics

An innovative combination of polyethylene glycols and polyfunctional methylolamide crosslinking agents has produced fabrics that are capable of thermal storage and release. Vigo and coworkers have shown that polyethylene glycols of low molecular weight (300-1,000) can be durably bound to cotton and other fibers by a pad-dry-cure method utilizing a polyfunctional crosslinking agent (*128, 129*). The treated fabrics release heat when the temperature drops and absorb heat when the temperature rises. The thermal activity and the temperature at which the modified fabrics are thermally active are dependent on the molecular weight of the polyol and on the type and concentration of crosslinking agent and catalyst. The resultant fabrics, in addition to thermal properties, exhibit other multifunctional properties, such as antistatic behavior, water absorbency, resiliency, soil release, and pilling and abrasion resistance.

Summary

The Southern Regional Research Center has provided the U. S. textile industry and the U. S. consumer with value added products from cotton, one of our nation's most important renewable resources. Basic research has established mechanisms for reactions of cellulose that have paved the way for product development. New markets for cotton have been realized and old markets have been retained as challenges from new fibers were met through research. The annual yardage of treated cotton in the apparel and bedding markets alone is estimated to be enough to reach from planet earth to the moon and back more than four times!

Literature Cited

1. Public Law No. 430, 75th U. S. Congress.
2. Scott, W. M. *Am. Dyest. Rep.* **1941**, *30*(22), 604-606, 619.
3. Goldthwait, C. F. U.S. Patent 2 379 574, 1945.
4. Goldthwait, C. F.; Kettering, J. H.; Moore, M., Jr.; *Surgery* **1945**, *18*(4), 507-510.
5. Sloan, W. G.;Moore, H. B.; Hoffman, M. J.; Cooper, A. S., Jr. *Text. Bulletin* **1965**, *90*(10), 106-110.
6. Sloan, W. G.; Hoffman, M. H.; Robinson, H. M.; Moore, H. B.; Cooper, A. S. *Am. Dyest. Rep.* **1963**, *52*(11), 26-33.
7. Fisher, C. J.; Robinson, H. M.; Jones, M. A. *South. Text. News* **1965**, *21*(24), 46, 48.
8. Cooper, A. S.; Voorhies, S. T., Jr.; Buras, E. M., Jr.; Goldthwait, C. F. *Text. Ind.* **1952**, *116*(1), 97-102, 194-195.
9. Buras, E. M., Jr.; Cooper, A. S.; Keating, E. J.; Goldthwait, C. F. *Am. Dyest. Rep.* **1954**, *43*(7), 203-208.
10. Gardner, H. K., Jr.; Buras, E. M., Jr.; Decossas, K. M.; Cooper, A. S., Jr.; Keating, E. J.; McMillan, O. J., Jr. *Text. Ind.* 1958, *122*(9), 100-104.
11. Dean, J. D.; Strickland, W. B.; Berard, W. N. *Am. Dyest. Rep.* 1946, *35*(14), 346-348.
12. Reeves, W. A.; Berard, W. N.; Leonard, E. K.; Brysson, R. J. *Canvas Prod. Rev.* **1959**, *35*(4), 32-33.
13. Donaldson, D. J.; Guice, W. A.; Drake, G. L., Jr. *Text. Ind.* **1970**, *134*(3), 171, 173, 180, 185.

14. Connor, C. J.; Cooper, A. S., Jr.; *Canvas Prod. Rev.* **1962**, *39*(7), 18, 26.
15. Connor, C. J.; Cooper, A. S., Jr.; Reeves, W. A.; Trask, B. A. *Text. Res. J.* **1964**, *34*(4), 347-357.
16. Connor, C. J.; *Proc., Cotton Util. Conf.* **1971**, 72.
17. Brysson, R. J.; Berard, W. N.; Bailey, J. V.; Dupre, A. M. *Canvas Prod. Rev.* **1957**, *32*(12), 36-37, 40-42.
18. Berard, W. N.; Leonard, E. K.; Reeves, W. A. In *Develop. Ind. Microbiol., Vol. 2* Plenum Press, Inc., New York, 1961, 79-91.
19. Vigo, T. L.; Danna, G. F.; Welch, C. M. *Text. Chem. Color.* **1977**, *9*(4), 77-79.
20. Danna, G. F.; Vigo, T. L.; Welch, C. M. *Text.Res. J.* **1978**, *48*(3), 173-177.
21. Goldthwait, C. F.; Smith, H. O. *Text. World* **1945**, *97*(7), 105, 107, 196, 198.
22. Goldthwait, C. F.; Sloan, W. G. *Text. Res. J.* **1961**, *31*(5), 434-446.
23. Schuyten, J. A.; Reid, J. D.; Weaver, J. W.; Frick, J. G., Jr. *Text. Res. J.* **1948**, *18*(7), 396-415.
24. Schuyten, J. A.; Weaver, J. G.; Frick, J. G., Jr.; Reid, J. D. *Text. Res. J.* **1952**, *22*(6), 424-432.
25. Schuyten, J. A.; Weaver, J. W.; Reid, J. D. *Am. Dyest. Rep.* **1949**, *38*(9), 364-368.
26. Weaver, J. W.; Schuyten, J. G.; Frick, J. G.; Reid, J. D. *J. Org. Chem.* **1951**, *16*(7), 1111-1116.
27. Phillips, F. J.; Segal, L.; Loeb, L. *Text. Res. J.* **1957**, *27*(5), 369-378.
28. Moreau, J. P.; Ellzey, S. E. *Am. Dyest. Rep.* **1967**, *56*(4), 38-42.
29. Connick, W. J.; Ellzey, S. E. *Am. Dyest. Rep.* **1968**, *57*(3), 71-73.
30. Moreau, J. P.; Drake, G. L. *Am. Dyest. Rep.* **1969**, *58*(4), 21-26.
31. Berni, R. J.; Benerito, R. R.; and Phillips, F. J. *Text. Res. J.* **1960**, *30*(8), 576-586.
32. Bullock, J. B.; Welch, C. M. *Text. Res. J.* **1965**, *35*(5), 459-471.
33. Schuyten, H. A.; Weaver, J. W.; Reid, J. D. In *Advances in Chemistry Series, No. 9* 1954, 7-20.
34. Reeves, W. A.; Guthrie, J. D. *Text. World* **1954**, *104*(2), 101, 176, 178, 180, 182.
35. Reeves, W. A.; Guthrie, J. S. *Ind. & Eng. Chem.* **1976**, *48*(1), 64-67.
36. Reeves, W. A.; Guthrie, J. D. U. S. Patent 2 722 188, 1956.
37. Drake, G. L.; Reeves, G. L.; Perkins, R. M. *Am. Dyest. Rep.* **1963**, *52*(16), 41-44.
38. Vail, S. L; Daigle, D. J.; Frank. A. W. *Text. Res. J.* **1982**, *52*(11), 671-677.
39. Frank. A. W.; Daigle, D. J.; Vail, S. L. *Text. Res. J.* **1982**, *52*(11), 678-693.
40. Frank. A. W.; Daigle, D. J.; Vail, S. L. *Text. Res. J.* **1982**, *52*(12), 738-750.
41. Daigle, D. J.; Frank, A. W. *Text. Res. J.* **1982**, *52*(12), 752--755.
42. Daigle, D. J.; Reeves, W. A.; Beninate, J. V. *J. Fire Flammability* **1970**, *1(7)*, 178-182.
43. Knoepfler, N. B. *Proc., Flame Ret. Cotton Batting, ARS 72-72* **1969**, 40-49.
44. Knoepfler, N. B. *Proc. 18th Cottonseed Process. Clinic, ARS 72-74*, **1969**, 74-85.
45. Knoepfler, N. B. *Proc. Europ. Conf. on Flammability and Fire Retard. 1980*, Bhatnagar, ed., **1980**, 115-125.
46. Madasci J. P.; Neumeyer, J. P. *J. Consum. Prod. Flammability* **1982**, *9*, 3-10.

47. Frick, J. G., Jr.; Andrews, B. A. Kottes; Reid, J. D. *Text. Res. J.* **1960**, *30*(7), 495-504.
48. Frick, J. G., Jr.; Kottes, B. A.; Reid, J. D. *Text. Res. J.* **1959**, *29*(4), 314-322.
49. Rao, J. M.; Roberts, E. J.; Rowland, S. P. *Poly. Letters* **1971**, *9*(9), 647-650.
50. Reeves, W. A.; Vail, S. L.; Frick, J. G., Jr. *Text. Res. J.* **1962**, *32*(4), 305-312.
51. Andrews, B. A. Kottes; Arceneaux, R. L.; Frick, J. G., Jr. *Text. Res. J.* **1962**, *32*(6), 489-496.
52. Vail, S. L. *Textile Res. J.* **1969**, *39*(8), 774-780.
53. Petersen, H. *Textilveredlung* **1967**, *2*, 744-757.
54. Petersen, H. *Textilveredlung* **1967**, *3*, 51-62.
55. Petersen, H. *Textilveredlung* **1968**, *38*, 156-176.
56. Petersen, H. *Textilveredlung* **1970**, *5*, 437-453.
57. McKelvey, J. B.; Berni, R. J.; Benerito, R. R. *Text. Res. J.* **1964**, *34*(12), 1102-1104.
58. McKelvey, J. B.; Berni, R. J.; Benerito, R. R. *Text. Res. J.* **1965**, *35*(4), 365-376.
59. Bullock, J. B.; Welch, C. M. *Text. Res. J.* **1965**, *35*(5), 459-471.
60. Frick, J. G., Jr.; Arceneaux, R. L.; Reid, J. D. U. S. Patent 3 144 299, 1964.
61. Frick, J. G., Jr.; Arceneaux, R. L.; Reid, J. D. U. S. Patent 3 219 632, 1965.
62. Andrews, B. A. Kottes; Reinhardt, R. M. *Book Pap. 1985 AATCC Intern. Conf. & Exhibit.* 1985, 174-179.
63. Andrews, B. K.; Reinhardt, R. M. U. S. Patent 4 888 878, 1984.
64. Andrews, B. A. Kottes; Harper, R. J.; Reed, J. W.; Smith, R. D. *Text. Chem. Color.* **1980**, *12*(11), 287-291.
65. Arceneaux, R. L.; Fujimoto, R. A.; Reid, J. D.; Reinhardt, R. M. *Am. Dyest. Rep.* **1962**, *51*(15), 45-52.
66. Guthrie, J. D. U. S. Patent 3 154 373, 1964.
67. Reeves, W. A.; Perkins, R.M.; Chance, L. H. *Text. Res. J.* **1960**, *30*(3), 179-192.
68. Reinhardt, R. M.; Cashen, N. A.; Reid, J. D. *Text. Chem. Color.* **1969**, *1*(20), 415-422.
69. Berni, R. J.; Gonzales, E. J.; Benerito, R. R. *Text. Res. J.* **1970**, *40*(4), 377-385.
70. Pierce, A. G., Jr.; Reinhardt, R. M.; Kullman, R. M. H. *Text. Res. J.* **1976**, *46*(6), 420-428.
71. Pierce, A. G., Jr.; Frick, J. G., Jr.; U. S. Patent 3 565 824, 1971.
72. Reinhardt, R. M.; Mazzeno, L. W.; Reid, J. D. U. S. Patent 2 917 412, 1959.
73. Blanchard, E. J.; Harper, R. J., Gautreaux, G. A.; Reid, J. D. *Text. Ind.* **1976**, *131*(1), 116-119, 122, 143.
74. Harper, R. J., Jr.; Blanchard, E. J.; Reid, J. D. *Text. Ind.* **1967**, *131*(5), 172, 174, 177, 181, 184, 186, 233, 234.
75. Blanchard, E. J.; Harper, R. J., Jr.; Bruno, J. S.; Reid, J. D. *Text. Ind.* **1967**, *131*(10), 98, 100, 102, 104, 114, 116, 194.
76. Blanchard, E. J.; Kullman, R. M. H.; Reid, J. D. *Text. Ind.* **1966**, *130*(5), 169-173.
77. Vail, S. L.; Young, H. P; Verburg, G. B.; Reid, J. D.; Reeves, W. A. *Text. Ind.* **1967**, *131*(12), 184, 186, 189, 191-191, 194-195.

78. Bertoniere, N. R.; Martin, L. F.; Blouin, F. A.; Rowland, S. P *Text. Res. J.* **1972**, *42*(12), 734-740.
79. Arthur, J. C., Jr. *J. Macromolec. Science-Chem.* **1976**, *A10*(4), 653-670.
80. Harris, J. A.; Reinhardt, R. M. *Text. Res. J.* **1981**, *51*(2), 73-80.
81. Benerito, R. R.; Webre, B. G.; McKelvey *Text. Res. J.* **1961**, *31*(9), 757-769.
82. Welch, C. M.; Bullock, J. B.; Margavio, M. F. *Text. Res. J.* **1967**, *37*(4), 324-333.
83. Vail, S. L.; Murphy, P. J. U. S. Patent 3 112 156, 1963.
84. Rowland, S. P.; Welch, C. M.; Brannan, M. A. F. U. S. Patent 3 526 048, 1970.
85. Welch, C. M.; Andrews, B. K. U. S. Patent 4 820 307, 1989.
86. Welch, C. M.; Andrews, B. K. U. S. Patent 4 936 865, 1990.
87. Welch, C. M.; Andrews, B. K. U. S. Patent 4 975 209, 1990.
88. Andrews, B. K.; Morris, N. M.; Donaldson, D. J.; Welch, C. M. U. S. Patent 5 221 285, 1993.
89. Andrews, B. A. Kottes, *Text. Chem. Color.* **1990**, *22*(9), 63-67.
90. Andrews, B. A. Kottes, "A Comparison of Formaldehyde-Free DP Finishes from Etherification and Esterification Crosslinking," *Am. Dyest. Rep.*, in press.
91. Steinwand, P. J., U. S. Patent Application 621 649, 1990.
92. Caulfield, D. F. *Tappi J.* **1994**, *77*(3), 205-212.
93. Herron, C. M.; Cooper, D. J. U. S. Patent 5 137 537, 1992.
94. Herron, C. M.; Hanser, T. R.; Cooper, D. J.; Hersko, B. S. European Patent Application 0 427 317 A2, 1990.
95. Calamari, T. A.; Schreiber, S. P.; Cooper, A. S., Jr.; Reeves, W. A. *Text. Manuf.* **1971**, *97*(1159), 295.
96. Tripp, V. W.; Moore, A. T.; Porter, B. R.; Rollins, M. L. *Text. Res. J.* **1958**, *28*(6), 447-452.
97. Mazzeno, L. W.; Kullman, R. M. H.; Reinhardt, R. M.; Moore, H. B.; Reid, J. D.; *Am. Dyest. Rep.* **1958**, *47*(9), 299-302.
98. Beninate, J. V.; Kelly, E. L.; Drake, G. L., Jr.; Reeves *Am. Dyest. Rep.* **1966**, *55*(2), 25-29.
99. Reeves, W. A., Beninate, J. V.; Perkins, R. M.; Drake, G. L., Jr *Am. Dyest. Rep.* **1968**, *57*(26), 35-38.
100. Moreau, J. P.; Drake, G. L., Jr. *Am. Dyest. Rep.* **1969**, *58*(4), 21-26.
101. Connick, W. J., Jr.; Ellzey, S. E., Jr. *Text. Res. J.* **1970**, *40*(2), 185-190.
102. Warren, J.; Reid, J. D.; Hamalainen, C. *Text. Res. J.* **1952**, *22*(9), 584-590.
103. Guthrie, J. D., *Am. Dyest. Rep.* **1952**, *41*(1), 13-14, 30.
104. Harper, R. J., Jr.; Blanchard, E. J.; Lofton, J. T.; Gautreaux, G. A. U. S. Patent 3 800 375, 1974.
105. Harper, R. J., Jr.; Blanchard, E. J.; Lofton, J. T.; Bruno, J. S.; Gautreaux, G. A. *Text. Chem. Color* **1974**, *6*(9), 201-205.
106. Harper, R. J., Jr.; Blanchard, E. J.; Lofton, J. T.; Gautreaux, G. A. U. S. Patent 3 960 477, 1976.
107. Blanchard, E. J.; Bruno, J. S.; Gautreaux, G. A. *Am. Dyest. Rep.* **1976**, *65*(7), 26, 32, 34, 65.
108. Blanchard, E. J.; Harper, R. J., Jr. *Book Pap., 1983 AATCC Nat. Tech. Conf.* **1983**, 299-305.

109. Harper, R. J., Jr.; Gautreaux, G. A.; Bruno, J. S.; Donoghue, M. J., U. S. Patent 3 788 804, 1974.
110. Harper, R. J., Jr.; Gautreaux, G. A.; Blanchard, E. J. U. S. Patent 3 795 480, 1974.
111. Harper, R. J., Jr.; Gautreaux, G. A.; Blanchard, E. J. U. S. Patent 3 807 946, 1974.
112. Harper, R. J., Jr.; Gautreaux, G. A.; Bruno, J. S. U. S. Patent 3 975 154, 1974.
113. Harper, R. J., Jr.; Blanchard, E. J.; Lofton, J. T.; Gautreaux, G. A. U. S. Patent 3 853 459, 1974.
114. Reinhardt, R. M.; Blanchard, E. J. *Am. Dyest. Rep.* **1988**, *77*(1), 29-30, 32-34, 53.
115. Harper, R. J., Jr. U. S. Patent 4 629 470, 1986.
116. Stone, R. L.; Harper, R. J., Jr. *Book Pap., 1986 AATCC Intern. Conf. & Exhib.* 1986, 214-224.
117. Harper, R. J., Jr., *et al. Book Pap., 1987 AATCC Intern. Conf. & Exhib.* 1987, 31-49.
118. Blanchard, E. J.; Reinhardt, R. M. *I&E Chem. Res.* **1989**, *28*(4), 490-492.
119. Reinhardt, R. M.; Blanchard, E. J. *Am. Dyest. Rep.* **1990**, *79*(6), 15-16, 18, 20, 22, 52.
120. Reinhardt, R. M.; Blanchard, E. J. *Pol. Preprints* **1990**, *31*(2), 586-587.
121. Blanchard, E. J.; Reinhardt, R. M.; Andrews, B. A. Kottes *Text. Chem. Color.* **1991**, *23*(5), 25-28.
122. Blanchard, E. J.; Reinhardt, R. M.; Graves, E. E.; Andrews, B. A. Kottes *I&E Chem. Res.* **1994**, *33*(4), 1030-1034.
123. Reid, J. D.; Daul, G. C. *Text. Res. J.* **1947**, *17*(10), 554-561.
124. Reid, J. D.; Daul, G. C. *Text. Res. J.* **1948**, *18*(9), 551-556.
125. Daul, G. C.; Reinhardt, R. M.; Reid, J. D. *Text. Res. J.* **1952**, *22*(12), 787-792.
126. Daul, G. C.; Reinhardt, R. M.; Reid, J. D. *Text. Res. J.* **1952**, *22*(12), 792-797.
127. Reinhardt, R. M.; Reid, J. D.; Fenner, T. W. *J. I&E Chem.* **1958**, *50*(1), 83-86.
128. Vigo, T. L.; Bruno, J. S.; *Text. Res. J.* **1987**, *57*(7), 427-429.
129. Vigo, T. L.; Bruno, J. S.; *Book Pap. Intern. Nonwoven Conf.* 1988, 427-435.

Chapter 4

Solid Substrate Fermentation in the Production of Enzyme Complex for Improved Jute Processing

B. J. B. Wood[1], B. L. Ghosh[2], and S. N. Sinha[2]

[1]Department of Bioscience and Biotechnology, University of Strathclyde, Royal College Building, 204 George Street, Glasgow G1 1XW, Scotland
[2]Indian Jute Industries Research Association, P.O. Box 16745, Calcutta 700 027, India

Plant residues adhering to jute fibers after "retting" (the removal of plant tissue from around the fibers by immersion in water, which causes and promotes microbial growth on the more easily biodegradable material, but to which the tough fibers are resistant) cause delays in spinning and weaving, reduce the quality rating of the finished product, and may damage machines used in the processes. For weaving, the spun thread is sized with Tamarind Kernel Powder (TKP), which is difficult to disperse in water. Enzymes extracted from Aspergillus grown on bran by SSF will digest the plant gums which bind plant debris to the jute fibers, so easing their removal. Similarly, digesting the polysaccharides in TKP gives faster dispersion at lower temperature. Production and use of the enzymes is described, other possible applications are discussed, and areas requiring further research are identified.

Jute is a fiber whose importance in the developed world has declined as a result of the plastics revolution during the past 40 years. In some developing countries however it retains much of its traditional importance. This report describes using enzymes to assist in producing jute thread and in weaving the thread into cloth under a program funded by the United Nations Industrial Development Organization.

Jute Jute (*Corchoros* spp.) is a family of fast-growing plants which is believed to have originated in the Middle-East. It thrives in warm, humid regions with good soil and abundant rainfall, and is therefore particularly well suited to cultivation in West Bengal, Bangladesh (the former East Bengal) and parts of Pakistan and China, where it is an important subsistence crop. Although its cultivation has a long history, jute became a crop of world economic importance during the time

0097–6156/96/0647–0046$15.00/0

when the Indian sub-continent was ruled by the British. Previously its principal uses had been in such applications as ropes, but its potential for spinning into a strong thread with many uses resulted in the development of a very substantial industry during the 19th and early 20th centuries. However, the development of cheap plastics after World War II caused jute's eclipse as a major fiber, but there are now some signs that it may be showing recovery as new ideas about using renewable materials gain increased importance.

Applications. Jute's use in rope-making has already been mentioned. String and twine for agricultural and industrial applications are another traditional application, but to the West its most important application was in the form of woven materials such as sacking, carpet backing and the base for linoleum manufacture. It has always maintained importance in the quality carpet industry, but the strength, uniformity and freedom from rotting which synthetic materials enjoy have greatly reduced jute's other markets in thedeveloped world, although its importance for sacking and linoleum continues in developing countries. Of recent years there has been a revival in interest in its wider uses in the West, and the reasons for this development have some lessons to offer us. Perhaps most surprising is the resurgence in the linoleum market. This was once a thriving industry to which certain towns, such as Kircaldy in Scotland, owed their prosperity. By the 1970s this industry was moribund, but about four years ago Kircaldy's only remaining linoleum maker produced its first complete list of linoleum products for about two decades. Modern linoleum is a thick, hard-wearing material able to compete on equal terms with the synthetic floor coverings which had previously displaced it, with the added attraction that the basic materials required for its manufacture are all renewable (jute, cork. linseed oil).

Geotextiles. Here the revival is at least in part due to the very property of fairly easy biodegradation which is a disadvantage in other applications. Geotextiles are materials used in the construction branches of the building and civil engineering industries. One application is in temporary strengthening and stabilization of surfaces over which heavy equipment will have to move. Since jute geotextiles will rot away after a while, they are excellently suited to temporary applications as there is no need to remove them at the end of the job, thus saving a cost associated with plastic geotextiles. In hot climates, such as the Arabian Gulf, where soil temperatures may easily exceed 40°C, plastic geotextiles may stretch under heavy loading, whereas jute does not. Plastic nets applied to stabilizing banks, the sides of cuttings and other earthworks while a plant cover is established, can result in the soil panning and forming an impervious layer which disrupts drainage and deprives plant roots of deep water. This is due to iron salts and other materials lodging on the plastic. In contrast to plastics jute will rot down, adding to the humus in the soil and attracting earthworms whose activities will tend to disrupt any pan which may be forming.

Newer Developments . Jute sacking is now being combined with plastic linings to give strong, reasonably priced containers. Innovative developments include jute-reinforced plastics which can replace plywood in applications such as tea-chests, and which are also being used for other purposes in which their easy shaping has advantages, an example being separators for shipping high quality fruit (*1*). Over the years there have been attempts to make carpets entirely with jute, instead of using it just as the backing material onto which other textiles are woven. Recently some jute mills have installed Belgian machines which are producing attractive industrial carpeting entirely from jute, and these are winning export orders to markets such as Singapore. Thus it may be concluded that after decades of decline, jute is taking on a new lease of life, and although it may never regain the pre-eminence which it once enjoyed, it has a much brighter future than would have seemed possible a few years ago. Ranganathan (*2*) provides a valuable survey of innovations in its use, and the examples quoted here are drawn from this reference.

Problems with Jute. These are very much bound up with the traditional processes for separating the fiber from the plant, particularly the "retting" stage in which plants undergo a controlled rotting while immersed in water; this topic will be covered in the next section. Jute's greatest disadvantage is its color; the characteristic brown places severe limitations on the extent to which it can be dyed. Exposure to sunlight results in the fiber undergoing further darkening. There are also technical problems associated with spinning and weaving, but these must await an outline of the processes involved in obtaining, spinning and weaving the textile. IJIRA have developed methods of separating the fiber from other plant materials which could overcome many of these problems, but jute production is mainly an activity for subsistence farmers who are not in a position to make the necessary investment in new equipment. Traditional processing is a very polluting activity which robs water supplies of oxygen and makes the water unfit for other uses. On these grounds alone traditional "retting" should be done away with, but the economic situation makes it very unlikely that this essential change will happen, at least in the near future.

Jute Production and Processing

The jute industry has declined greatly from the time when the city of Dundee (Scotland) was driven to its present size and importance on the industry generated by the processing and weaving into cloth of jute imported from India. Consequently there are no recent books dealing with jute processing. The following outline of the processes employed in traditional jute processing is based on Atkinson's 1964 text (*3*). See also Figure 1.

The jute plant is raised from seed which the farmer has saved from a previous crop. Because the farmers are so poor this is the best way to proceed from their point of view, but it does mean that there is no incentive to develop better strains of the plant, although it is difficult to escape from the view that modern breeding techniques could have a drastic effect on the properties of a plant group which has been farmed in an unchanged way for (probably) many centuries.

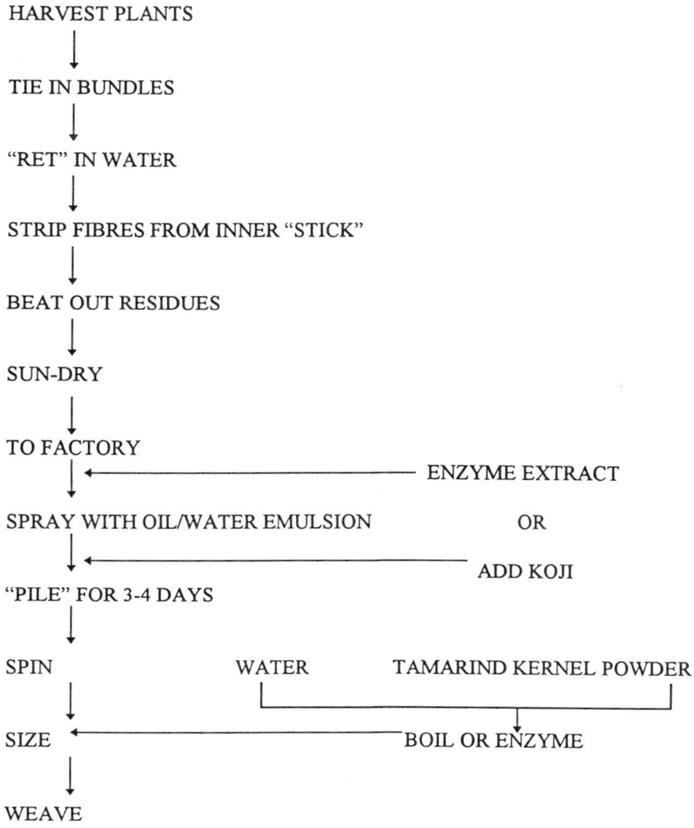

Figure 1. Jute Processing

The plant grows with remarkable speed, and is ready for harvest after a few months, when it is up to five meters tall and is about to flower. A few plants are left to set seed, but most of them are cut down near to ground level. The plant has an interesting structure. There is a central stem (called the "stick" in the jute industry) which has a rather open texture, rather like that of balsa wood, and which would certainly not be strong enough to support the plant unaided. The stick is surrounded with a criss-crossing mesh of fibers, which will be the eventual jute crop. This construction is very strong and yet pliable, enabling the plant to stand upright and withstand strong winds and tropical rain downpours without serious damage. The fibers are held together by other plant polysaccharides and are surrounded by the conductive tissues, cambium, etc.

Retting. After harvest the plants are tied into bundles, weighted with stones and immersed in water. Running water is preferred because this replenishes the oxygen which is rapidly depleted by the activities of the many bacteria which develop on the plant material, and whose growth and metabolic activities are intense in the warm water. However this is a serious pollution hazard for the streams in which the plant bundles are immersed, and has been outlawed in many Western countries where similar processes are used in the production of temperate climate fibres such as linen. This process is called "retting" and its purpose is to degrade the plant materials which bind the fibers to the stick and to each other. The jute fiber is lignocellulosic and the cellulose is of a highly crystalline type; consequently the fiber is more resistant to microbial attack than are the components which surround it. After an appropriate time the plant bundles are recovered from the water and the fibers are separated from the stick and other plant material by hand, a most laborious process of beating the bundles and drawing the fibers out from the residues. It is believed that the fibers accumulate iron and other mineral salts from the water during retting and that these contribute to the fibers' color and also catalyse the oxidative darkening which the fibers undergo on exposure to sunlight.

Preparation for Spinning. The fibers are spread out to dry in the sun and air, and are then bundled for shipment to the mill. At the mill the bundles are separated, graded, then fed through machines which hammer them to promote the separation of the fibers from each other, and also to loosen plant debris still attached to the fibers; this debris is referred to as "bark" in the jute industry, and constitutes something of a problem during subsequent processing stages. At the lower end of the plant the fibers tend to form bundles held together by polysaccharides; these bundles, termed "root ends", are somewhat resistant to retting and to dispersion by mechanical hammering, and are also a nuisance during subsequent processing. During this mechanical pre-treatment the jute is also sprayed with a mixture of water and mineral oil to moisten and lubricate the fibers. The jute is then formed into piles and left for three to four days. No study seems to have been undertaken of events during this "piling" stage of processing, but there must be some microbial activity taking place in the warm, humid compost heaps of the piles.

Spinning. The material from the piles is then fed into machines which align the fibers and draw them out into ever thinner bundles until they are finally spun into thread. Root ends and bark particles interfere with both of these processes, resulting in stoppages, equipment jamming, thread breaking, etc. The thread may be sold as it is, but most of it is woven into cloth. Root ends and bark will disrupt this process, and finished product with excessive bark particles will be down-graded, with reduction in the price which it commands. This reduction in quality is especially important in cloth destined for the carpet industry, which demands the highest quality jute textile.

Tamarind Kernel Powder and Weaving. As in other weaving processes, the thread needs to be "sized" by coating with a suitable material before weaving. Sizing is with an aqueous solution of a plant polysaccharide such as starch, which gives the thread bulk to help its handling by the weaving machines, holds moisture thus reducing the build-up of static electrical charges, and makes the thread smoother thereby helping its passage through the looms. Starch is the traditional size in most weaving, but during the Second World War Tamarind Kernel Powder (TKP) replaced it in the Indian jute industry, and has remained in use ever since. This material is obtained from tamarind seeds, themselves a by-product from production of tamarind pulp for the food industry. The seeds are parched to make the hulls brittle, then after dehulling the kernels are milled to produce the kernel powder, which is quite rich in protein and contains a most unusual polysaccharide which has a cellulose "backbone" with numerous carbohydrate side-groups attached to it. This structure makes it fairly easily dispersable in water to give a viscous solution excellent for sizing. TKP is also used as a food thickener, but was not regarded as possessing any nutritional value other than the protein associated with it. However it may be appropriate to re-evaluate the polysaccharide in the light of current interest in the so-called soluble fiber as a valuable component of the human diet. Its chemistry has been the subject of several reports (*4-10*).

Although the TKP polysaccharide is fairly easily dispersable in water as compared with cellulose, it does take rather a long time to obtain a suitable dispersion by the conventional method, which involves passing steam into a mixture of TKP and water through a steam lance. As will be shown subsequently, enzyme extracts originally devised for processing the jute fibers have also found application in shortening and simplifying the production of aqueous TKP dispersion.

Application of Enzymes

Groups at IJIRA (*11*) and in Bangladesh (*12*) have developed methods for producing and applying enzymes in jute processing. As will be seen, these methods claim to ease processing, upgrade the fiber quality and reduce the number of stoppages during spinning and weaving. Somewhat similar claims have been made for purely chemical additions during processing, and we shall attempt to reconcile these two sets of claims.

The rationale underlying the application of enzymes in processing the jute fibers is that of aiding dispersion of the plant gums which bind bark to fibers and bind fibers together into root ends. Additionally, mills using the process report that fiber which has been so treated is "brighter" than otherwise comparable jute which has not received the enzyme treatment. This latter claim is difficult to quantify in the absence of objective standards for jute's appearance, or indeed any clear definition as to the meaning of "brightness" in this context, although this term is widely used in the industry, and would be readily understood by anyone working in it. Nevertheless, this is an opinion expressed independently by a number of craftsmen with lifetime experience of jute, and on occasion, even the author with no experience of jute (BJBW) has been able to detect a difference in appearance between conventional and enzyme-treated fibers. The term "biosoftening" is not entirely satisfactory as a description of the full range of functions claimed for the enzyme complex, but it has become established within IJIRA, who also promote its use in the jute industry, and so will be employed here.

Biosoftening. In outline, the enzyme biosoftening process involves producing an enzyme complex by growing a fungus (*Aspergillus terreus* , as a strain isolated at IJIRA, is currently employed) on wheat bran as a solid substrate fermentation (conveniently described by the Japanese term "koji"). The mold-covered bran can be applied directly, or enzymes can be extracted with water and this extract, as obtained or after concentration, used. In either case the enzyme-containing material is mixed with the jute immediately prior to the "piling" process. Aqueous extracts can be applied with the water/oil dispersion which is sprayed onto the jute during the preparatory mechanical treatment, or sprayed on during the preparation of the piles. If dried koji is being used it is sprinkled onto the piles as each layer of jute is added.

The process favored by the chemists at IJIRA is to spray a dilute solution of urea onto the jute during piling. Their rationale for this, that the urea causes dispersion of the polysaccharide gums, is unacceptable at the very low urea concentrations which they employ. A more reasonable explanation is that the urea, by supplying readily assimilated nitrogen, will enhance microbial growth in the piled fiber, which will probably contain very little of this essential nutrient after retting, which would be expected to remove much or all of the (probably limited) nitrogen in the freshly harvested jute plants. Unfortunately no microbiological studies of the piling process have been reported. It can reasonably be argued that the koji and extracts from it would also supply nutrients to any microbes active in the piles, and so a similar logic may be applied in assessing the true biochemical basis for the enzyme biosoftening process.

On the other hand the enzyme explanation for the actions exerted by koji extract in aiding dispersion of TKP would seem to be the only reasonable one, given the short time period used and the probability that the initial microbial load of the TKP will be low.

Enzyme Production

The strain of *Aspergillus terreus* which was isolated by IJIRA staff is maintained as a pure culture. From this stock culture subcultures are prepared as inocula for scale-up. Cultures on sterile wheat bran are incubated under controlled conditions in the laboratory until growth is complete and extensive sporulation has occurred. This koji is then dried and stored until required as an inoculum for enzyme production (Figure 2).

"In-house" Production at Mills . Some jute mills prefer to produce their own koji. This may meet the whole or part of their requirement for enzyme. They may maintain their own inoculum line or use dried koji from IJIRA. Despite the fact that mills could not possibly afford to employ a microbiologist each, the standard achieved by the people producing koji at mills is remarkably high. Typically such a producer is one of the mill's engineers who has offered to take on the task of koji production and has received only the most basic training in sterile technique. They seem to enjoy the unusual challenge, and bring some unexpected insights to the task. Koji vessels used in the mills are two-part metal canisters comprising a base and a rather closely fitting lid. One producer had noticed that loose-fitting lids gave better growth of the mold, and when this was explained to him in terms of oxygen transfer rates, he quickly devised a plan to modify all his koji vessels. The conditions under which bran preparation and sterilization, inoculation, incubation and harvest are carried out are very primitive if judged by the standards appropriate to the microbiological industry as a whole, but are perfectly adequate as a response to the particular requirements which are being addressed. The high ambient temperature and humidity normally found in Jute producing areas mean that there is no real need for special provision by way of incubators; there is normally steam available from the mill for sterilization, and while some of the pressure vessels would not meet safety standards applied in industrialised countries, no significant accidents have been recorded at mills producing their own koji. When fungal growth is complete the koji is air-dried and stored until required. The mill workers will either extract enzymes with water and use this in piling or use the dried koji directly, spraying extra water onto the piles as they are formed.

Production at the IJIRA facility. United Nations Industrial Development Organization funding has enabled IJIRA to construct a specially designed koji factory at a jute mill on the outskirts of Calcutta. This factory comprises facilities for storing bran and dried koji, an autoclave, a ribbon mixer, incubation rooms, a cross-flow drier, small extraction towers and a Millipore filter press to concentrate the enzyme extract. Laboratory facilities aid in standardising processes and products (*13-15*).

The operation is on a larger scale than at any of the mills and so can operate under the supervision of a qualified microbiologist assisted by skilled and semi-skilled workers. It supplies mills with dry koji, aqueous extract or concentrated aqueous extract as specified by the customer.

One advantage of preparing extract "in house" is that the extracted bran can be re-used as substrate for further mold growth. There seems to be an optimum mixture of new and re-used bran for enzyme production. The flour milling plant from which the bran comes is rather old and badly worn, and the bran contains more associated endosperm than is desirable. Therefore it seems reasonable to suggest that used bran has had the more easily utilised starch, etc, removed, so that the mold which grows on it the second time around is forced to expend more effort on producing enzymes to hydrolyzse the less easily digested polysaccharides present in the bran, and these are the types of polysaccharideswhich need to be dispersed for improved jute processing. On the other hand if recycled bran is used without supplementation with some fresh bran then fungal growth is poor, although no work has been done to determine if this results entirely from nutrient depletion or corresponds in some measure to inhibitor production by the growing or sporulating mold which colonised the bran first time around.

Application of Enzyme Preparation

As the program develops it is intended that the enzyme extracts will be standardized with respect to the concentrations of the enzymes principally involved in biosoftening, but at present application rates for the koji or extract tend to be rather *ad hoc* . A particular advantage of enzyme biosoftening treatment is that it does not involve any significant alteration or disruption to the operating routines followed in the mill where raw jute is being processed. In TKP dispersion preparation there is a considerable saving in time, energy and labour if enzyme extract is used. TKP typically has a composition comprising protein 16.7%, carbohydrate 63.4%, fiber 5.4%, sugars 3.1%, oil 6.9%, tannins, etc. 1.7%, ash 2.7%, dry weight basis, with the powder as supplied typically containing about 10% moisture. As mentioned earlier the principal component is a polysaccharide, or series of related polysaccharides, basically a cellulose-like backbone of beta 1-4 linked glucose molecules with pentose and other sugars forming short side-chains. This polysaccharide is thought to be the principal sizing agent, but it is believed that the protein present in the extract can act to bind jute fibers together. The experience of mills is that, when TKP dispersions produced with koji extract are in use, there is a considerable reduction in the frequency with which spinning and weaving are halted due to thread breakage, giving an overall production increase around 4%. This increase in production efficiency is attributed to the protein component of the TKP, and it is argued that the much milder conditions used to enzymically disperse the TKP in water, plus perhaps some modifications in the protein(s) structure resulting from the activity of proteases present in the koji, condition the protein to act as a glue for the fibers, although it is not clear that these suggestions have a firm scientific basis, and there is a need for a good scientific study of the problem. IJIRA recommends that the mixture of TKP, water and enzyme extract be incubated for 15 to 20 minutes at 40 to 50°C, then the mixture is heated to 70°C to denature the enzyme and also to complete dispersion of the TKP in water.

While the enzyme composition of the extract from *A. terraeus* has not been exhaustively examined, the principal activities believed to be present are listed in Table I. Sinha *et al*. (*16*) report evidence that the mold produces a substance which inhibits cellulase attack on intact cellulose, but has no effect on hydrolysis of partly degraded cellulose. They demonstrated that the dialysed extracts from the koji will degrade intact cellulose, and they also showed that the intact extract will selectively inhibit the degradation of intact cellulose (the so-called cotton cellulase activity) by other enzyme systems. The active component seems to be a simple, thermostable molecule, and it is unfortunate that no effort has been made to establish its identity unequivocally, as this property (which their work indicates is not confined to action on A. terraeus cellulase) could have practical use in a number of applications and could also yield interesting information on the binding of enzyme and substrate. There is also scope for further work on identifying the components of the cellulase complex which are actually present in the *A. terraeus* extracts.

Future Developments

While IJIRA's wish to continue to work with the *A. terraeus* which has served them so well to date in this project is understandable, it must be acknowledged that it is very improbable that extract from a single organism represents the best possible balance of enzymes for either of the tasks for which the enzyme preparation is employed. A detailed study of the contributions made by each of the enzymes present in the extract, buttressed by a more complete understanding of the polysaccharides principally responsible for the problems caused by bark and root ends, is clearly desirable and should provide a framework for studies leading to a standardised mixture of enzymes giving enhanced performance in biosoftening. A second blend should optimise TKP dispersion. The need to identify the agent protecting undegraded cellulose against cellulase has been noted above, but bears repetition here because of its importance.

If jute is to realize its full economic potential in modern markets, the problems associated with its color and the darkening which occurs on exposure to sunlight need to be resolved. Is tannase present in the enzyme complex? This could account for some of the beneficial effects on jute color from the biosoftening treatment. In the longer term it is clearly very desirable to get away from traditional retting with its serious polluting effects on the water in which it is done, and this measure is also expected to reduce the color and darkening problems, as noted above. It may also be that fibers which have not undergone retting may be stronger than their retted counterparts. Enzymatic cleaning will surely be essential

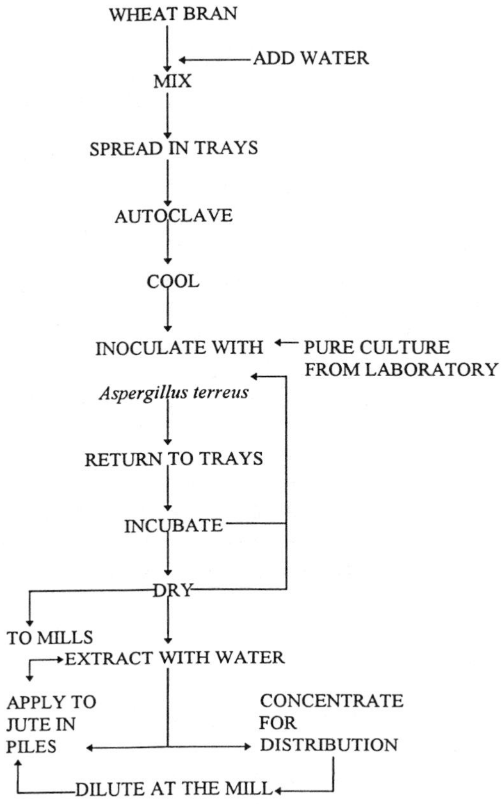

Figure 2. Enzyme production for the jute industry

Table I. Enzymes present in *Aspergillus terreus* Koji

FOR JUTE PROCESSING

PECTINASES

HEMI-CELLULASES

β-GLUCANASES

CELLULASE COMPLEX

INHIBITOR OF "COTTON CELLULASE" ACTIVITY

FOR TKP PROCESSING

β-GLUCANASES

PROTEINASES

in processing fiber which has ben mechanically stripped from freshly harvested jute plants, thus increasing the importance of the IJIRA enzyme production program.

In the longer term genetic engineering of the molds used for enzyme production may prove to be the best way to obtain an optimised and highly active enzyme blend, but in the shorter term the easiest way forward will probably be to grow mold strains selected from cultivars already available in such places as the Central Food Technological Research Institute in Mysore, and then blend extracts from them into balanced enzyme mixtures appropriate for each application.

Microbiology of Piling. The events which occur during the "piling" stage of jute processing remain unknown, but must surely involve considerable microbial activity. Research leading to accurate knowledge of the microbes concerned and any succession which they exhibit, could suggest ways of beneficially controlling the process and promoting or enhancing the growth of useful species, possibly even leading to deliberate inoculation with selected organisms. Is it possible that a combination of enzyme extract treatment in association with measures to enhance the growth of selected microorganisms could be more beneficial than either treatment alone? The suggestion that the mold extract may have effects other than those attributable to the enzymes which it contains receives some support from a rather unexpected place, studies on animal nutrition. Newbold *et al* . (*17*) report that adding *Aspergillus oryzae* fermentation extract to the diet of a ruminant animal, specifically the sheep, increased the numbers of cellulolytic (by 94%) and total (by 30%) bacteria in the animals' rumens, while leaving the numbers of fungi and protozoa unchanged. The activity was not destroyed by autoclaving the fungal extract prior to feeding it to the sheep, so the results cannot be accounted for by enzyme activities in the extract, even though *A. oryzae* is valued in food fermentation for its very active production of extracellular enzymes. The work followed from an observation that including the extract in the sheeps' diet produced increases in milk yield and weight gain, demonstrating that the increase in micrbial counts was a real effect with influence on the rumen's performance, and not just a statistical phenomenon.

Advantages of SSF. In a developing country such as India the advantages of Solid Substrate Fermentation over Stirred Tank Reactor processes are considerable. SSF requires smaller investment in plant and can be conducted with a less skilled labour force. The product can be stabilized pending use or further processing by simply drying it. Product extraction gives a concentrated extract and a moist solid residue rather than the large volumes of highly polluting liquid which remain after extraction from a liquid fermentation. Even in developed areas such as Europe and North America there is a revival of interest in SSF; India and Japan, with their considerable experience in SSF, are in a strong position to capitalise on the knowledge thus gained. An example of this is the text edited by Pandey (*18*) reporting the proceedings of a very useful conference on applications of SSF held in Trivandrum in 1994. This offers an overview of the current resurgence of interest in SSF and the reasons why it is particularly attractive for developing economies. It then reviews methods for conducting and monitoring SSF; production of food, feed and fuel by SSF; enzyme production for the food, leather, textile and other industries; finally there is a mixture of minor products and consideration of matters such as safety aspects in SSF operation. In addition, the Indian leather industry's Central Leather Research Institute in Madras is developing SSF enzyme production for dehairing and bateing hides.

In conclusion, the authors believe that this program demonstrates that SSF has a part to play in the modern world, and that if the difficult lessons taught by this experience can be properly assimilated and applied, SSF has much to offer other agro-technological industries in the developing and even the developed countries of the world.

Acknowledgments

We are grateful to the United Nations Industrial Development Organisation for their support of the jute biosoftening project, and for their consenting to our publishing this report of the work. We thank IJIRA's Director for permission to publish this paper. We thank Mr Edmund P. Phillips, a student with BJBW, for assistance with surveying the literature on the composition of tamarind kernel powder. We thank Forbo-Nairn Ltd, Kircaldy, Fife, Scotland for information and illustrative material on the production of linoleum.

Literature Cited

1. IJIRA Technical Note (undated). *Poly-Jute Packaging for Practical Solutions.*
2. Ranganathan, S.R. *Statement on Status of Indian Jute Diversified Products.* IJIRA, Calcutta, India, **1990.**
3. Atkinson, R.R. *Jute; Fibre to Yarn.* Temple Press Books: London, England, **1964,** p p 212.
4. Ghose, T.P.; Krishna, S. *Current Science,* **1945**, 14, 299-300.
5. Khan, N.A.; Mukherjee, B.D. *Chem. & Ind.* **1959** , 1413-1414.
6. Kooiman, P. *Recueil des Travaux Chimiques des Pays-Bas* **1961**, 80, 849-865.
7. Savur, G.R. J. *Chem. Soc.* **1956**, 2600-2603
8. Savur, G.R. *Chem. & Ind.* **1956** , 212-214.
9. Srivastava, H.C.; Singh, P.P. *Carb. Res.* **1967**, 4, 326-346.
10. White, E.V.; Rao, P.S. *J. Amer. Chem. Soc.* **1953**, 75, 2617-2619.
11. IJIRA Technical Note, *Enzyme unlocks the Inherent Potential of T.K.P. by Biomodification,* **1988.**
12. Anon. *Jute (Newsletter of the International Jute Organisation, Dhaka, Bangladesh)* **1991,** 6(1), 6.
13. Wood, B.J.B. *Technical Report: Biosoftening of Jute (First mission).* United Nations Industrial Development Organisation:Vienna, Austria, **1990**.

14. Wood, B.J.B. *Technical Report: Biosoftening of Jute (Second mission).* United Nations Industrial Development Organisation:Vienna, Austria, **1990.**

15. Wood, B.J.B. *Technical Report: Biosoftening of Jute (Third mission).* United Nations Industrial Development Organisation:Vienna, Austria, **1992.**

16. Sinha, S.N.; Ghosh, B.L.; Ghose, S.N. *Can. J. Microbiol.* **1982,** 27, 1334-1340.

17. Newbold, C.J.; Brock, R.; Wallace, R.J. *Lett. in App. Microbiol.* **1992**, 15 , 109-112.

18. Pandey, A., Ed.; *Solid State Fermentation* . Wiley Eastern Ltd, New Age International Publishers: New Delhi & London; **1994**.

Chapter 5

Preservation and Tanning of Animal Hides

The Making of Leather and Its Investigation by the U.S. Department of Agriculture

William N. Marmer

Eastern Regional Research Center, Agricultural Research Service, U.S. Department of Agriculture, Wyndmoor, PA 19038

Animal hides are the most valuable co-product of the packing industry. They are a renewable resource of great export value, raw or tanned. Much of the current research in hides and leather researach at the Eastern Regional Research Center (ERRC) addresses environmental issues of the packing and tanning industries. Hides are traditionally preserved in NaCl brine. Research has focused on damage from halophiles and on alternatives to NaCl brine, *e.g.*, KCl brine or electron-beam irradiation. Pretanning steps, chemical and enzymatic, essentially leave a matrix of collagen-I. New methodology allows assay of bating enzyme activity during one such step. About 90% of all hides are tanned with Cr-III salts, and though the resulting product is not regarded as toxic, the toxicity of Cr-VI species raises public concern over all Cr-containing material. This has spurred a quest for alternatives both to chrome tanning and to disposal of huge amounts of solid, chrome-containing tannery waste. We have developed a molecular model of collagen-I to assist our understanding of the mechanism of chrome tanning and organic crosslinking, and we now can process the solid waste into chrome-free protein fractions and recyclable chromium oxide. The final steps of leather production involve physical and chemical finishing. Chemical and physical softening has been studied by acoustic emission. UV-cured finishes were demonstrated as alternatives to solvent-based spray finishes.

It is no surprise that the United States produces substantial numbers of cattle for beef and dairy needs, so it should not be surprising that we also produce substantial numbers of cattle hides. These hides are, in fact, the most valuable co-product of the meat packing industry. The domestic tanning industry only consumes somewhat less than half of our hide production, so hides represent a tremendous export product, worth about two billion dollars annually. The demarcation between packer and tanner is sometimes difficult to distinguish, and there is growing vertical integration; at least one of the three giants in the packing industry now processes some of its hides to what is called bluestock, i.e., hides processed through the first stages of tanning.

The USDA's Agricultural Research Service maintains a substantial research interest in the processing of animal hides, and all of its research program on hide utilization is consolidated at its facility outside Philadelphia, the Eastern Regional Research Center (ERRC). Among the assortment of laboratories and testing rooms at ERRC is a very unique facility, housed in its own building, the research tannery. This pilot plant is the only such public facility in the United States; besides supporting the research program, the tannery is used by the industry as a training facility during short courses held every six months.

To understand the significance of the ERRC program in leather research, a systematic review is provided of the traditional steps necessary to convert an animal hide into finished leather. The steps are outlined in Table I and discussed below:

Overall, the hide must be preserved while it is awaiting tanning, and this is done traditionally by soaking the hide in sodium chloride brine (**hide curing**). The hides are then shipped to the tannery and stored. Leathermaking continues at the tannery, where the hide is **trimmed**, the salt and blood are removed (**soaking**), adhering adipose tissue is separated (**fleshing**), and hair, epidermis and soluble proteins are stripped away (**liming and unhairing**, using lime and sodium sulfide at pH 12.5, which also hydrolyzes the amide groups of the amino acids glutamine and asparagine). Thick hides, sometimes 8 mm thick or more, are then **split** into two layers, with the bottom layer sometimes diverted for collagen and gelatin by-products. For leather, lime must be neutralized and removed (**deliming**), and this is usually done with ammonium sulfate. Next comes an enzymatic treatment called **bating**, in which proteolytic enzymes remove hair follicles and other noncollagenous materials.

The next step on the way to leather is crosslinking; this is accomplished for about 90% of all hides by salts of chromium III. The hides are acidified to pH 2.5 (**pickling**) and treated with the chrome salt (**tanning**). The chrome is fixed by raising the pH. The hide is now preserved against putrefaction. Its thermal stability has been raised considerably, a necessary change for withstanding subsequent manufacturing steps to finished leather and leather goods, and for durability under user applications. The resultant bluestock material now can be called leather, and as noted before, the packing industry is moving toward selling its hides in bluestock form. The bluestock is **wrung** of excess moisture, **trimmed** of uneven perimeter areas, and **split**, with the bottom layer destined for suede and the top layer for top grain leather. Uneven bottom surfaces are **shaven**, generating tens of thousands of tons of landfill waste annually in the USA alone.

Now, this bluestock is finished into the leather the consumer knows. **Retanning** stabilizes chrome from leaching, imparts softness and body, and in some cases bleaches the blue color imparted by chrome. **Coloring** is -- of course -- dyeing, and **fatliquoring** conveys softness by lubrication of internal fibers. The grain layer is smoothed and excess liquids are expressed during **setting out.** Conventional **drying**

Table I. The Tanning Process

Tanning Step	*Explanation*
Hide Curing	Temporary protective treatment for pelts; traditionally by NaCl
Trimming	Removes odd-shaped, unworkable areas of the hide
Soaking	Removes salt, blood; restoration of moisture
Fleshing	Removes attached adipose tissue (also done prior to curing)
Liming/Unhairing	Removes hair, epidermis, some soluble proteins; uses $Ca(OH)_2$ and Na_2S
Splitting	Diverts bottom layer into food use (*e.g.*, sausage casings)
Deliming	Removes alkali (using $(NH_4)_2SO_4$)
Bating	Removes noncollagenous materials enzymatically
Pickling	Lowers pH (using salt, acid) for reception of tanning chemicals
Tanning	Preserves from putrescence; imparts thermal stability; 90% of hides tanned with salts of Cr-III
Wringing	Removes excess moisture
Trimming	Removes unusable perimeters; generates Cr-containing waste
Splitting	Lower layer destined to suede; top layer to grain leather
Shaving	Adjusts thickness; generates Cr-containing waste
Retanning	(Minerals, vegtans, syntans) Adds body, softness; bleaches color
Coloring	Dyeing (with acid, metallized, direct and basic dyes)
Fatliquoring	Lubricates for flexibility, softness (lipid derivatives; emulsifiers)
Setting Out	Removes excess moisture; smoothing of grain surface.
Drying	Removes all but equilibrium moisture
Conditioning	Introduces controlled amounts of moisture for softness
Staking	Mechanical flexing for softness
Buffing	Sanding of grain surface; generates Cr-containing waste
Finishing	Impregnates and coats with polymeric materials for abrasion and stain resistance, color effects, gloss and handle properties

uses hot air, more recently supplemented by microwave or radiofrequency drying. **Conditioning** makes final adjustments to the moisture content. **Staking** mechanically massages the leather for additional softness. **Buffing** smooths out any rough surfaces and - like **shaving** - generates much solid waste destined to landfills. Finally, polymeric impregnants, undercoats and topcoats are applied during **finishing** to add body, abrasion and stain resistance, color coats, and gloss (as with patent leather), and to alter tactile properties. Most domestic leather is targeted to the footwear, leathergoods and upholstery industries, and little to the garment industry. An excellent review of the leathermaking process was recently issued in revised form (1).

KCl as a Substitute for NaCl in Hide Curing

ERRC researchers have given considerable attention to hide preservation, especially because of the tremendous environmental impact of disposing of tons of sodium chloride both at the packing plant in the brine curing step and at the tannery during soaking. Kalium, Ltd., Regina, Saskatchewan, is a Canadian miner of potassium chloride. Representatives from Kalium and a meat packer, Lakeside Packers of Brooks, Alberta, posed the question, "Can we substitute KCl for NaCl in hide curing?" The added expense would be offset by the utilization of the waste KCl in fertilizer, thus saving a major cost of disposal of NaCl. But what about the quality of leather achievable from KCl-cured hides? Under a cooperative agreement, ERRC scientists investigated the fate of these hides through all stages of tanning (2, 3).

Under the agreement with Kalium, the effects of KCl during curing were investigated. At Lakeside Packers, 100 hides were cut along the backbone and one set of matched sides was cured with KCl and the other set with NaCl. These sets were then chrome tanned to bluestock at Dominion Tanners in Winnipeg. Lakeside also shipped 150 KCl-cured hides and 400 NaCl-cured hides to Teh Chang Tannery in Taiwan to test their fate during long-distance shipping. In the USA, IBP in Joslin, IL, prepared 900 hides in KCl and 900 in NaCl and shipped them to three American tanneries (Blueside in St. Joseph, MO, Garden State in Williamsport, MD, and Pfister & Vogel in Milwaukee, WI).

Samples from each set were brought back to ERRC in Philadelphia for evaluation. No discernable differences were found in tensile strength, extensibility, or area yield (Table II):

Table II. KCl *vs* NaCl Brine Curing: Mechanical Properties of Resultant Leather

	KCl-Cured	*NaCl-Cured*
Tensile Strength (psi):	3078 ± 547	2952 ± 587
Extension at Break (in):	0.340 ± .030	0.340 ± .040
Area Yield (ft^2)	21.88 ± 1.66	21.74 ± 1.55

Simplistically, one might expect no changes in treatment protocol or results when switching from NaCl to KCl, but attention is called to one major difference between the two salts, their solubilities as a function of temperature. The NaCl curve is flat, but KCl's solubility falls with falling temperature. It is important to maintain high salt concentration in the brine, and attention has to be given to the raceway during the winter season. It is also conceivable that under extremely cold conditions the concentration of KCl brine within cured hides might fall below the minimum concentration for adequate bacterial protection.

The approach of substituting sodium with potassium has broader potential than just for brine curing. Sodium-containing agents throughout the tanning process might be substituted due to the ever-increasing environmental concerns of industry.

Halophilic Bacteria and Curing Brines

Although one might think that traditional brine curing relieves concern over bacterial contamination, this is not quite true. Halophiles are a unique type of bacteria that only grow in concentrated salt solutions. The condition labelled "Red Heat" has been known by tanners and salt fish dealers for centuries. Red Heat is now known to be the manifestation of pigmented halophilic bacteria. Tanners associate Red Heat with hide damage even though up to now no one has positively demonstrated a causal relationship.

ERRC researchers produced leather from experimental brine-cured hides that were inoculated with halophilic bacteria isolated from commercial brine-cured hides. The samples held at 106°F (41°C) for seven weeks showed serious erosion of the grain surface of the hide. No such damage was seen in samples that had been stored at room temperature (4). Scanning electron microscopy of damaged cross-sections showed an unraveling of the fiber structure on the grain surface of the hide.

Electron Beam Curing of Hides

Another approach to hide preservation completely eliminates salt curing. Microbial contamination is destroyed by electron beam irradiation. Although the electron-beam apparatus is just a larger version of the electron gun found in an ordinary TV picture tube, the electrons in this case have much higher energies: from 3 to 10 million electron volts. Facilities to house e-beam irradiators are substantial in bulk, protective shielding, and - needless to say - capital cost. Nevertheless, electron beam irradiation is an established industrial process used in many plants all over the world. Most bandages and other soft medical supplies have been electron-beam sterilized inside the package after manufacture. The process is also used to crosslink rubber and various plastics. We have even experimented with electron-beams for curing polymeric finishes on leather.

In cooperation with Ionizing Energy Corp. of Canada, ERRC staff investigated the use of electron beam irradiation as an alternative to brine curing for the preservation of cattle hides (3). In the earliest work, sterile packaging was required, using only a small amount of bactericide for safety. Preserved samples that were packaged and treated in 1986 are still on display at ERRC.

Although preservation by electron-beam irradiation is good, there is some loss of tensile strength of the resultant leather as the irradiation dose is increased. Using lower doses coupled with refrigeration and bactericidal treatment seems to be the best protocol, and packaging and sealing each hide individually are no longer required. The hides need only to be covered in groups to prevent dehydration. Three- or 10-MeV irradiators are employed with either double-passing each side at a dosage of 0.6 Mrad or single-passing at 1.2 Mrad; dosage is regulated by cart speed through the apparatus. The aim is to keep the hides in their "green state" while they are shipped overseas or to domestic tanneries. This method of preservation would be most affordable for a large packing company, and it would be a better investment for the packer were the process also usable for preservation of cuts of meat.

Colorimetric Assay of Bate Enzymes

Bating, the proteolytic process to remove noncollagenous materials, used to be done in manure pits until Rohm and Haas Company cofounder, Otto Rohm, discovered that processed hog pancreases (mixed with wood flour) could effect the same results in a socially more acceptable manner. This material was introduced commercially prior to World War I as Rohm and Haas's very first product, Oropon.

During hide processing, bate activity should be controlled. The goal is effective removal of undesirable material while avoiding damage to the essential structural proteins of leather. A colorimetric bate assay recently developed at ERRC provides the information needed for such control (5).

The substrate for the assay is hide powder azure, a denatured and insoluble collagen to which a blue dye is covalently bonded. Samples of well-dried hide powder azure are stored in tubes under silica gel. The test sample -- either bate enzyme or the bath in which hides are undergoing bating -- is incubated with the hide powder azure substrate. Proteolysis of the substrate produces soluble fragments still bearing bonded blue dye. Since the substrate is insoluble, vigorous mixing is required during incubation. Trichloroacetic acid is added to quench the reaction and filtration is accomplished rapidly through three filters: a Vacutainer (a coarse filter) and then medium and fine porosity syringe filters.

Absorbance of the solubilized dye is measured at 594 nm. Concentration correlates linearly with the enzymatic activity of the bate. Serendipitously, it was found that ammonium sulfate, which is used in the prior deliming step, inhibits bate activity. The tanning industry is now taking care to rinse the hides thoroughly of ammonium sulfate prior to bating. Another approach to measuring enzymatic activity is under development, whereby the degree of collagen hydrolysis may be determined by viscometry instead of colorimetry.

Molecular Modeling of Collagen-I

One of the ERRC projects targets tanning and, as discussed below, "untanning." Chrome tanning accounts for 90% of all tanning, and for most purposes there is nothing known to effectively compete with chrome in its role as a hide preservative, as a hide stabilizer to temperature extremes, and as a well-fixed, non-leaching tanning agent. A molecular model of collagen-I, the major constituent of prepared hides, should be a good tool for developing an understanding of the mechanism of chrome crosslinking and the design of alternative crosslinkers. Building a molecular model of collagen-I has been a major undertaking.

Collagen-I consists of a triple helix of 1014 amino acid residues per chain of repeating (Gly-X-Y) units, linked to the next helical segment by a telopeptide of 20-30 residues. Two of the three chains are identical. Modeling work on the crystalline segment was accomplished using Sybil software from Tripos.

Initially, a model of a single helix of glycine-proline-proline (Gly-Pro-Pro) was developed, and then hydroxyproline was substituted for the second proline. Three of these modified helices (Gly-Pro-Hyp) were combined to form a triple helix. Finally, the true amino acid sequence was introduced to produce the Smith model of a collagen microfibril, a bundle of five triple helices (Figures 1 and 2) (6-8).

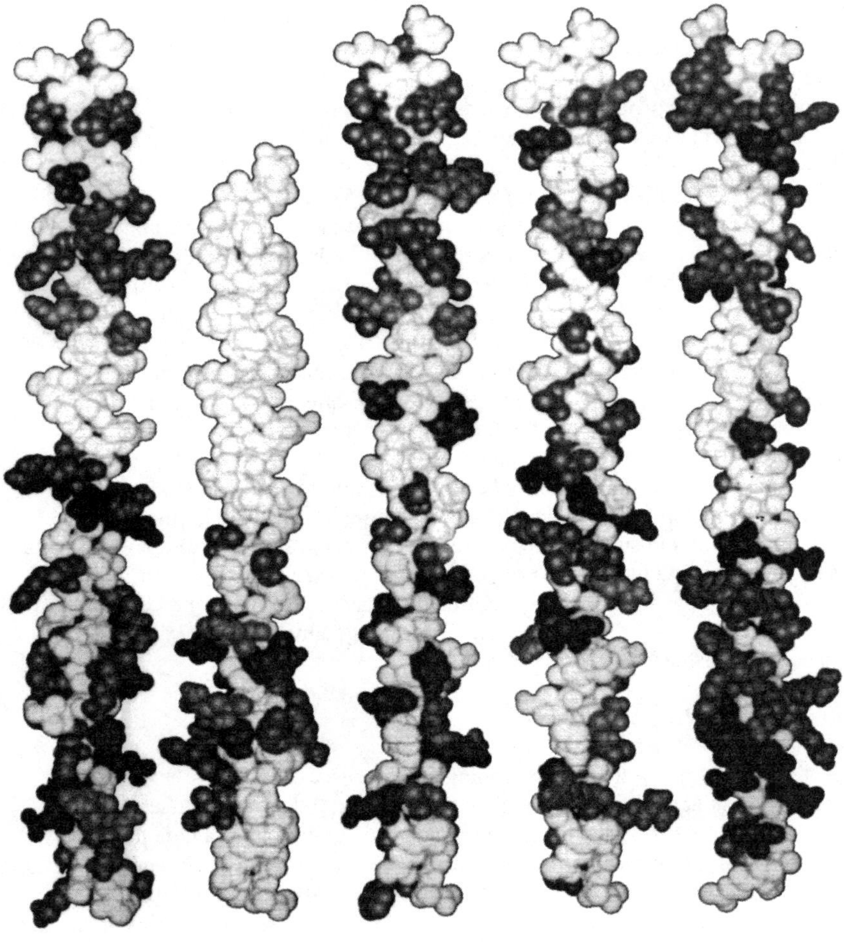

Figure 1. Space-filling molecular model for fragments from the five triple helices of Type I Collagen. The gap region is represented by the abbreviated length of the second triple helix. Four classes of sidechains are depicted from light to dark shading: structural (Gly, Pro, Hyp), nonpolar (Phe, *e.g.*) and neutral polar (Ser, *e.g.*), basic (Lys, Arg, His), and acidic (Glu, Asp). A color representation is depicted in Figure 3 of Reference 6.

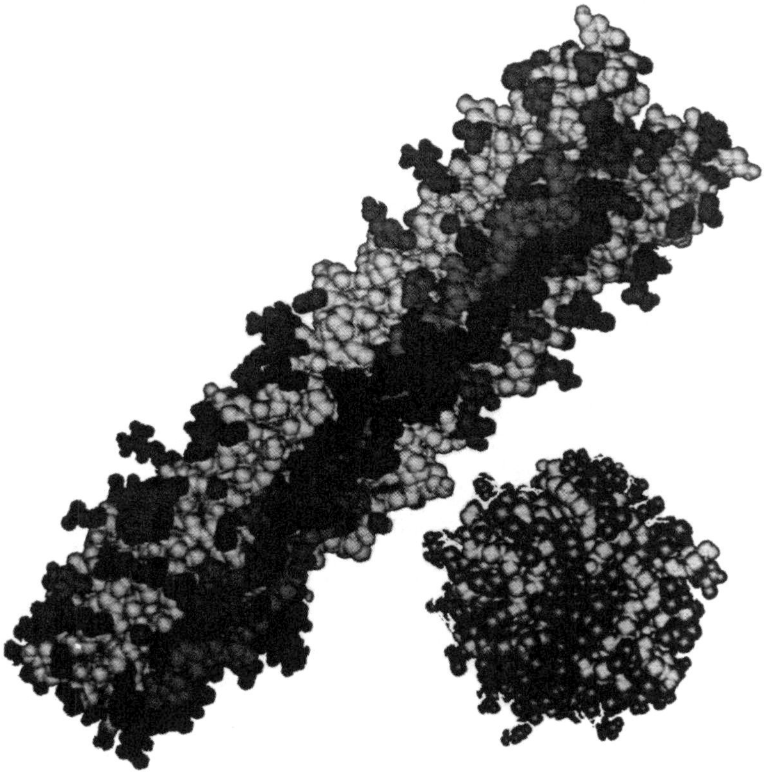

Figure 2. Lateral and cross-sectional views of the space-filling model for a fragment of the Collagen I microfibril constructed from the triple-helical fragments of Figure 1. The gap region is visible at the top of the lateral view. Shading is as described for Figure 1, except that the three polypeptide chains of one triple helix are highlighted in three dark shades to emphasize the path of this fragment across the lateral view of the microfibril ending in the gap region. A color representation is depicted in Figure 2 of Reference 7.

At the C-terminal end of each collagen chain is a 5 mer repeat of the Gly-Pro-Hyp sequence, which serves to nucleate helix formation. In the microfibril, the C-terminal nucleating region of one triple helix is separated from the N-terminus of the next triple helix by a gap region containing both N- and C- terminal telopeptides. These gaps are seen as dark areas in electron micrographs. The characteristic staining patterns observed by electron microscopy can be visualized on the model by assigning colors based on the ionic character of the amino acid side chain. New regions concentrated in $-CO_2^-$ appear when amide-bearing residues (Asn, Gln) are hydrolyzed (to Asp and Glu, respectively). This change is thought to facilitate chromium uptake during subsequent chrome tanning. Regions with local concentrations of hydrophobic sites (e.g., Phe) are seen, as are neutral-polar regions (Ser, Thr).

We are just beginning to experiment with the molecular model to study crosslinking during tanning of hides. In one example, we studied the spacial requirements of dicarboxylic acid crosslinks, using 6, 8, and 10-carbon straight-chain acids bonded to lysinyl amino groups in the model via salt bridges. Chain-length 6 is too short; C-8 is better; C-10 is best. These results corroborate with chemical experiments, which show a similar pattern in the ability of such binding to raise the thermal stability of the collagen samples.

Enzymic Digestion of Chrome Shavings

The shaving and trimming of bluestock leather generate a huge amount of chromium-containing solid waste, most of which ends up in landfills. Chromium in leather is Cr (III), not the toxic Cr (VI), and EPA exempts the industry from handling its chromium-containing waste as toxic material. Nevertheless, some <u>local</u> environmental authorities do not make a distinction between the two valence states. The tanning industry challenged us to develop a means of diverting this waste from landfills. Even if this waste is not regarded as toxic, it represents an economic loss of protein and chromium.

The magnitude of this waste in this country alone is staggering; almost 60 thousand metric tons per year of bluestock waste are landfilled. Worldwide, this waste is an order of magnitude larger. Half the waste is chrome shavings and the rest is trimmings and buffing dust. (Further, huge amounts of <u>finished</u> leather also end up in landfills, and finished leather waste is another significant challenge.) ERRC researchers discovered a cost-effective means of digesting the bluestock waste with alkali and alkaline protease enzyme; valuable chrome-free protein fractions are recovered, as is a cake of chromium (III) oxide, recyclable back into the tanning process (9, 10).

The flow diagram shown in Figure 3 illustrates what now is a patented process (11, 12). The solid waste is first treated under alkaline conditions (magnesium oxide, alone or combined with sodium hydroxide or carbonate, is most effective); this allows the isolation of a valuable high-molecular-weight product (13) -- a gelable, chrome-free protein of molecular weight *ca.* 100 K -- potentially usable in adhesives, as films, and for encapsulation. The remaining sludge is then treated with an alkaline protease enzyme to recover a protein hydrolysate and a precipitate of chromium (III) oxide. The hydrolysate with the enzyme is recycled several times to digest more sludge.

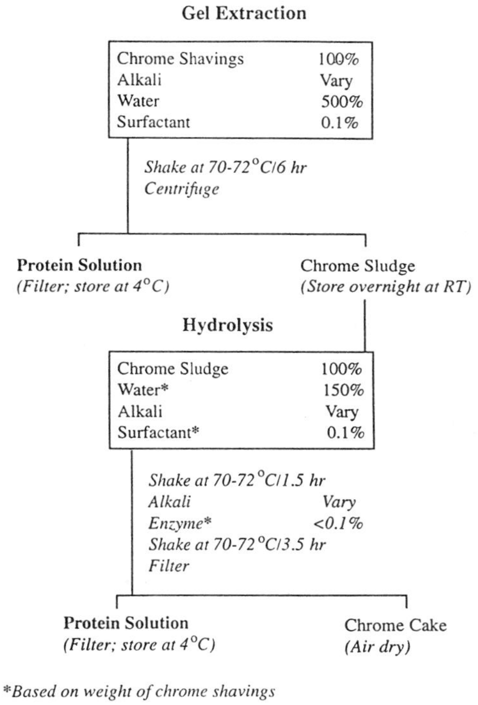

Figure 3. Recycling Solid, Chrome-Containing Tannery Waste per U.S. Patent 5,271,912 (Reference 12).

Chromium (III) oxide is recovered by filtration. The hydrolysate has a molecular weight of *ca.* 20 K and measurable gel strength.

Numerous tanneries around the world are evaluating this process. Alternatives have been developed whereby alkaline digestion unassisted by enzymes is used to prepare a lower quality protein hydrolysate, usable as a fertilizer component.

Acoustic Emission of Leather

The stiff product that results from tanning must be softened. This is done chemically by application of fatliquors (sulfated castor oil is a classic example). Softening imparted by fatliquoring has been related to the breakdown of 50 μm-thick fibers into 5 μm-thick fibers, as seen by electron microscopy. Softening also is imparted mechanically by staking, a process that can be envisioned as hundreds of fingers massaging the leather as it passes through a machine.

Drying techniques also influence softness. Drying is a very sensitive step done on drying racks typically in convection or drying ovens. Overdrying makes a stiff product. Preserving area yield is critical, since leather is marketed by its area, so the leather is toggled or pasted onto the drying racks.

To study the mechanisms of fatliquoring, drying and staking, mechanical testing is carried out. Stress-strain testing uses the Instron tensile tester. Also performed are studies of viscoelasticity, the time-dependent response to stress, so important in shoe lasting and for upholstery and garment leather.

A unique way to study softening is to mount a sample on an Instron tester and attach a piezoelectric transducer to the leather surface, enabling the sounds emitted during stretching to be detected and recorded. ERRC scientists have applied this technique, acoustic emission, to their leather studies (14). When thick bundles of fibrils split, short acoustic pulses are emitted. Stretching again after relaxation is a "quieter" process. The finer bundles of fatliquored leather are more deformable.

Figure 4 shows a record of extension of a piece of leather that had not been fatliquored. The stress-stain curve is shown as a scatter plot (left ordinate). Its initial shape indicates stiffness, with later relaxation, and finally failure at about 25% extension. The bold, bottom trace is the acoustic pulse frequency curve (right ordinate). It rises promptly as stretching is initiated and continues to record sound until failure. The upper acoustic trace is the pulse energy curve (also right ordinate), calculated from the pulse amplitude and pulse decay time. It looks similar to the pulse curve until the leather has been stretched about two-thirds of the way to failure. Then large energies appear irregularly, indicating the actual breakage of fibers prior to failure.

The curves generated from a piece of fatliquored leather are quite different (Figure 5). The initial slope of the stress-strain curve is smaller, indicating less stiffness. Failure does not occur till almost 50% extension. The acoustic pulse and energy curves show no early acoustic activity, and the extension through to failure is a relatively quiet process. Although occasional high energy pulses are seen, the low activity on the frequency curve downgrades their significance. The fatliquor has reduced interfiber adhesion, resulting in fibers that are pulled apart only near maximum extension, but with little breakage.

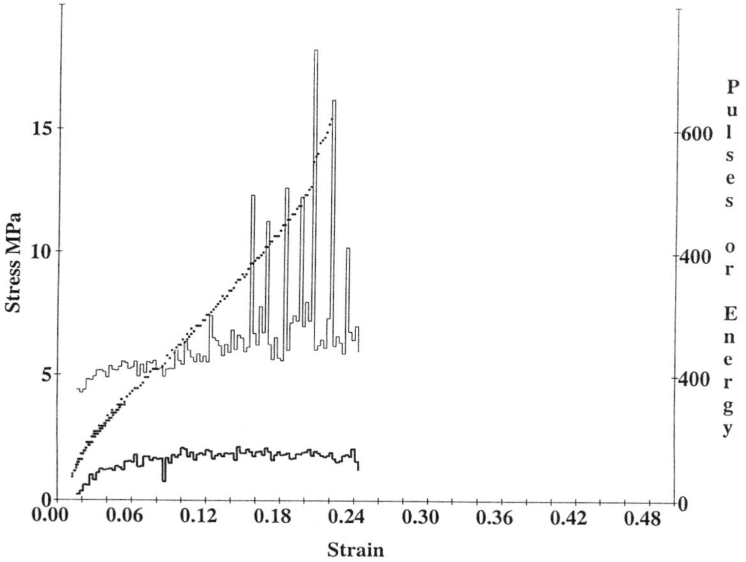

Figure 4. Stress-strain curve (left ordinate) and acoustic emission (right ordinate) during enlongation of crust leather not fatliquored (or staked). Thin line: energy (arbitrary units) per pulse (over elongation interval of 1%); thick line: number of pulses per interval; scatter plot: stress (Mpa) *vs.* strain (mm extension per mm sample length). (adapted from Reference 14).

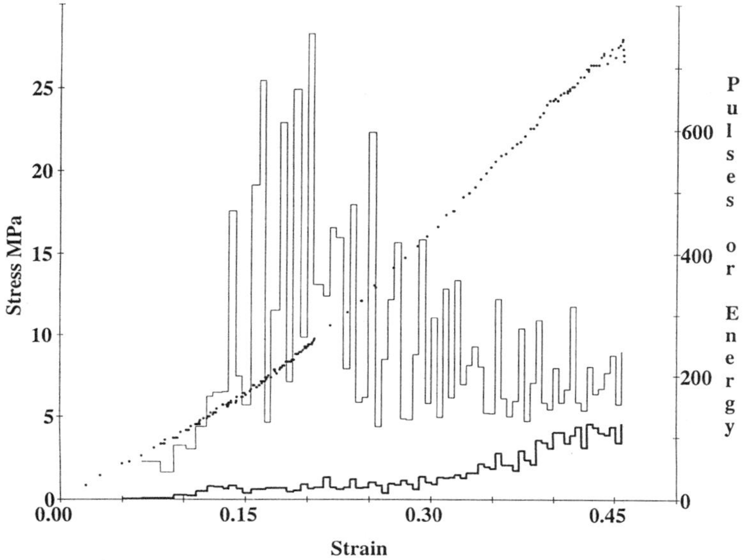

Figure 5. Stress-strain curve (left ordinate) and acoustic emission (right ordinate) during elongation of leather dried from fatliquor (unstaked). Lines and units as per Figure 4 (adapted from Reference 15).

Acoustic emission studies are being applied to study not only the mechanisms of fatliquoring, staking and drying, but also in the design of novel fatliquoring agents and water repellant treatments.

UV-Cured Finishes for Leather

The final step in manufacturing leather is the application of polymeric finishes. Traditionally, these finishes (dispersions of polyacrylates, polybutadienes or polyurethanes) have been formulated in carriers of organic solvents. Recently, the industry has been offered so-called aqueous-formulated finishes (mainly dispersions of crosslinked polyurethanes), though they do contain small amounts of hydrocarbon solvents. ERRC scientists explored another type of finishing, UV-curing (15), which is in widespread use in the coatings industry for other materials, including paper and metal.

In one typical formulation they developed specifically for leather, an acrylated urethane oligomer (33.3 parts) was combined with *N*-vinylpyrrolidone (33.3 parts) to add hardness; *iso*-decyl acrylate (33.3 parts) for flexibility and softness; trimethylolpropane triacrylate (TMPTA; 5.0 parts) as a cross-linker; and diethoxyacetophenone (DEAP; 2 parts) as a photoinitiator. All these components are commercially available and used in the USA in large quantities. Newer monomers are now available and more testing is warranted.

There are disadvantages to this process. Some people suffer allergic reactions to acrylates. There is a capital investment for acquisition of the equipment, and the chemicals can be costly. Application to a nonuniform surface can be tricky, though roller coating has worked well in plant trials. The use of high intensity UV lamps demands safety precautions. There also can be some problems with effectiveness of the cure, depending on thickness of oligomer application, formulation, and style of finish (particularly with patent leather finishes). Most critically, however, UV curing of leather is ineffective in application of impregnants; photoinitiation of impregnants is blocked by the collagen matrix, unless one switches from UV curing to electron-beam curing (2 Mev). Nevertheless, UV curing offers the leather industry some great opportunities for application of intermediate coats and topcoats. The advantages are in the complete consumption of all applied chemicals. There are no losses from spraying, there are no volatile emissions of solvent hydrocarbons, there is no water to evaporate, the apparatus consumes little power and its footprint is a tremendous saver of floorspace over conventional spray booths and dryers. Finishing is accomplished rapidly and the finishes produced are good quality.

Pilot plant runs in the ERRC tannery were accomplished using a small apparatus, essentially a moving belt (30 ft/min) with two UV lamps (mercury or argon, 300 W/in, 210-270 nm). Different styles of finishes were applied successfully, including color coats and patent leather finishes. Scuff resistance, measured using a Taber Abraser, and flexibility, assessed using a Bally Flexometer, were good. UV-cured leather finishing thus is a promising alternative to conventional finishing, and a combination of economics and environmental constraints may lead the industry to adopt this concept in the near future.

As long as beef and dairy cattle are raised and consumed, there will be a replenishable resource in the hide co-product. Despite attempts by industry to produce a synthetic substitute for leather, nothing yet matches the "real thing." Our program at ERRC aims to add value to the agricultural co-product and to reduce the environmental impact of all operations that process hides through to finished leather. This long-standing research program is likely the only public and integrated effort in the United States that targets the broad area of converting hides into leather. Its research staff, using its laboratories, testing facilities and tannery - itself a unique facility in the USA - keep abreast of the problems faced by the industry and continue to develop and promote "cleaner" processes and better products for the consumer.

Literature Cited

1. Anon. *Leather Facts;* Third Edition; New England Tanners Club: Peabody, MA, 1994.
2. Bailey, D. G. *J. Am. Leather Chem. Assoc.* **1995,** 90, 13-21.
3. Bailey, D. G. *World Leather* May **1995**, 43-47.
4. Bailey, D. G.; Birbir, M. *J. Am. Leather Chem. Assoc.* **1993,** 88, 291-299.
5. Mozersky, S. M.; Bailey, D. G. *J. Am. Leather Chem. Assoc.* **1992,** 87, 287-295.
6. Chen, J. M.; Feairheller, S. H.; Brown, E. M. *J. Am. Leather Chem. Assoc.* **1991,** 86, 475-486.
7. Chen, J. M.; Feairheller, S. H.; Brown, E. M. *J. Am. Leather Chem. Assoc.* **1991,** 86, 487-497.
8. Brown, E. M.; Chen, J. M.; Feairheller, S. H. *XXII IULTCS Congress Proceedings;* Associaçao Brasileira dos Químicos e Técnicos da Indústria do Couro: Porto Alegre, Brazil, **1993**; Vol. 1, pp 69-74.
9. Taylor, M. M.; Diefendorf, E. J.; Thompson, C. J.; Brown, E. M.; Marmer, W. N. In *Polymers from Agricultural Products;* Fishman, M. L.; Friedman, R. B.; Huang, S. J., Eds.; American Chemical Society: Washington, DC, **1994**; pp. 171-187.
10. Taylor, M. M.; Marmer, W. N.; Diefendorf, E. J.; Brown, E. M. *J. Am. Leather Chem. Assoc.* **1994,** 89, 221-228.
11. Taylor, M. M.; Diefendorf, E. J.; Marmer, W. N.; Na, G. *U.S. Patent 5,094,946,* March 10, 1992.
12. Taylor, M. M.; Marmer, W. N.; Diefendorf, E. J.; Brown, E. M. *U.S. Patent 5,271,912*, December 21, 1993.
13. Brown, E. M.; Thompson, C. J.; Taylor, M. M. *J. Am. Leather Chem. Assoc.* **1994,** 89, 215-220.
14. Kronick, P. L.; Page, A.; Komanowsky, M. *J. Am. Leather Chem. Assoc.* **1993,** 88, 178-186.
15. Scholnick, F. *J. Am. Leather Chem. Assoc.* **1990,** 85, 36-41.

Applications in Polymers

Chapter 6

Biopolymers from Fermentation

W. F. Fett, S. F. Osman, M. L. Fishman, and K. Ayyad

Eastern Regional Research Center, Agricultural Research Service, U.S. Department of Agriculture, 600 East Mermaid Lane, Philadelphia, PA 19118

Biopolymers are used for a variety of food and nonfood applications, and many can be produced by microbial fermentation of agricultural commodities. Commonly used fermentation medium components include corn starch, corn gluten meal, and corn steep liquor. Commercialized bacterial exopolysaccharides produced by fermentation include dextran, xanthan, gellan, cellulose and curdlan. Bacterial polyhydroxyalkanoates recently have found use in the manufacture of biodegradable plastics. New, potential products obtained by microbial fermentation include yeast glucans and several fungal and bacterial exopolysaccharides. We initiated a screening program in our laboratory, resulting in the identification of several high-yielding alginate-producing bacterial strains, as well as several producers of novel exopolysaccharides. Bacterial alginates may substitute for algal alginates in certain applications. The complete structures of most of the novel bacterial exopolysaccharides were determined.

Hydrocolloids or water-soluble gums are used as thickeners, stabilizers, emulsifiers and gelling agents in food and nonfood industries. Plant hydrocolloids have been employed for such purposes for thousands of years (1). However, it was not until the 1960s that the first microbial hydrocolloid, the bacterial exopolysaccharide (EPS) xanthan gum, was available for commercial use. Bacterial EPSs are found outside of the cell either in the form of a tightly-held capsule or a loosely-held slime layer. They are usually anionic and of high molecular weight. Xanthan gum is produced by the bacterium *Xanthomonas campestris,* causal agent of black rot disease of cruciferous crops. It was first isolated by Lilly and coworkers (2) in the late 1950s and then extensively studied for possible commercial applications at the USDA, ARS, Northern Regional Research Center (now the National Center for Agricultural Utilization Research, NCAUR) located in Peoria, IL. This polysaccharide was approved by the U.S. Food and Drug Administration (FDA) for general food use in 1969 and first sold by the Kelco Company (now a division of Monsanto Chemical Company). Xanthan gum is an acidic, high molecular weight heteropolysaccharide with a pentasaccharide repeating unit consisting of a cellulosic backbone with trisaccharide side chains on alternate glucose residues (3). The side chains are substituted with pyruvate and acetate. Due to its unique physical properties xanthan is used as a thickener, stabilizer and suspending agent for several food and nonfood applications (4). Xanthan forms highly viscous aqueous solutions at low concentrations, stable over a wide range of pH, temperature and salt concentrations. In addition, these solutions exhibit high pseudoplasticity. Most recently, xanthan is being employed as a shortening replacer for food use. Annually, about 20,000 metric tons of xanthan gum are used for commercial purposes in the U.S.A. (5).

Due at least in part to high development costs, it would be twenty-four years (in 1993) before another microbial EPS, gellan gum, was approved by the FDA for general food use in the U.S.A. Gellan gum is produced by the saprophytic bacterium *Pseudomonas (Auromonas) elodea*. It consists of a repeating unit of an acidic, linear tetrasaccharide containing the sugars glucose, glucuronic acid and rhamnose, as well as acetate and glycerate (6). The deacylated form produces a firm, brittle gel in the presence of monovalent or divalent cations (4). The polysaccharide was sold beginning in 1990 as an agar substitute, especially useful for plant tissue culture applications.

Curdlan is a water-insoluble, linear β-(1,3)-linked D-glucan produced as an EPS by *Alcaligenes faecalis* var. *myxogenes* (7). This polymer forms gels when heated to 55 C or above and subsequently cooled. Curdlan also gels when alkaline solutions are neutralized. In Japan curdlan is sold by Takeda Chemical Ind., Ltd. for food use and by Wako Pure Chemical Ind., Ltd. for nonfood use. Certain modified curdlans are reported to have anticancer activity (7). At the present time, curdlan is not approved for food use in the U.S.A.

A bacterial EPS that has many nonfood applications is dextran (8). Commercial dextran is obtained primarily from *Leuconostoc mesenteroides* strain NRRL B-512 or its derivative B-512(F). This strain was isolated at the NCAUR in the 1940s (9). The polysaccharide produced by this strain consists solely of α-(1,6)-linked glucan with variable degrees of branching to the O-3 of the backbone moieties (8). The polymer is synthesized extracellularly from sucrose by the action of the enzyme dextransucrase. It is possible to obtain dextrans of narrow molecular weight ranges for specific applications by controlling culture conditions. For example, a 70 kD dextran is used as a blood expander and a 40 kD dextran is used in organ perfusion solutions. Modified dextran is also used for the manufacture of a variety of chromatographic media (8). Annual consumption of dextran for nonfood uses in the U.S.A. is approximately 2000 metric tons (5). Currently, dextran is not approved for food use in the U.S.A.

A fifth bacterial EPS, cellulose, has found food and nonfood commercial applications. Cellulose is a linear, non-water soluble polymer composed solely of β-(1,4)-linked glucose. The source of bacterial cellulose for commercial use is *Acetobacter xylinum*. Bacterial cellulose, in contrast to plant cellulose, can be obtained in highly purified form free of lignin. This material has a high degree of crystallinity, a much larger surface area than plant cellulose, and has excellent mechanical strength and absorptive capacity (10). Bacterial cellulose is used to prepare a dessert called "nata" in the Philippines (11), and is also used by Sony

Corporation to manufacture high quality audio speaker systems. Several other potential applications are under study (10).

Additional bacterial EPSs such as emulsan and alternan and fungal polymers such as pullulan, scleroglucan and yeast glucans (12,13) are currently either not produced commercially or are produced in small amounts. These polymers are being studied for a variety of potential commercial uses.

Intracellular bacterial polymers have several potential as well as realized commercial uses. Much recent industrial and academic interest has centered around using bacterial polyesters for the production of biodegradable plastics. Under conditions of adequate carbon availability and nutrient stress (14) many bacteria synthesize polyhydroxyalkanoic acids (PHAs) as intracellular energy reserve materials, stored as inclusion bodies. The PHAs consist of hydroxyalkanoic acids with chain lengths between 3 and 14 carbon atoms. The specific PHAs produced are dependent on the bacterium and the carbon source available in the fermentation broth. The bacterium *Alcaligenes eutrophus* has been the most studied PHA producer. Zeneca BioProducts utilizes this bacterium to produce a biodegradable resin (BioPol) made from polyhydroxybutyrate- hydroxyvalerate copolymer. The resin is used to make biodegradable films, coatings and containers.

All commercially available bacterial EPSs are currently produced by batch fermentation using complex media. Agricultural byproducts commonly used as carbon and energy sources for fermentation include glycerol, molasses, corn starch and starch hydrolysates, glucose, sucrose and whey (65% lactose). Commonly used nitrogen sources include corn steep liquor, corn gluten meal, dried distillers solubles, yeast extract, fish meal, soybean meal and cottonseed flour (5,15). Several fermentation variables affect polymer yield and quality (e.g., molecular weight, degree of substitution with non-sugar moieties, level of various contaminating macromolecules). These include the carbon source, nitrogen source, the carbon to nitrogen ratio, the levels of various cations and anions, available oxygen levels, temperature and stage of microbial growth at time of harvest. Polymer molecular weight can be indirectly affected by the effects of culture conditions on EPS depolymerases. In general, the nature of the EPS produced is independent of carbon source, but there are exceptions (5).

Acidic Exopolysaccharides of Group I Pseudomonads

Alginates. The term alginate encompasses a group of structurally related, linear, acidic polysaccharides containing varying amounts of O-acetylated β-1,4-linked D-mannuronic acid and its C-5 epimer L-guluronic acid (16). These two uronic acids are present as homopolymeric or heteropolymeric block structures. The number of homopolymeric blocks of guluronate in the polymer determines its gelling properties. A high amount of polyguluronate leads to the formation of firm, brittle gels in the presence of divalent cations, particularly calcium, while low polyguluronate alginates form more elastic gels (16). Alginates comprise a major structural polymer in brown algae, and these algae, harvested from the ocean, are currently the sole source of alginates for commercial use. Algal alginates have many applications in food and nonfood industries as thickeners and gelling agents. They were first produced commercially in California in 1929. Annual worldwide consumption of alginates is estimated at approximately 23,000 metric tons (5). The algal species, age at harvest, geographic location of harvest and plant part extracted all can have a significant effect on both the level of guluronic acid present and the cost of alginate (17). The production of alginate by bacterial fermentation of agricultural commodities would allow for a product with a constant composition and availability.

Several years ago we initiated a study of EPSs produced by the rRNA homology group I pseudomonads (18). This group of bacteria consists primarily of pseudomonads which produce fluorescent iron-binding pigments (siderophores) when grown on media low in iron. Members of this important group of bacteria include human pathogens, saprophytes, and plant pathogens. They are the subject of intensive study due to their pathogenic potential, their extreme metabolic capabilities useful for bioremediation applications and their potential as biocontrol agents for a variety of serious plant diseases. The type species of this group is the opportunistic human and plant pathogen *P. aeruginosa*. At the time the project was initiated, very little was known about the EPSs produced by this group of pseudomonads. *Pseudomonas aeruginosa* was previously shown to produce alginate (19), and alginate-producing variants had been obtained by laboratory manipulations of *P. fluorescens, P. putida* and *P. mendocina* (20,21). Alginates produced by pseudomonads were shown to be similar to algal alginates except for two important properties. First, there were no homopolymeric blocks of guluronic acid present, which affects their gelling properties in the presence of cations and, secondly, some mannuronic acid residues were mono and/or disubstituted with acetate (13).

We began our studies using the soybean pathogen *P. syringae pv. glycinea*. Several strains of the bacterium were demonstrated to produce alginate in glucose-containing broth media and in infected plant leaves (22). The alginates were acetylated and contained up to 20% guluronic acid. The guluronic acid content of the alginates produced by individual strains was dependent on the particular environmental conditions at the time of synthesis. This differed from alginates produced by *P. aeruginosa,* whose mannuronic acid to guluronic acid ratio appeared to be independent of growth conditions for each particular strain (23). Early studies in our laboratory utilizing ^{1}H-NMR and enzymatic assays using a guluronate-specific alginate lyase indicated the possible presence of low levels of homopolymeric guluronic acid block structures in some *Pseudomonas* alginates. However, further studies by others indicated that pseudomonad alginates are devoid of such sequences (5,16). Results of high-performance size exclusion chromatography indicated that the alginates produced in infected plants were higher in molecular weight (2.0×10^4 to 4.7×10^4 D) than corresponding alginates produced *in vitro*. The relatively low molecular weights (3.8×10^3 to 5.2×10^3 D) of the *in vitro* produced alginates were most likely due to the action of alginases synthesized by the producing bacteria (Fett, unpublished). Subsequently, we found that alginates were produced by additional plant pathogenic and saprophytic group I pseudomonads, but not by pseudomonads which did not belong to group I (24,25,26).

The bacterium *P. aeruginosa* is not a viable source of alginates due its pathogenicity towards humans. The other well-studied alginate-producing bacterium, *Azotobacter vinelandii,* is also not used currently because it has a high oxidation rate (27). Thus, we conducted further investigations to determine whether any of the other group I pseudomonads included in our studies might be useful for the production of alginates (28). A total of 115 strains of fluorescent *Pseudomonas* species (*P. cichorii, P. fluorescens, P. syringae* and *P. viridiflava)* were tested for yields of alginates when grown in batch culture in a proprietary liquid medium (PLM) (Kelco). The PLM contained either fructose or glucose (both at 5%, w/v) as the primary carbon and energy source. Selected strains were also grown in a modified Vogel and Bonner medium (MVBM) containing gluconate (5%, w/v). This medium was previously formulated by others to support maximal alginate production by the human pathogen *P. aeruginosa* (29). Cultures were incubated for five days at 24 C with shaking (250 to 300 r.p.m.). The yields of alginate present in isopropanol-precipitated material were estimated by use of a colorimetric assay for uronic acid content (30). Maximal yields of alginates were 5 g/L for PLM with fructose, 3 g/L for PLM with glucose and 9 g/L for MVBM. Culture fluids containing such high amounts of alginate were highly viscous to gel-like. Starch was not tested as a carbon and energy source since none of the pseudomonads tested could utilize this feedstock.

The yields of alginates obtained represent a good starting point for further development of selected *Pseudomonas* strains for commercial production of bacterial alginates. Bacterial alginates could be used for applications which do not require the formation of strong gels, e.g., as viscosifiers for textile dyes. A knowledge of the environmental factors affecting alginate synthesis would be useful to develop fermentation conditions which would allow for increased yields. To this end, we determined the effect of various stresses and other culture parameters on the yields of alginates produced by representative group I pseudomonads other than *P. aeruginosa*. Results indicated that many fluorescent pseudomonads respond to various environmental stresses by increasing production of alginates (31). Hyperosmotic conditions brought on by addition of sodium chloride (0.2 to 0.5 M) to a complex liquid medium led to a significant increase in alginate production; up to 8-fold on a mg alginate/L culture media basis and up to a 22-fold increase on a mg alginate/cell dry weight basis. Similarly, addition of ethanol (1 to 3%) to the medium caused up to a 12-fold increase in alginate production on a mg alginate/L culture media basis. However, strains which did not produce alginate in control cultures did not produce alginates under stress conditions. The results were very similar to those found for *P. aeruginosa* (32). The data indicated that environmental stress can increase constitutive alginate synthesis, but by itself cannot induce alginate production. A very recent report, however, demonstrated that inclusion of copper into a solid medium induced alginate production by several copper-resistant strains of *P. syringae* (33). We noted that constitutive alginate production was quite low (a maximum of 49 mg alginate/L) (31) so that even a 22-fold increase did not represent an acceptable commercial yield. Addition of sodium chloride or ethanol to culture media which supported high yields of alginates (3 to 9 g/L) did not lead to further increases. The primary limiting factor in these cultures may have been oxygen availability due to the highly viscous nature of the culture fluids. The effect of various nitrogen sources (ammonium sulfate, potassium nitrate, sodium nitrate and sodium glutamate) and divalent cations on alginate production by a strain of the plant pathogen *P. viridiflava* strain 671m cultured in a completely defined medium also was determined. Ammonium sulfate supported the greatest production of alginate on a mg/g cell dry weight basis. Inclusion of Mn^{+2} (1 mM) and Mg^{+2} (10 mM) in the growth medium greatly stimulated alginate production, while inclusion of Fe^{+2} was inhibitory (Singh, et al., unpublished). Similar results have been reported for *P. aeruginosa* (29,32,34).

How environmental signals are sensed by alginate-producing bacteria and how the message is transduced to the cytoplasm has been the subject of extensive molecular studies on *P. aeruginosa* and additional group I pseudomonads. The regulation of alginate production is extremely complex, with both global and specific two-component signal-transducing regulatory systems interacting with each other. Primary regulation of alginate biosynthesis by pseudomonads is by specific activation of promoters of two alginate biosynthetic genes, *algC* and algD. A number of positive and negative regulatory genes have been identified (32,35,36,37). A recently described regulatory gene, designated *repA,* was cloned from *P. viridiflava* (38). RepA is a global regulator, controlling not only alginate synthesis, but also synthesis of extracellular enzymes required by *P. viridiflava* to attack its host plants.

Novel exopolysaccharides. Based on our early experiments with group I pseudomonads, we hypothesized that all of these bacteria produce alginate as an acidic EPS, but further studies proved this hypothesis to be incorrect. The group I pseudomonads turned out to be a rich source of novel acidic EPS structures. As a species, *Pseudomonas marginalis,* a pectolytic plant pathogen which causes soft-rot of harvested fruits and vegetables, produces three different acidic EPSs (39,40,41). Their acidity comes from pyruvate, succinate or lactate substituents, not from uronic acid as usually found for acidic bacterial EPSs. The first was a galactoglucan which we named marginalan (type A, Fig. 1). At the time of publication the exact location of the succinate substituent was not known. Recent data indicate that the galactose moiety is disubstituted with succinate and pyruvate whereas glucose is not substituted (42). Subsequently, we found marginalan to be produced by certain strains of *P. fluorescens* and *P. putida* (25,26). Two additional acidic EPSs are produced by strains of *P. marginalis* (types B and C, Fig. 1), but so far their production appears to be specific to a very limited number of strains of this species (40,41). A fourth novel acidic EPS is produced by the type strain (ATCC 17588) of the group I, nonfluorescent, opportunistic human pathogen *P. stutzeri* (Fig. 1) (43). Again, this polymer is acidic solely due to the presence of a lactic acid substituent. Genetically-manipulated *P. stutzeri* was reported previously to produce alginate as an acidic EPS (44). As a species, *P. stutzeri* is quite diverse (18) so it is not surprising that more than one acidic EPS is produced. An acidic EPS other than alginate is also produced by the mushroom pathogen *P.* "gingeri" (Fig. 1) (45). This was the first EPS of the group I pseudomonads, other than alginate, shown to contain uronic acid. It is of interest that the *P.* "gingeri" EPS is identical in structure to the EPS produced by *Escherichia coli* strain K55 (46) and differs only in the location of the acetate group from *Klebsiella* type 5 capsular EPS (47).

Most recently, we have been studying acidic EPSs produced by three additional group I pseudomonads: saprophytic *P fluorescens* strain H13 (26), saprophytic *P. chlororaphis* strain B2075 and *P. flavescens* strain B62, a recently described walnut pathogen (48). The composition of the *P. fluorescens* strain H13 polymer is unusual for the group I pseudomonads as it contains glucosamine (26). The extracellular material produced by *P. chlororaphis* is a mixture of a hexosamine-containing polymer and alginate. The EPS produced by *P. flavescens* strain B62 also appears to

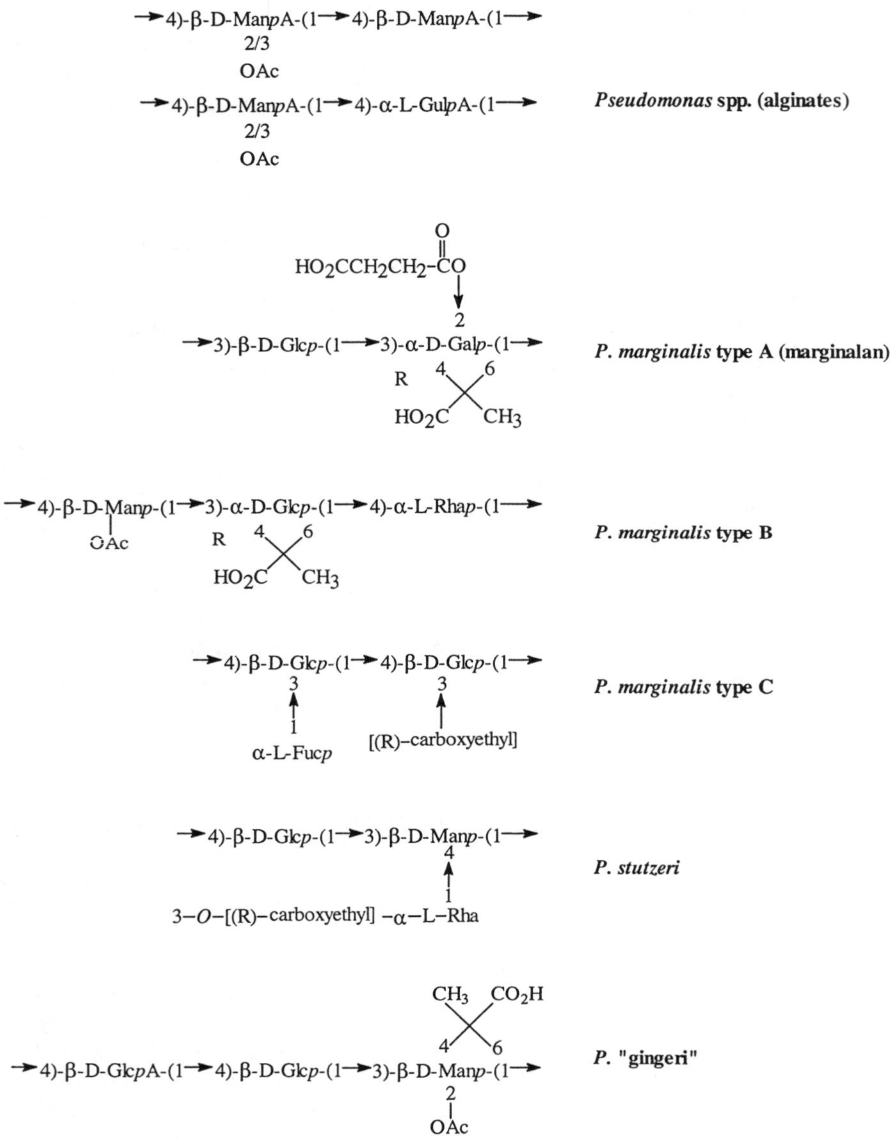

Figure 1.

Structures of exopolysaccharides produced by group I pseudomonads. ManA, mannuronic acid; GulA, guluronic acid; OAc, O-acetyl group; Glc, glucose; Gal, galactose; Man, mannose; Rha, rhamnose; Fuc, fucose; GlcA, glucuronic acid; *p*, pyranose form of the sugar.

be unusual for this group as it contains a uronic acid substituted with lactic acid (Cescutti *et al.*, unpublished).

Physical properties. In order to determine if any of the EPSs isolated from the group I pseudomonads might have commercial applications, a variety of physical measurements as well as utility tests were run. All EPS samples examined in these studies had undergone three steps of precipitation with isopropanol and dialysis against distilled water, 0.1 M NaCl and then distilled water once again. Samples were then digested with nucleases and a protease to remove any contaminating nucleic acid and protein. The lack of these contaminants was confirmed by UV spectroscopy. Finally, any contaminating lipopolysaccharides were removed by ultracentrifugation (100,000 x g, 4 h). Gel formation was tested for several of the EPSs *(P. flavescens* EPS, the *P. marginalis* type A, B, and C EPSs, and the *P. stutzeri* EPS) when applied in sugar-acid and calcium-type jellies according to standard A.O.A.C. procedures (49). For the sugar-acid type jelly test, total soluble solids were above 60% with a pH below 3.8 after boiling the polysaccharide (0.41% w/v) solution. For the calcium-type jelly test, calcium orthophosphate was added to the polysaccharide solution (0.08% w/v) to give 0.01% (w/v) at a total soluble solids level below 30% at room temperature. Bloom strength was quantified with a TA-XT2 Analyzer (Stable Micro Systems, U.K.) adjusted to 4 mm depression/second using a 0.5 inch plunger. Results indicated that none of the EPS samples tested performed as well as pectin in these tests (Table I).

Table I. Jelly Strength of *Pseudomonas* Exopolysaccharides When Applied in Sugar-Acid and Ca-Type Jellies

Polysaccharide Type or Source	*Sugar-Acid Jelly(g)*	*Calcium-Jelly(g)*
Pectin	150[a]	140[a]
P. flavescens EPS	24.7 ± 0.2[b]	3.7 ± 0.1
P. marginalis Type A	33.5 ± 0.1	3.5 ± 0.2
P. marginalis Type B	45.1 ± 0.2	4.1 ± 0.1
P. marginalis Type C	38.5 ± 0.5	6.8 ± 0.1
P. stutzeri ATCC17588	21.0 ± 0.3	3.5 ± 0.1

[a]IFT values for a pectin graded as standard.
[b]Values are averages of three replicate trials ± standard deviation.

The ability to form gels in the presence of calcium ions and after heating was determined with samples prepared at 5.0 mg/ml (for calcium-mediated gellation) or 10 and 20 mg/ml in Milli-Q purified water. To test for calcium-mediated gellation, the samples were dialyzed against 100 mM $CaCl_2$ with a commercial sample of algal alginate as a positive control. After overnight dialysis, only the algal alginate exhibited gel formation. For heat- mediated gelation, all samples were placed in a constant temperature water bath held at 95 C for 30 minutes. After gradual cooling to room temperature only the commercial sample of curdlan, which was included as a positive control, gelled.

The weight average molar mass (Mw), Z average root mean square radius (Rgz) and the weight average intrinsic viscocity (IVw) of selected EPS samples were determined (Table II). Measurements were made by high-performance size

Table II. Physical Properties of *Pseudomonas* Exopolysaccharides

Polysaccharide Type or Source	*Mw ($x10^{-6}$)*	*Rgz (nm)*	*IVw (dL/g)*
P. fluorescens H13	2.85 ± 0.20[a]	123 ± 5	12.6 ± 0.3
P. "gingeri" Pf9	1.06 ± 0.04	108 ± 6	30.6
P. marginalis Type A	1.68 ± 0.03	108 ± 6	8.1 ± 0.1
P. marginalis Type B	1.02 ± 0.03	97 ± 7	35.3 ± 6.0
P. marginalis Type C	1.49 ± 0.03	112 ± 3	25.5 ± 1.0

[a]Values are averages of three replicate trials ± standard deviation except for *P.* "gingeri" Pf9 EPS (1 trial only).

exclusion chromatography with appropriate detectors. Freeze-dried samples were dissolved in 0.05 M $NaNO_3$, passed through a 0.2 µm Nucleopore filter (Costar Corp., Cambridge, MA) and equilibrated at 45 C for 20 minutes. Sample injection volume was 100 µl, and the mobile phase was 0.05 M $NaNO_3$. The mobile phase was degassed and filtered off line by passing it through a 0.4 µm Nucleopore filter. Detection of concentration was by differential refractive index.

For the Mw and Rgz measurements, solutions were pumped through three serially placed chromatography columns (all 8 mm i.d., 300 mm long): Shodex OH-pak SB-806 SB-805, and SB-803 (JM Science Inc., Buffalo, NY). The exclusion limits for these columns as specified by the manufacturer for pullulans were 4 x 10^7, 2 x $10^6$and 1 x 10^5 g/mole, respectively. The Mw and Rgz were measured with a Dawn F MALLS photometer fitted with a helium-neon laser
(A= 632.8 nm) and a K-5 flow cell (Wyatt Technology, Santa Barbara, CA). The nominal flow rate was 0.7 mL/min and the sample concentration was 2 mg/mL.

For IVw measurements, solutions were pumped through four serially placed gel permeation chromatography columns (all 4.6 mm i.d., 250 mm long): two Synchropak GPC 4000 columns and one each of Synchropak GPC 1000 and GPC 100 (SynChrome, Inc., Lafayette, IN). The fractionation range for this column set approximates that of the Shodex columns used in the measurement of Mw and Rgz. The IVw was measured with a model 100 differential viscometer (Viskotek Co., Houston, TX) using a nominal flow rate of 0.45 mL/min and a sample concentration of 1 mg/mL.

The weight average molar masses of the exopolysaccharides in Table II ranged from 1.02 to 2.85 x 10^6, the Z average radii from about 97 to 123 nm and the weight average intrinsic viscosities from about 8.1 to 35.3 dL/g. By way of comparison, a xanthan gum with a Mw of 1 x 10^6 has an IVw of about 24 dL/g in 0.5 M NaCl (50). From this comparison, one may conclude that the chain conformation of *P. marginalis* B and C, and *P.* "gingeri" EPS's may be comparable in extension to that of xanthan. To be sure of this, the physical properties of these three EPS's need to be measured at 0.5 M ionic strength rather than 0.05 as used in our study. The IVw's of EPS's from *P. marginalis* (type A) and *P. fluorescens* appear to be in about the same range as carboxymethyl or hydroxyethel cellulose, but higher than pectin, amylose, hydroxyethyl starch, pullulan, guar gum, and locust bean gum.

Conclusions

The search for new, industrially useful biopolymers obtained by microbial fermentation of renewable agricultural resources such as corn byproducts continues in earnest. The potential for the development of products with unique rheological properties as well as the biodegradable nature of bacterial polymers motivate such research. Although limited at present, a major market for biodegradable plastics made from bacterial polyalkanoates may develop. However, this will be dependent on the economics of the fermentation and manufacturing processes as well as the regulatory climate in the U.S.A.

Our studies with the group I pseudomonads, which comprise a rather limited number of bacterial species, demonstrated the unlimited potential for the discovery of novel biopolymers. Undoubtably, thousands of additional bacterial EPSs and other microbial polymers remain to be discovered, many of which will have unique properties for industrial applications. Studies of biopolymers produced by microbes found in extreme environments may be particularly fruitful.

Literature Cited

1. Glicksman, M. In *Food Hydrocolloids,* Glicksman, M., Ed.; CRC Press, Boca Raton, FL, 1982, Vol. 1; pp 3-18.
2. Lilly, V. G.; Wilson, H. A.; Leach, J. G. *Appl. Microbiol.* **1958,** *6,* 105-108.
3. Jansson, P-E.; Kenne, L.; Lindberg, B. *Carbohydr. Res.,* **1975,** *45,* 275-282.
4. Kang, K. S.; Pettitt, D. J. In *Industrial Gums: Polysaccharides and Their Derivatives,* Whistler, R. L. and BeMiller, J. N., Eds.; Academic Press, Inc., New York, NY, 1993, pp 341-397.
5. Sutherland, I. W. *Biotechnology of Microbial Polysaccharides,* Cambridge University Press, New York, NY, 1990.
6. Kuo, M-S.; Mort, A. J. *Carbohydr. Res.,* **1986,** *156,* 173-187.
7. Harada, T.; Terasaki, M.; Harada, A. In *Industrial Gums: Polysaccharides and Their Derivatives,* Whistler, R. L. and BeMiller, J. N., Eds.; Academic Press, Inc., New York, NY, 1993, pp 427-445.
8. DeBelder, A. N. In *Industrial Gums: Polysaccharides and Their Derivatives,* Whistler, R. L., and BeMiller, J. N., Eds.; Academic Press, Inc. New York, NY, 1993, pp 399-425.
9. Glicksman, M. In *Food Hydrocolloids,* Glicksman, M. Ed.; CRC Press, Boca Raton, FL, 1982, Vol 1; pp 158-166.
10. Ross, P.; Mayer, R.; Benziman, M. *Microbiol. Rev.,* **1991,** *55,* 35-58.

11. Embuscado, M. E.; Marks, J. S.; BeMiller, J. N. *Food Hydrocoll.*, **1994,** *8,* 407-418.
12. Leathers, T. D.; Hayman, G. T.; Cote, G. L. *Curr. Microbiol.*, **1995,** *31,* 19-22.
13. Roller, S.; Dea, I. C. M. *Crit. Rev. Biotechnol.*, **1992,** *12,* 261-277.
14. Doi, Y. *Microbial Polyesters.* VCH Publishers, Inc. New York, NY, 1990.
15. Greasham, R.; Inamine, E. In *Manual of Industrial Microbiology and Biotechnology,* Demain, A. L. and Solomon, N. A., Eds.; American Society for Microbiology, Washington, D. C., 1986, pp 41-48.
16. Gacesa, P.; Russell, N. J. In *Pseudomonas Infection and Alginates: Biochemistry, Genetics and Pathology,* Gacesa, P. and Russell, N. J., Eds.; Chapman and Hall, London, 1990, pp 29-49.
17. King, A. H. In *Food Hydocolloids, Vol. II,* Glicksman, M., Ed.; CRC Press, Boca Raton, FL, 1983, pp 115-188.
18. Palleroni, N. J. In *Bergey's Manual of Systematic Bacteriology,* Krieg, N. R., Ed.; The Williams and Wilkens Co., Baltimore, MD, 1984, Vol. 1; pp 141-199.
19. Evans, L. R.; Linker, A. *J. Bacteriol.*, **1973,** *116,* 915-924.
20. Govan, J. R. W.; Fyfe, J. A. M.; Jarman, T. R. *J. Gen. Microbiol.*, **1981,** *125,* 217-220.
21. Hacking, A. J.; Taylor, I. W. F.; Jarman, T. R.; Govan, J. R.W. *J. Gen. Microbiol.*, **1983,** *129,* 3473-3480.
22. Osman, S. F.; Fett, W. F.; Fishman, M. L. *J. Bacteriol.*, **1986,** *166,* 66-71.
23. May, T.; Chakrabarty, A. M. *Trends Microbiol.*, **1994,** *2,* 151-157.
24. Fett, W. F.; Osman, S. F.; Fishman, M. L.; Siebles, T. S. III. *Appl. Environ. Microbiol.*, **1986,** *52,* 466-473.
25. Fett, W. F.; Osman, S. F.; Dunn. M. F. *Appl. Environ. Microbiol.*, **1989,** *55,* 579-583.
26. Fett, W. F.; Wells, J. M.; Cescutti, P.; Wijey, C. *Appl. Environ. Microbiol.*, **1995,** *61,* 513-517.
27. Gacesa, P. *Carbohydr. Polym.*, **1988,** *8,* 161-182.
28. Fett, W. F.; Wijey, C. *J. Industr. Microbiol.*, **1995,** *14,* 412-415.
29. Chan, R.; Lam, J.; Lam, K.; Costerton, J. W. *J. Clin. Microbiol.*, **1984,** *19,* 8-16.
30. Blumenkrantz, N.; Asboe-Hansen, G. *Anal. Biochem.*, **1973,** *54,* 484-489.
31. Singh, S.; Koehler, B.; Fett, W. F. *Curr. Microbiol.*, **1992,** *25,* 335-339.
32. May, T. B.; Shinabarger, D.; Maharaj, R.; Kato, J.; Chu, L.; DeVault, J. D.; Roychaudhury, S.; Zielinski, N. A.; Berry, A.; Rothmel. R. K.; Misra, T. K.; Chakrabarty, A. M. *Clin. Microbiol. Rev.*, **1991,** *4,* 191-206.
33. Kidambi, S. P., Sundin, G. W., Palmer, D. A., Chakrabarty, A. M., Bender, C. L. *Appl. Environ. Microbiol.*, **1995,** *61,* 2172-2179.
34. Martins, L. E.; Brito, L. C.; Sa-Correia, I. *Enzyme Microb. Technol.*, **1990,** *12,* 794-799.
35. Fett, W. F.; Wijey, C.; Lifson, E. R. *FEMS Microbiol. Lett,* **1992,** *99,* 151-158.
36. Deretic, V.; Schurr, M. J., Boucher, J. C.; Martin, D. W. *J. Bacteriol.*, **1994,** *176,* 2773-2780.
37. Fialho, A. M.; Zielinski, N. A.; Fett, W. F.; Chakrabarty, A. M.; Berry, A. *Appl. Environ. Microbiol.*, **1990,** *56,* 436-443.
38. Liao, C.-H.; McCallus, D. E.; Fett, W. F. *Molec. Plant-Microbe Interact.*, **1994,** *7,* 391-400.

39. Osman, S. F.; Fett, W. F. *J. Bacteriol.*, **1989,** *171,* 1760-1762.
40. Osman, S. F.; Fett, W. F. *Carbohydr. Res.*, **1990,** *199,* 77-82.
41. Osman, S. F.; Fett, W. F. *Carbohydr. Res.*, **1993,** *242*, 271-275.
42. Matulova, M.; Navarini, L.; Osman, S. F.; Fett, W. F. *Carbohydr. Res.*, **1996,** *283,* 195-205.
43. Osman, S. F.; Fett, W. F.; Dudley, R. L. *Carbohydr. Res.*, **1994,** *265,* 319-322.
44. Goldberg, J. B.; Gorman, W. L., Flynn, J. L.; Ohman, D. E. *J. Bacteriol.*, **1993,** *175,* 1303-1308.
45. Cescutti, P.; Osman, S. F.; Fett, W. F.; Weisleder, D. *Carbohydr. Res.*, **1995,** *275,* 371-379.
46. Anderson, A. N.; Parolis, H. *Carbohydr. Res.*, **1989**, *188,* 157-168.
47. Dutton, G. G. S.; Yang, M.-T. *Can. J. Chem.*, **1973,** *51,* 1826-1832.
48. Hildebrand, D. C.; Palleroni, N. J.; Hendson, M.; Toth, J.; Johnson, J. L. *Int. J. Syst. Bacteriol.*, **1994,** *44,* 410-415.
49. Anonymous. Methods for determining the bloom strengths. *J. Offic. Meth. Anal.*, **1990**, *73,* 929.
50. Yalpani, M. *Polysaccharides: Syntheses, Modifications and Structure/Property Relations.* Elsevier Science Publishers, Amsterdam, The Netherlands, 1988.

Chapter 7

Microcellular Starch-Based Foams

G. M. Glenn, R. E. Miller, and D. W. Irving

Cereal Products Utilization Research Unit, Western Regional Research Center, Agricultural Research Service, U.S. Department of Agriculture, 800 Buchanan Street, Albany, CA 94710

Interest has been growing in the development of a new class of foams called microcellular foams. These foams have extremely small cells (<10 μm) and possess unique and useful properties. This chapter discusses some of the important methods for making macro- and microcellular polymeric foams and discusses their structural, physical and mechanical properties. Particular emphasis is focused on starch-based foams. Starch can be extrusion processed into foams having relatively large cells. Microcellular starch-based foams can be made using methods originally developed for other polymeric materials. These foams are white, opaque, have densities ranging from 0.10 g/cm^3 to 0.32 g/cm^3, compressive strengths of 0.22 MPa to 0.97 MPa, and moduli of elasticity from 5.1 MPa to 33 MPa. The microcellular starch-based foams have low thermal conductivity (0.024 W/mK to 0.036 W/mK), large pore volume (0.17 to 0.63 cm^3/g) and large total surface areas (50 to145 m^2/g).

Starch is the principal carbohydrate storage biopolymer of higher plants (biopolymers). It is the principal constituent of wheat endosperm (64% to 74%) and other cereal grains (*1*) and is a valuable source of food. Starch is a mixture of two glucan polymers, amylose and amylopectin. Amylose is a straight chain polymer of α-D-glucopyranosyl units linked by (1→4) bonds and amylopectin is a polymer of α-D-glucopyranosyl units linked by (1→4) bonds with (1→6) branches (*2-4*). Typical starches isolated from wheat and corn consist of approximately 28% amylose and 72% amylopectin (*5-7*). However, breeding programs have developed commercial varieties of corn that produce starches with amylose contents ranging from 0.8% to 80% (*8*).

Starch is the lowest priced and most abundant worldwide food commodity (*9*). The low cost and availability of starch have been consistent over a span of many years which makes it attractive as an industrial raw material. Approximately 4.5 billion pounds of starch, primarily corn starch, are used annually in the U.S. for industrial applications (*8*). Nearly 75% of the corn starch used in industrial applications is converted into adhesives for the paper, paperboard, and related industries (*8*). Most of the wheat starch used for industrial applications is consumed by the paper industry as a wet-end adhesive, surface coating, or as an adhesive for corrugated board. It is also used for textile sizing and carbonless paper (*10*). Other applications for wheat and corn starch include chemicals derived from starch fermentation, starch adhesives, starch-based sizing products, soil conditioners, and starch-based plastics (*8-14*)

The market for single-use, disposable products which is currently being met in large part by petroleum-based plastics is one of the most promising markets for further growth in starch utilization. The market for single-use, disposable plastics was nearly 16 billion pounds in 1992 (*15-16)*. Although petroleum-based plastics perform well for this application, they have some drawbacks. For instance, they are derived from a non-renewable resource and they degrade very slowly, especially in landfills where they comprise 7% of the municipal solid waste by weight and 18% to 20% by volume (*16*). Biodegradable products containing starch continue to be researched and developed for the disposable products market (*8, 17)*. These new biodegradable products are attractive because they are derived from renewable resources, they are compostable, and they burn clean which is important for countries that burn their waste (*17*).

Polymeric foams, many of which are made from plastic polymers, help comprise a number of potential niche markets for starch-based products. Polymeric foams consist of a solid polymer matrix formed into discrete elements or empty cells and a gas phase that fills the void space (*18, 19*). Polymeric foams may be categorized into two broad groups, macrocellular and microcellular, based on their cell size. Most commercially available, low density, macrocellular foams are made from organic polymers and have large (100 to 1000 μm) closed cells with cell wall thicknesses ranging from 10 to 100 μm (*20*). Microcellular foams have cell diameters under 10 μm and have physical and mechanical properties that are unique and of commercial interest (*20-23*). However, microcellular foams are generally more costly to produce and have been primarily made on a small scale (*20*).

This chapter reviews various processes for making macro- and microcellular foams and discusses some of the structural, physical, and mechanical properties they possess. Particular emphasis is given to foams made from starch.

Formation of Polymeric Foams

Macrocellular Foams. The most widely used system for producing polymeric foams consists of simultaneously generating a gas and nucleating gas-filled bubbles in a fluid polymer phase (*18-19*). The bubbles expand and are stabilized as the polymer solidifies to make the resultant foam (*19*). A gas can be incorporated throughout a polymer melt by mechanical agitation or by blowing agents. Blowing agents commonly used in extrusion processing include compressed gases, typically CO_2 or N_2, and liquids with low boiling points. These agents form gases which expand as the polymer melt leaves the die (*18*). Chemical blowing agents that are used in making foams include compounds that react when mixed or heated to produce a gas within the polymer matrix (*24*).

One of the first commercial foams to achieve prominence was made of cellulose acetate. Cellulose and amylose have the same empirical formula but cellulose is a straight-chain polymer of β-D-glucopyranosyl units linked by (1→4) bonds (*25*). Cellulose acetate foams were first widely used in aircraft manufacture and used later for specialty applications (*26*). The foams were made by extruding cellulose acetate mixed with an acetone/ethanol liquid. At elevated temperatures, the liquid mixture became a solvent and plasticizer for the cellulose acetate as well as a foaming agent.

Macrocellular foams made from petroleum-based polymers such as polystyrene are commercially important. These foams are made either by extrusion or expandable bead processing. In extrusion processing of polystyrene, a blowing agent is introduced in the extruder barrel where it is thoroughly mixed with the polymer melt and extruded. A foam is produced when the blowing agent expands and forms bubbles within the melt upon exiting the die of the extruder. The foam is stabilized as the melt cools and solidifies.

Expandable bead polystyrene is made by incorporating a blowing agent within small spheres of polystyrene. The beads are then heated to initiate melting and to induce gaseous bubbles within the bead. After cooling and curing for several hours, the beads are poured into molds and again heated to form a polymer melt and to induce further expansion. The process causes the beads to become tacky and adhere to each other which aids the molding process. There were more than 1.6 billion pounds of polystyrene processed into various products in 1992 of which nearly 600 million pounds was made into foamed articles (27).

Other foams produced commercially are made of polyethylene, phenolics, phenylene oxide, polyurethane, polyvinyl chloride, and silicone (*28*). The markets that might be suitable candidates for biodegradable starch-based foams are the polyethylene and polyurethane foam markets used in packaging applications. The polyethylene foams are produced by extrusion processing. Polyurethane foams are produced by the reaction between isocyanate and hydroxyl compounds and a second reaction between isocyanate and water which liberates CO_2 gas as a blowing agent (*24*). Alternatively, the second reaction can be minimized and a blowing agent that has more desirable physical properties such as low thermal conductivity may be utilized (*24*).

Starch-Based Macrocellular Foams. The first macrocellular foams containing starch were baked foods made using leavening agents. The use of leavening agents dates back to the time of the ancient Egyptian some 6,000 years ago (*29*). Leavening agents create large numbers of thin walled cells by generating gases within the dough mixtures. These gases may be generated by microbial activity (fermentation) or by the reaction of chemical additives (*29-30*). The dough matrix is stabilized by the baking process resulting in a foamed product containing large cells.

Extrusion processing is another important method for making starch-containing foams. It has become a rapid, efficient way to make snack foods and ready to eat cereals as well as some non-food products. The fundamental studies that provided the basis for commercial development of starch-containing foods date back to the 1970's (*31*). Extrusion of starch involves mixing, heating and conveying the starch through a heated barrel using a single or twin screw configuration. The two most important parameters affecting extrudate expansion are moisture content and extrusion temperature (*31*). The starting material typically has moisture contents ranging from 10-25%. However, studies using corn starch show that maximum expansion occurs in the range of 14 to 15% moisture content and barrel temperatures of 170 °C to 230 °C (*31-32*).

Extrusion processing of starch begins by feeding the starch into the barrel of an extruder where it is compacted by the screw and quickly heated to form a melt. The moisture within the barrel is thoroughly dispersed throughout the melt and becomes superheated under the conditions of high heat and pressure that are generated. As the starch melt exits the barrel through the die, the pressure drops to atmospheric converting superheated moisture to steam and causing the melt to expand into a foam (*33-34*). This foam process occurs in three steps the first of which is the nucleation step where vapor filled bubbles form within the melt (*33*). Next, the bubbles grow as water vapor diffuses from the melt and finally, the starch matrix cools and dries as it stabilizes. Water has a dual role in starch foam extrusion. It acts as both a plasticizer that reduces melt viscosity and as a blowing agent (*33*).

Starch foams can also be made by puffing starch. Puffing occurs in popcorn when temperatures reach approximately 177 °C (*35*). At these temperatures, the water within the popcorn kernel becomes superheated and pressure within the kernel approaches 135 psi (*35*). Once the pericarp that encases the kernel ruptures, the superheated water within the kernel forms steam and causes the product to expand or puff. Few other cereal grains are able to puff without puffing equipment that artificially creates the necessary conditions for puffing. Puffing machines are used commercially to manufacture a variety of food products (*36*) but the process is poorly suited for high volume, non-food markets.

Microcellular Foams. A category of foams distinct from the foams previously mentioned is the microcellular foams. Microcellular foams have been made from various polymeric materials and are classified as foams with cell sizes smaller than 10 µm, cell densities greater than 10^9 cells/cm^3, and a specific density reduction of 5 to 98% of the solid parent material (*21, 23*). Microcellular foams have been successfully made using replication of removable pore formers, polymerization of inverse monomer emulsions, blowing/nucleation techniques, and phase separation of polymer solutions (*22, 37*).

The first method mentioned, foams made by replicating removable pore formers, utilizes small solid particles around which monomer solutions are polymerized. The solid particles are then removed by dissolution, pyrolization or some other means leaving a distribution of small voids throughout the polymer matrix (*22*). The second method mentioned for making microcellular foams, polymerization of inverse monomer emulsions, consists of polymerizing monomers that are miscible in oil around surfactant-stabilized water droplets that are dispersed in the oil.

The third method discussed for making microcellular foams, blowing/nucleation techniques, is being used to make microcellular foams from plastics (*38*).The typical approach used to make microcellular plastics is to inject CO_2 under high pressures into the barrel of an extruder during plastic processing (21, 23, 39-40). The liquified gas is mixed thoroughly with the polymer and expands as the polymer leaves the die. The amount of CO_2 injected into the barrel must be accurately gauged to obtain a constant and optimum gas:polymer weight ratio (*40*). An excess of gas leads to large cells (*23*). Microcellular foams with pores that are uniform in size and under 10 µm can be made by maintaining a gas/polymer mixture with a constant gas:polymer weight ratio (*21*).

Starch foams blown by liquid CO_2 injections have been reported (*34*). Ferdinand et al. (*34*) used a temperature profile that kept the starch extrudate below 100 °C and injected CO_2 into the extruder barrel. The starch extrudate had a higher density using CO_2 versus water/steam as the blowing agent. The low density in water\steam blown extrudate may be attributed to the starch melt being better able to stabilize bubbles at the higher temperatures achieved in a water\steam blown extrudate. The cells of the CO_2 blown extrudate were fairly uniform in size but had thick walls and were much larger than 10 µm (*34*). Small cell size may be possible by better defining the critical processing parameters and improving processing methods.

Another important method for making microcellular foams is phase separation of polymer solutions. Pekala et al. (*37*) have divided this technique into three categories consisting of phase separation by rapid cooling, chemical induction, and by gelation and crystallization. Phase separation by rapid cooling occurs when a homogenous solution containing high molecular weight polymers is cooled forcing separation of the solvent from the polymer and forming a continuous polymer matrix in which are dispersed discrete cells containing solvent (*22*). Removal of the solvent droplets from the matrix results in an open cell structure where the voids are replicas of the solvent droplets (*41*). One of the first reported microcellullar foams made using this technique was made using solutions of dextran in water (*42*). The solutions were quickly frozen and freeze-dried. The dextran foams had low densities (0.05 g/cm^3) and pore sizes under 1 µm. The cooling method has been used to develop microcellular foams from various other polymeric materials but has been limited by the inability to make foam samples thicker than 1 cm because of the difficulty in achieving rapid and uniform freezing rates (*43*).

Phase separation by chemical induction utilizes multifunctional monomers that are polymerized and cross-linked to form a three dimensional network that exists as a stable, rigid gel (*22*). This method of fabricating microcellular foams is called "sol-gel" processing and is used to make transparent, low density, microcellular foams known as the aerogels (20, *22, 37, 44-45*). Aerogels were first described in 1931 (*46*) and were produced from firm aqueous gels (aquagels) by displacing water within the gel matrix with air (*46-47*). The aerogels could not be prepared by simply air-drying aquagels because of surface tension. Surface tension was created as menisci formed by water within the pores of the gel retreated into the gel body forcing the gel structure to collapse (*48*). Kistler protected the aquagel structure from surface tension by transforming the liquid within the gel into a supercritical fluid using a heated autoclave (*46-47*). A supercritical fluid does not form menisci and can be easily removed from the gel without creating surface tension. Using this method, aerogels have been produced from materials that include alumina, tungsten, ferric and stannic oxide, nickel tartrate, cellulose, cellulose nitrate, silica, egg albumin, resorcinol, and gelatin (*46-47*).

Phase separation by gelation and crystallization occurs when a high molecular weight, semicrystalline polymer is dissolved in an organic solvent and cooled to induce phase separation and produce a gel (*22*). A microcellular foam can be produced from the gel if it is rigid enough to withstand the compressive forces associated with solvent evaporation.

Starch-Based Microcellular Foams. Starch is known to swell and hydrate in the presence of water in a process known as gelatinization (*49*). As the gelatinized starch is cooled, a gelation process occurs as the starch molecules reassociate (*49*). Starches containing higher amounts of amylose set quickly into gels because the straight-chain amylose molecules tend to associate more freely than the branched amylopectin molecules (*49*).

Starch-based microcellular foams can be made from rigid starch gels. For instance, 8% (w/w) solutions of unmodified wheat and corn starches can be gelatinized, poured into molds, and cooled to form rigid aquagels (*50*). The rigid aquagels collapse into dense materials, however, when simply dried in air. Low density starch foams can be made if the water phase is removed by freeze-drying as described earlier for dextran solutions (*42, 50*). Large pore sizes result when the aquagels are frozen slowly and freeze-dried (*50*). Microcellular foams may be formed using freeze-drying methods provided the freezing is done rapidly and there are many nucleating sites to deter the formation of large ice crystals (*43*).

An alternative method of forming starch-based microcellular foams from rigid starch aquagels is to exchange the water that fills the matrix of the aquagel with a liquid possessing lower surface tension. Glenn and Irving (*50*) demonstrated that microcellular foams could be made from rigid wheat and corn starch aquagels by equilibrating the aquagels in ethanol before air-drying. Since the surface tension created at the air/ethanol interface was only one third that of air/water, the gels did not collapse. High amylose corn starch aquagels had less compressive strength than the unmodified wheat and corn starch aquagels and collapsed when exchanged in ethanol then air dried. The high amylose gels formed low density foams when exchanged with ethanol followed by liquid CO_2 before air-drying. This drying process reduced surface tension by over seventy times that of air/water (*50*). Filling the gel matrix with supercritical CO_2 before air drying (critical point drying or CPD) completely eliminates surface tension and also results in low density foams but is the most expensive drying method of the three investigated (*50*).

Cellular Morphology of Foams

Macrocellular Foams. Polymeric foams consist of a solid polymer matrix and a gaseous phase and are formed into discrete elements or cells (*22, 51-52*). The cell structure within the foam may be open or closed (Figure 1a-c). Open cells provide a continuous gaseous phase throughout the foam. Closed-cell foams have continuous cell walls that seal intracellular air thus making the gaseous phase discontinuous (*19, 22, 54*). As mentioned previously, foams can be made in various ways but the most common method utilizes gas expansion (*19*). Expanding gases form bubbles within the polymer melt which are then stabilized and result in closed-cell or open cell foams (*19, 53*). The closed-cell foams may have spherical cells with thick walls as in the case of thermoplastic structural foams. By contrast, low density, closed-cell foams have thin walls and cells composed of pentagonal dodecahedrons (twelve faces per cell with each face consisting of a pentagon, see figure 2 a, d) (*53-54*).

a

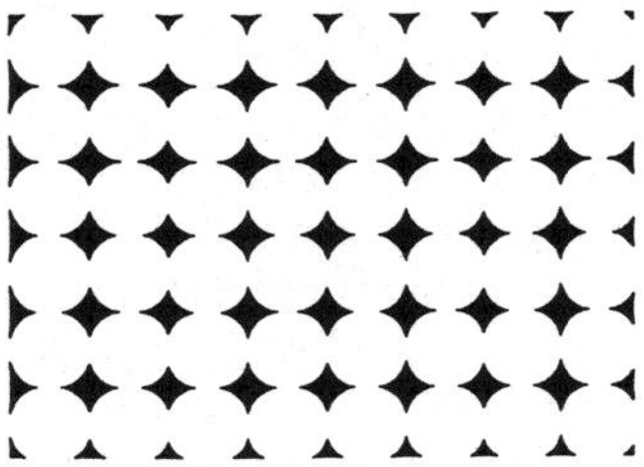

b

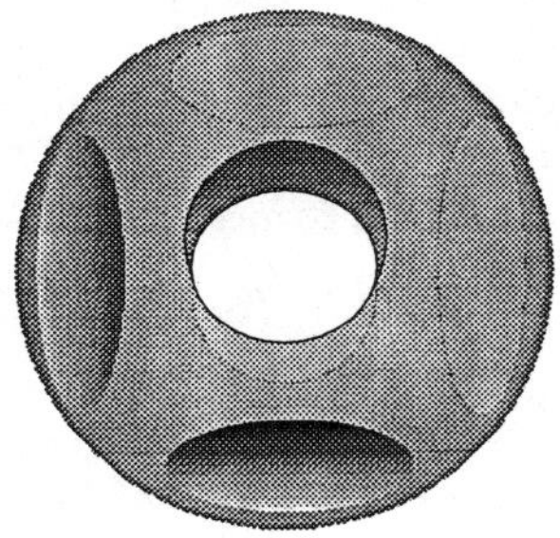

c

Figure 1. Drawings are graphical representations of open and closed cell foams. Foams with closed cells have a continuous solid phase but a discontinuous gaseous phase (a). Foams with open cells have both a continuous solid and a gaseous phase (b,c). The cell (c) represents and individual cell illustrated in b.

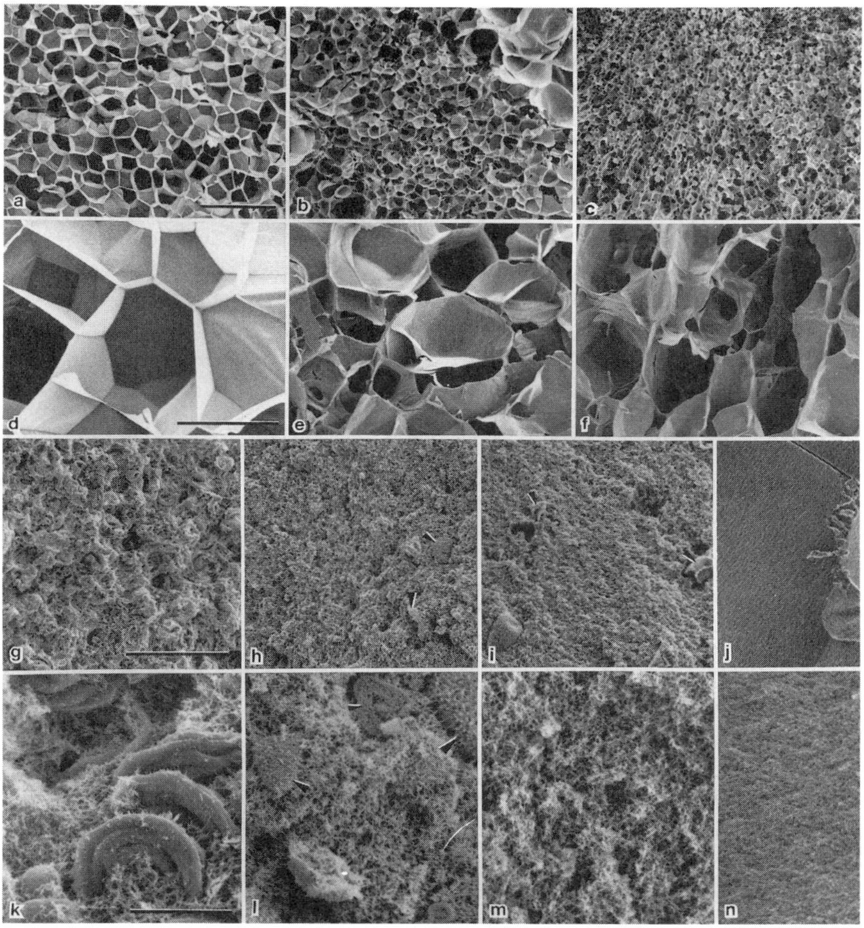

Figure 2a-f. Scanning electron micrographs of beaded polystyrene (a,d), puffed wheat (b,e), and freeze-dried foam made from 8% wheat starch aquagel (c,f). The foams all contained large cells with sheet-like cell walls. Note that the cells of the beaded polystyrene foam appear to be of different sizes because the cells are not cut along the same axis. Figure 2g-n are scanning electron micrographs of wheat starch foams (g,k), regular corn starch foams (h,l), high amylose corn starch foams (i,m), and transparent silica aerogel (j,n). Note that figures g-j are the same magnification as figures d-f. The wheat starch, corn starch, and high amylose corn starch foams had much smaller pore sizes than the samples in d-f. The pore size of the silica aerogel was much too small to discern even at high magnification (n). The wheat starch foams contained numerous starch granule remnants and some pores as large as 2μm (g,k). The corn starch foams also contained numerous starch granule remnants (arrows-h,l) and pores as large as 1μm. The high amylose corn starch foams had fewer starch granule remnants than the other starch foams (arrows-i) and a pore size similar to that of corn starch foams. Magnification bar for a-c (in fig a) = 500μm; bar for d-f (in fig d) = 100μm; bar for g-j (in fig g) = 100μm; bar for k-n (in fig k) = 10μm.
(Reproduced with permission from ref. 50. Figure not copyrightable.)

Open cells may develop in some foams where individual cells within the foam matrix rupture due to excessive gas expansion (*53*). A foam may have a mixture of both closed and open cells or be all open cell. The classic example of open cell foams has been reported for polyurethane foams. These foams have an interconnecting network of struts and have cell faces that are pentagonally shaped (*53*). The expanding cells in typical foams initially are closed but can open when the wall membranes are chemically dissolved or mechanically ruptured.

The range in cell size distribution of a stabilized foam is largely dependent upon the amount of blowing agent used and the uniformity of its distribution throughout the polymer when in a fluid state (*23*). Commercial foams often have a nonuniform cell size distribution because of the methods used to mix the blowing agent and polymer (*51*). Polyurethane foams are somewhat unique for a blown foam since, depending on processing parameters, they can be open or closed-cell, rigid or flexible, and may have large (600-900 μm) or small cells (10-100 μm) (*22*).

Starch-Based Macrocellular Foams. Low density starch-based foams that use water/steam as a blowing agent (33) typically develop closed cells (Figure 2 b, e). In a study of the mechanism of popping popcorn, Hoseney et al (*35*) showed that popped corn had closed-cells with a wide range in cell size and wall thickness. The cells typically were pentagonal in shape and were larger than 20 μm. The cell size and wall thickness of extruded starch-based foams are affected by various processing parameters such as moisture content, temperature, die pressure, and die diameter (*31*, 55). Extruded starch foams have a closed cell structure (*56-57*). Cell size varies in extruded starch foams. Low density starch-based foams typically have cell diameters that range from 0.1 mm to 1.5 mm (Figure 2 b, e). Cells with diameters smaller than 0.1 mm may have thick cell walls (*57*).

Starch foams have been made by a thermally induced phase inversion process better known as freeze-drying (*41*). Frozen gels envelop nucleated ice particles and form foams containing both open and closed cells (Figure 2 c, f). The cell size is a direct replica of nucleated ice particles (*43*). Small crystals are possible with rapid freezing techniques and the use of proper ice nucleators (*42-43*). The walls of cells from freeze-dried starch gels are continuous and sheet-like just as was obtained with starch foams prepared by extrusion or puffing (compare figure 2 e and f) (*50*).

Microcellular Foams. Some microcellular foams have a similar morphology to the foams previously described. Baldwin et al. (*38*) showed that microcellular plastics made using liquid CO_2 typically have a uniform cell geometry. The cells were closed and appeared to be polyhedral with relatively thick cell walls. Starch-based foams produced by CO_2 injection during extrusion processing resulted in products of reduced density and cell size (*34*). The cells of these starch-based foams were spherical, variable in size, and had thick walls.

Microcellular foams have some morphologies that differ from the macrocellular foams previously described (*22*). Microcellular foams may not have easily recognizable "cells." (*22*). In these cases, the cells may be irregular voids surrounded by a polymer mass composed of irregular structures (*20*). Some morphologies that have been described include reticulate, spinodal, inverse emulsion, and aerogel (*20*). A number of morphologies made by phase separation have a complex microstructure and are difficult to describe. The reticulate structure is open-cell consisting of a strut-like network (*53*). Spinodal foams are open cell and consist of curved polymer structures that span throughout the foam matrix. Inverse emulsion foams may be open or closed cell (*20, 22*). The cellular morphology of aerogels has been recently reported (*20, 22, 58*). The aerogels are the microcellular foams with the smallest average cell size among the microcellular foams reported. Aerogels have an open-cell structure and have pore sizes under 0.1 μm (*22*). The microstructure is formed by small spherical particles (3-10 nm in diameter) that covalently bind to other particles forming chains. These chains intertangle or cross-link to form rigid gels. The structure as viewed under transmission electron microscopy resembles intermeshing chains of beads (*22, 58*).

Starch-Based Microcellular Foams. Starch gels that are critical point dried in CO_2 have a complex structure unlike that of extruded or freeze-dried starch (Figure 2 g-i, k-m). Their structure consists of a matrix of small intermeshing chains or fibers in which are embedded residual granules of starch. Since amylose chains diffuse out of the starch granules more readily than amylopectin (*49, 59*), the intermeshing fibers are likely to be composed mostly of amylose chains. The starch granule remnants are likely to be enriched with amylopectin. The number of the starch granule remnants are lower for high amylose corn starch compared to regular wheat and corn starch foams (*50*). The cells or pores appear open and have diameters estimated to be under 2 μm. It is interesting to note the differences in cell wall structure between the microcellular starch-based foams and other starch foams which typically consist of a smooth, continuous film (compare figure 2 e, f and figure 2 k-m). The differences in cell wall structure have a profound impact on the mechanical properties of the foams (*50*) The pore sizes of the starch-based microcellular foams (Figure 2 k-m) was larger than that of a typical silica aerogel (Figure 2 j, n).

Physical and Mechanical Properties of Polymeric Foams

Macrocellular Foams. The number of applications for polymeric foam products continues to expand because of their useful physical and mechanical properties. Polymeric foams are generally light weight, have high strength:weight ratio, provide thermal insulation and effectively attenuate shock, vibration, and sound. The selection of a polymer suitable for industrial foam applications depends upon its properties, ease of processing, and cost. A sustained growth in demand for polymeric foams has been achieved by utilizing foams in a myriad of products that include furniture, vehicles, bedding, carpet pads, packaging, food and drink containers, and shoes (*19*).

Macrocellular Starch-Based Foams. The physical and mechanical properties of starch-based foams have been studied less than most polymeric foams, especially from a non-food perspective. Non-foamed, starch-based products should bear some common properties with their foamed counterparts. Non-foamed starch products have been evaluated as potential substitutes for petroleum-based plastics (*60-63*). These products have achieved limited commercial success because of moisture sensitivity and the tendency to embrittle during aging (*64*). Embrittlement has been attributed to moisture loss and free volume relaxation (*64*).

Embrittlement during aging is less well documented for starch-based foams than for starch plastics. However, the effect of moisture content on the physical and mechanical properties of starch foams is well documented (*65*). Katz and Labuza (*65*) showed that the mechanical properties of various foamed food products are highly sensitive to moisture content and are affected by the way the products are stored. Hutchinson et al. (*66*) investigated the effect of moisture on extruded food foams including starch foams. They showed that the post-extrusion environment affects foam properties as much as the extrusion conditions themselves. Compressive strength, impact strength and flexural modulus have all been correlated with bulk density. Bulk density of extruded starch foams varies with changes in relative humidity (*66*). Starch-based foams have the greatest flexural modulus at low moisture content whereas compressive strength and impact strength are greatest for foams containing intermediate amount of moisture (*66, 67*).

The mechanical properties of extruded foams relate to bulk density in foamed plastics (*68*) and food foams (*66-67, 69*). Flexural strength, impact strength and flexural modulus increase with density in extruded wheat starch foams (*66*). Hayter et al. (*69*) reported a correlation between decreasing pore size and decreasing bulk density of starch foams. The opposite is reported for plastic foams where decreasing bulk density is correlated with increasing pore size. Hayter et al. also showed that extruded starch foams have less than one-fourth the impact resistance of a plastic foam with a similar bulk density (*69*).

The expansion ratio of starch foams affects bulk density and, consequently, an array of other mechanical properties. Chinnaswamy and Hanna (*70-71*) found that among the factors affecting the expansion ratio were the amylose content and moisture content of the starch. The maximum expansion was achieved with starch mixtures containing 50% amylose on a dry weight basis and 13-14% moisture content (*71*). The bulk density of these starch-based foams was inversely related to the expansion ratio and ranged from 0.163 to 0.348 g/cm^3. The shear strength of the foams increased 4 to 5 fold with increasing amylose content but was inversely related to the expansion ratio.

Extruded starch has been used commercially for several years for making snack-foods and crispbreads (*66, 70*). The most important non-food, starch-based foam reported to date is currently used as a biodegradable, loose-fill packaging material. Lacourse and Altieri (*56*) exploited the fact that starches high in amylose produced foams with low bulk density (*71*). They determined the bulk densities, percent resilience (ability to recover after deformation), and a measure of compressive strength for different starch foams and a polystyrene foam used commercially for loose-fill packaging. Extruded high amylose corn starches were made with densities similar to the polystyrene foam but with lower resiliency and compressive strength (*56*). Extruded starch mixtures containing propylene oxide (5%) and polyvinyl alcohol (8%) had over three times the bulk density of polystyrene foams but a similar resiliency and compressive strength.

Microcellular Foams. Microcellular plastic foams became of interest in recent years as a way to conserve petroleum resources and reduce material costs. It was reasoned that plastic foams with cell sizes smaller than the size of pre-existing flaws would reduce material costs without compromising the mechanical properties (*23, 39-40*). Studies indicate that microcellular plastic foams typically exhibit improved impact strength, toughness, stiffness to weight ratio, and reduced thermal conductivity compared to unfoamed plastic (*37-40*). The major obstacle to the commercialization of microcellular plastic foams is the need to reduce processing costs for making high quality products.

Aerogels are another type of microcellular foam that has been extensively studied (*22, 44, 48*). Some of the unique properties they possess are low density (0.004-0.300 g/cm^3), extremely small cell size (0.01-25 µm), low thermal conductivity, high surface area (400-1000 m^2/g), porosities as high as 99.8% and they are often translucent or transparent (*22, 37, 48*). The widest practical application for aerogels to date has been for use in Cerenkov radiation detectors. However, other potential uses include insulation for windows, refrigerators, and passive solar collectors (*44*). The greatest hurdle to further commercialization of aerogel foams is the fabrication cost. Aerogels require incubation periods to allow for solvent exchange and they are typically processed in expensive and potentially dangerous high pressure vessels (*44*).

Starch-Based Microcellular Foams. The physical and mechanical properties of relatively few microcellular foams made from starch polymers have been reported. Coudeville et al. (*42*) used freeze-drying techniques to create a low density (0.05 g/cm^3) foam coating of dextran on microspheres. The foams had open cells of uniform size (1 µm). Freeze-dried starch-based foams that were much larger (1.56 cm in diameter) and frozen slowly had large cell sizes due to the formation of large crystals (*50*). Many of the foams were of poor quality and broke easily in tensile tests. However, freeze-dried samples that had remained intact had bulk densities only 5-6 times higher than low density polystyrene foams and approximately twice the compressive and tensile strength (compare Tables I, II and III).

Microcellular foams prepared from rigid aquagels of unmodified wheat, corn, and high amylose corn starches have been studied and characterized (*50*). Of the various starches tested (Tables I-III), high amylose starch gels had the lowest bulk density (0.15 g/cm^3) which was eight fold higher than that of the polystyrene foam tested (Table III). The starch microcellular foams were hygroscopic (11.5% moisture content at 50% relative humidity) and had higher compressive strength and modulus of elasticity than the polystyrene foam sample but had similar tensile strength. (compare Tables I-III). The high amylose starch microcellular foams were white and opaque and had lower thermal conductivity than the polystyrene foam (Table III). The pore volume and total surface areas were higher for the high amylose starch foams than for the other starch foams tested (Tables I-III).

The densities and the compressive strengths of starch microcellular foams made from unmodified wheat and corn starches were higher than the densities of the polystyrene sample and the high amylose starch foams (Tables I-III). Even though the densities were much higher than that of the polystyrene, the unmodified wheat and

Table I. Physical and mechanical properties of unmodified wheat starch foams conditioned at least 48 hours at 50% relative humidity

Sample	D (g/cm^3)[a]	S_c (MPa)[b]	E (MPa)[c]	(W/m·K)[d]	S_u (Mpa)[e]	V_p(cm^3/g)[f]	A_s(m^2/g)[g]
Air-dry wheat	0.27a[h]	0.71a	23a	0.044a	-----	-----	-----
CO_2 Extracted	0.26a	0.57a	21a	0.037c	0.17a	-----	-----
CPD	0.26a	0.56a	21a	0.036c	-----	0.48	116
Freeze	0.12b	0.23b	2.9b	0.040b	0.46b	-----	-----

[a]D = Density; moisture content of samples was 11.5%
[b]S_c = Compressive Strength measured at 10% deformation.
[c]E = Compressive Modulus of Elasticity.
[d]Thermal conductivity at 22.8°C mean temperature. Thermal conductivity of still air is 0.0226 W/m·K.
[e]S_u = Tensile Strength measured at breaking point.
[f]V_p = Pore volume determined by BET analysis.
[g]A_s = Total surface area determined by BET analysis.
[h]Values within columns followed by a different letter are significantly different at the 95% confidence level (Fisher's protected lsd).
SOURCE: Adapted from ref. 50.

Table II. Physical and mechanical properties of unmodified corn starch foams conditioned at least 48 hours at 50% relative humidity

Sample	D (g/cm^3)[a]	S_c (MPa)[b]	E (MPa)[c]	(W/m·K)[d]	S_u (Mpa)[e]	V_p(cm^3/g)[f]	A_s(m^2/g)[g]
Air-dry	0.31a[h]	1.14a	36a	0.037a	-----	-----	-----
CO_2 Extracted	0.29a	1.00a	32a	0.033c	0.28a	-----	-----
CPD	0.29a	0.97a	33a	0.033c	-----	0.17	50
Freeze	0.12b	0.19b	2.9b	0.040b	30.48b	-----	-----

[a]D = Density; moisture content of samples was 11.5%
[b]S_c = Compressive Strength measured at 10% deformation.
[c]E = Compressive Modulus of Elasticity.
[d]Thermal conductivity at 22.8°C mean temperature. Thermal conductivity of still air is 0.0226 W/m·K.
[e]S_u = Tensile Strength measured at breaking point.
[f]V_p = Pore volume determined by BET analysis.
[g]A_s = Total surface area determined by BET analysis.
[h]Values within columns followed by a different letter are significantly different at the 95% confidence level (Fisher's protected lsd).
SOURCE: Adapted from ref. 50.

Table III. Physical and mechanical properties of high amylose corn starch and expanded-bead polystyrene (PS) conditioned at least 48 hours at 50% relative humidity

Sample	D (g/cm^3)[a]	S_c (MPa)[b]	E (MPa)[c]	(W/m·K)[d]	S_u (Mpa)[e]	V_p(cm^3/g)[f]	A_s(m^2/g)[g]
Air-dry	1.31a[h]	-----	------	-----	-----	-----	-----
CO_2 Extracted	0.16c	0.27a	8.1a	0.024a	0.23a	-----	-----
CPD	0.15c	0.22a	5.1a	0.024a	-----	0.63	145
Beaded Polystyrene	0.019b	0.098b	3.6a	0.036b	20.19a	-----	-----

[a]D = Density; moisture content of samples was 11.5%
[b]S_c = Compressive Strength measured at 10% deformation.
[c]E = Compressive Modulus of Elasticity.
[d]Thermal conductivity at 22.8°C mean temperature. Thermal conductivity of still air is 0.0226 W/m·K.
[e]S_u = Tensile Strength measured at breaking point.
[f]V_p = Pore volume determined by BET analysis.
[g]A_s = Total surface area determined by BET analysis.
[h]Values within columns followed by a different letter are significantly different at the 95% confidence level (Fisher's protected lsd).
SOURCE: Adapted from ref. 50.

corn foams all had thermal conductivities similar to polystyrene. The tensile strength was much lower than the compressive strength in the unmodified wheat and corn starch foams. The reason given for this difference was because of the number of starch granule remnants present in the matrix of the foams (Figure 2 k-i). These remnants could act as sites where cracks could initiate and cause structural failure (*50*). The high amylose starch foams, in contrast, did not have nearly the number of starch granule remnants and had similar strength in both tension and compression (Table III). The fibrous, open cell structure of the microcellular starch foams was not as strong under tension as the freeze-dried starch foams which had a continuous, film-like cell wall structure (compare Figure 2 f and k-i). Alternative methods of processing starch to form microcellular foams with continuous cell walls should result in improved tensile strength.

Microcellular foams in general have been produced on the laboratory scale and are still being evaluated for commercial potential (*22*). The potential uses of starch-based microcellular foams for industrial products must be considered as they continued to be developed. The starch-based microcellular foams have properties such as low thermal conductivity and high compressive strength that are superior to many commercial macrocellular polymeric foams. Their moisture sensitivity may limit their use for some applications. However, further research in developing starch-based microcellular foams with improved moisture resistance and tensile strength may result in new, important markets for starch.

Literature Cited

1. MacMasters, M. M.; Hinton, J. J. C.; Bradbury, D. In *Wheat Chemistry and Technology;* Pomeranz, Y., Ed.; Am. Assoc. Cereal Chem.: St. Paul, MN., 1971, pp 51-114.
2. Wolfrom, M. L.; El Khadem, H. In *Starch/Staerke Chemistry and Technology*; Whistler, R. L.; Paschall, E. F., Eds.; Academic Press: New York, NY, 1965, Vol. 1; pp 251-278.
3. Hough, L.; Jones, J. K. N. In *Starch and Its Derivatives*; Radley, J. A., Ed.; Chapman & Hall: London, 1953, Vol. 1; pp 25-57.
4. Greenwood, C. T. *Advan.Carbohydr. Chem.* **1956**, *11*, 335-393.
5. Medcalf, D. G.; Gilles, K. A. *Cereal Chem.* **1965**, *42*, 558-568.
6. Whistler, R. L.; Weatherwax, P. *Cereal Chem.* **1948**, *25*, 71-75.
7. Deatherage, W. L.; MacMasters, M. M.; Rist, C. E. *Trans. Am. Assoc. Cereal Chem.* **1955**, *13*, 31-38.
8. U.S. Congress, Office of Technology Assessment, *Biopolymers: Making Materials Nature's Way*, OTA-BP-E-102; U.S. Government Printing Office: Washington DC, 1993; pp 19-50.
9. Whistler, R. L. In *Starch Chemistry and Technology*; Whistler, R. L.; Bemiller, J. N.; Paschall, E. F., Eds.; Academic Press: New York, NY., 1984, pp 1-9.
10. Kirby, K. W. In *Developments in Carbohydrate Chemistry*; Alexander, R. J.; Zoebel, H. F., Eds.; American Association of Cereal Chemists: St. Paul, MN, 1992, pp 371-386.

11. Otey, F.H.; Doane, W. M. In *Starch Chemistry and Technology*; Whistler, R. L.; Bemiller, J. N.; Paschall, E. F., Eds.; Academic Press: New York, NY., 1984, pp 389-414.
12. Koch, H; Roper, H. *Starch/Staerke,* **1988**, *40,* 121-131.
13. Roper, H.; Koch, H. *Starch/Staerke,* **1990**, *42,* 123-130.
14. Bushuk, W. *Cereal Foods World,* **1986**, *31*, 218-226.
15. Anon. *Modern Plastics*, **1992**, *69*, 85-92.
16. Thayer, A. M. *Chem. Engin. News*, **June 25, 1990**, 7-14.
17. Colvin, R. *Modern Plastics*, **1995**, *72*, 17-19.
18. Hilyard, N. C.; Young, J. In *Mechanics of Cellular Plastics*, Hilyard, N. C., Ed.; Macmillan Publishing Co.: New York, NY., 1982, 1-26.
19. Saunders, J. H. In *Handbook of Polymeric Foams and Foam Technology;* Klempner, D.; Frisch, K. C., Eds.; Oxford University Press: New York, NY., 1991, 1-25.
20. Williams, J. M.; Wrobleski, D. A. *J. Mater. Sci. Letters*, **1989**, *24*, 4062-4067.
21. Baldwin, D. F.; Park, C. B.; Suh, N. P. In: *Cellular and Microcellular Materials;* Kumar, V.; Seeler, K. A., Eds.; ASME: New York, NY, 1994, Vol. 53; pp 85-107.
22. LeMay, J. D.; Hopper, R. W.; Hrubesh, L. W.; Pekala, R. W. *Mater. Res. Soc. Bull.* **1990**, *194*, 19-45.
23. Park, C. B.; Baldwin, D. F.; Suh, N. P. In: *Cellular and Microcellular Materials,* Kumar, V.; Seeler, K. A., Eds.; ASME: New York, NY, 1994, Vol. 53; pp 109-124.
24. Benning, C. J. In: *Plastic Foams;* Akin, R. B.; Mark, H. F.; Scavuzzo, J. J.; Stivala, S. S., Zukor, L. J., Eds.; Wiley and Sons: New York, NY, 1969, Vol. 1; pp 117-344.
25. Young, A. H. In: *Starch Chemistry and Technology*, Whistler, R. L.; Bemiller, J. N.; Paschall, E. F., Eds.; Academic Press, New York, NY, 1984; pp 249-283.
26. Ferrigno, T. H. In: *Rigid Plastics Foams*, Reinhold Publishing Corp., New York, N. Y., 1967; pp 357-358.
27. Narayan, R. In: *Polymers From Agricultural Coproducts*, Fishman, M. L.; Friedman, R. B.; Huang, S. J., Eds.; ACS Symposium Series 575; ACS: Washington DC, 1994; pp 9-12.
28. *Modern Plastics Encyclopedia '93;* McGraw Hill: New York, NY, 1993.
29. van Dam, H. W. In: *Chemistry and Physics of Baking;* Blanshard, J. M. V.; Frazier, P. J.; Galliard, T., Eds.; Royal Society of Chemistry Burlington House: London, UK, 1986; pp 117-130.
30. Handleman, A. R.; Conn, J. F.; Lyons, J. W. *Cereal Chem.* **1961**, *38*, 294-305.
31. Mercier, C.; Feillet, P. *Cereal Chem.* **1975,** *52*, 283-297.
32. El-Dash, A. A.; Gonzales, R.; Ciol, M. In: *Extrusion Cooking Technology;* Jowitt, R., Ed.; Elsevier Applied Science: London, UK, 1984; pp 51-74.
33. Smith, A. C. In: *Food Extrusion Science and Technology;* Kokini, J. L.; Ho, C.; Karwe, M. V., Eds.; Marcel Dekker: New York, NY, 1992; pp 573-618.
34. Ferdinand, J. M.; Lai-Fook, R. A.; Ollett, A. L.; Smith, A. C.; Clark, S. A. *J. Food Engin.* **1990,** *11*, 209-224.
35. Hoseney, R. C.; Zeleznak, K.; Abdelrahman, A. *J. Cereal Sci.* **1983**, *1*, 43-52.

36. Sullivan, J. F.; Craig, J. C. *Food Technol.* **1984**, *38*, 52-55, 131.
37. Pekala, R. W.; Alviso, C. T.; Hulsey, S. S.; Kong, F. M. *Cellular Polymers* **1992**, *38*, 129-135.
38. Baldwin, D. F.; Suh, N. P.; Shimbo, M. *Cellular Polymers*, **1992**, *38*, 109-128.
39. Martini-Vvedensky, J. E.; Suh, N. P.; Waldman, F. A., *Microcellular Closed Cell Foams and Their Method of Manufacture*; US Patent, #4473665, 1984.
40. Martini, J.; Waldman, F. A.; Suh, N. P. *Soc. Plastics Eng. Technol.* **1982**, *28*, 674-676.
41. Young, A. T.; Moreno; Marsters, R. G. *J. Vac. Sci. Technol.* **1982**, *20*, 1094-1097.
42. Coudeville, A.; Eyharts, P.; Perrine, J. P.; Rey, L.; Rouillard, R. *J. Vac. Sci. Technol.* **1981**, *18*, 1227-1231.
43. Charoenrein, S. *Ph. D. Dissertation*, 1989, University of California, Davis.
44. Fricke, J. *Scientific Amer.* **1988**, *258*, 92-97.
45. Rabinovich, E. M. In: *Sol Gel Technology for Thin Films, Fibers, Preforms, Electronics, and Specialty Shapes.* Klein, L. C., Ed.; Noyes Publishing: Park Ridge, NJ, 1988; pp 260-294.
46. Kistler, S. S. *Nature* **1931**, *127*, 741-742.
47. Kistler, S. S. *J. Phys. Chem.* **1932**. *36*, 52-55.
48. Fricke, J. In: *Aerogels*; Fricke, J., Ed.; Springer-Verlag: New York, 1986, pp 2-19.
49. Zoebel, H. F. In: *Starch Chemistry and Technology*. Whistler, R. L.; BeMiller, J. N.; Paschall, E. F., Eds.; Academic Press Inc., New York, NY, 1984; pp 285-309.
50. Glenn, G. M.; Irving, D. W. *Cereal Chem.* **1995**, *72*, 155-161.
51. Throne, J. In: *Science and Technology of Polymer Process;* Suh, N. P.; Sung, N., Eds.; MIT Press: Cambridge, MA, 1979; pp. 77-131.
52. Benning, C. J. In: *Plastic Foams;* Akin, R. B.; Mark, H. F.; Scavuzzo, J. J.; Stivala, S. S., Zukor, L. J., Eds.; Wiley and Sons: New York, NY, 1969, Vol. 2; pp 1-10.
53. Hilyard, N. C.; Young, J. In: *Mechanics of Cellular Plastics,* Hilyard, N. C., Ed.; Macmillan Publishing Co.: New York, NY, 1982; pp. 1-26.
54. Throne, J. L. In: *Mechanics of Cellular Plastics,* Hilyard, N. C., Ed.; Macmillan Publishing Co.: New York, NY, 1982, pp. 263-322.
55. Hayter, A. L.; Smith, A. C.; Richmond, P. *J. Mat. Sci.* **1986**. *21*, 3729-3736.
56. Lacourse, N. L.; Altieri, P. A., *Biodegradable Packaging Material and the Method of Preparation Thereof;* US Patent #4,863,655, 1989.
57. Hutchinson, R. J.; Siodlak, G. D. E.; Smith, A. C. *J. Mat. Sci.* **1987**. *22*, 3956-3962.
58. Pekala, R. W., *J. Mat. Sci.* **1989**, *24*, 3221-3229.
59. Shannon, J. C.; Garwood, D. L. In: *Starch Chemistry and Technology*. Whistler, R. L.; BeMiller, J. N.; Paschall, E. F., Eds.; Academic Press Inc.: New York, NY, 1984, pp 25-86.
60. Otey, F. H.; Westhoff, R. P.; Doane, W. M. *Ind. Eng. Chem. Res.*, **1987**, *26*, 1659-1663.

61. Doane, W. M.; Swanson, C. L.; Fanta, G. F. In: *Emerging Technologies for Materials and Chemicals from Biomass*, Rowell, R. M.; Schultz, T. P.; Narayan, R., Eds.; ACS Symposium Series 476, Washington, DC, 1992; pp. 197-230.
62. Gould, J. M.; Gordon, S. H.; Dexter, L. B.; Swanson, C. L. In: *Agricultural and Synthetic Polymers: Biodegradability and Utilization.*, Glass, J. E.; Swift, G., Eds.; ACS Symposium Series 433; ACS: Washington, D. C., 1990.
63. Wiedmann, W; Strobel, E., *Starch/Staerke*, **1991**, *43*, 138-145.
64. Shogren, R. L., *Carbohydr. Polymers*, **1992**, *19*, 83-90.
65. Katz, E. E.; Labuza, T. P. *J. Food Sci.* **1981**, *46*, 403-409.
66. Hutchinson, R. J.; Mantle, S. A.; Smith, A. C., *J. Mat. Sci.* **1989**, *24*, 3249-3253.
67. Hutchinson, R. J.; Siodlak, G. D. E.; Smith, A. C. *J. Mat. Sci.* **1987**, *22*, 3956-3962.
68. Baer, E., In: *Engineering Design for Plastics*; Baer, E., Ed.; Chapman and Hall; New York, NY, 1964.
69. Hayter, A. L.; Smith, A. C.; Richmond, P., *J. Mat. Sci.* **1986**, *21*, 3729-3736.
70. Chinnaswamy, R.; Hanna, M. A., *J. Food Sci.*, **1988**, *53*, 834-840.
71. Chinnaswamy, R.; Hanna, M. A., *Cereal Chem.* **1988**, *65*, 138-143.

Chapter 8

Edible Films for the Extension of Shelf Life of Lightly Processed Agricultural Products

A. E. Pavlath, D. S. W. Wong, J. Hudson, and G. H. Robertson

Western Regional Research Center, Agricultural Research Service, U.S. Department of Agriculture, 800 Buchanan Street, Albany, CA 94710

In order to extend the shelf-life of lightly processed agricultural produce and maintain the nutritional and esthetic values, various edible natural materials, i.e., proteins, carbohydrates and fats, can be used as a protective barrier to replace the natural skin or peel removed during processing. While none of these materials was effective alone, their emulsified combination allowed the development of a protective film on the cut surface. Various relative humidity levels were tested to determine the optimum conditions for minimum weight loss without any mold formation.

The outer tissues, skin or peel of fruits and vegetables protect against weight loss, discoloration, loss of texture and flavor, and other undesirable processes. However, fruit and vegetable preparation generally includes the removal of the skin, coring, slicing and/or dicing. The loss of the natural protective barrier limits the period of time during which these minimally processed items can be used.

Furthermore, these preparation processes can be tedious and time-consuming. In small-scale commercial practice, such as in a cafeteria, the cost of automatic processing machines can be high. On the other hand large scale commercial processing in a centrally located factory would provide both economical and environmental benefit since the disposal and/or utilization of the waste could be done more efficiently in a centralized location.

Transport of lightly processed fruits and vegetables leads to additional problems. Paramount of these, is the accelerated deterioration , e.g., enzymatic browning that occurs when the skin is removed and the flesh is exposed. While the protection provided by the skin can be enhanced by applying various waxy materials on the unprocessed produce, the direct application of any lipid to a cut, wet surface is unlikely to provide a strong and adhering protective coating. The processed pieces would have to be stored and transported under controlled atmospheres, a generally costly procedure, and even under these conditions the problems are not fully solved. The nature of the difficulties can be easily shown by the desiccation of an apple. At room temperature, a whole unwaxed apple will lose less than 0.5% moisture in a day, whereas a halved apple will lose up to 6% moisture under similar conditions. At the same time, oxidation of the cut surface contacting air causes browning. The discoloration may be slowed down by treating the pieces with sulfite, however, there have been serious concerns about possible health problems caused by sulfite. In addition, the removal of the skin or peel, may cause other problems, e.g. changes in respiration, physiological disorders, enzymatic breakdowns, etc.

One solution is the creation of an edible coating on the cut surface to provide protection similar to that given by the natural peel. This technology would enable economic central processing and long distance transport without deterioration. Development of such coatings could lead to increased consumption and expanded markets for apples and other agricultural produce. This paper describes a new approach to achieve this goal: edible films and coatings.

Possible components for edible coatings

The use of an edible coating to extend the shelf-life of food is not new. In 12th century China, citrus fruits were heavily coated with waxes to retain moisture *(1)*. Fruits and vegetables were coated with carnauba wax oil-water emulsion *(2)*. Today thin waxy layers are used on numerous agricultural produce to prevent moisture loss during storage. However, while numerous applications are described in the literature for commodities with intact natural surfaces, the application of coatings to cut surfaces is very rare. Dried fruit was coated with zein from alcohol solution which retarded moisture gain, but altered the flavor of the product *(3)*. Dry carboxymethyl cellulose (CMC) powder which was applied to cut surfaces, adsorbed the moisture within the pores of the surface, and the swollen CMC not only prevented the loss of moisture, but also provided a barrier to oxygen which is generally responsible for discoloration *(4)*.

Generating a protective layer can be commercially successful only if the layer does not have to be removed before consumption. This condition requires that the components of the layer must be fully approved for food use. In addition, the coating should have only minimal effect on the taste and other esthetic factors which determine consumer acceptance. Therefore, naturally occuring proteins, carbohydrates and fats would be the preferred components for these layers.

The epidermal layers of an unprocessed fruit or vegetable control the transmission of various molecules, such as oxygen, carbon dioxide, ethylene and water between the interior tissues and the environment. Each of these compounds has an important role in the postharvest life of fruits and vegetables and a replacement of the original protective layer would have to maintain a similar rate of their transmission.

The amount of the transmitted material, Q, can be calculated by the integrated form of Fick's Law:

$$Q=P.A.\Delta p.t/d,$$

where A is the surface area, Δp is the pressure gradient across the barrier, t is the time, d is the thickness of the film and P is the permeability constant characteristic to its structural composition. The variables are generally set by the conditions of the given application. The surface area is given by the desired size of the cleaned commodity. Unless we use a special environment for storage, Δp is set but it is not constant. The value of t in this equation represents the required shelf-life, which generally should be at least 10-14 days to allow for processing, packaging, transportation, distribution and some storage before reaching the consumer. Finally, the thickness is also limited by the requirement that the coating should have little or no effect on the original taste of the uncoated material. For this reason, the optimal thickness should be not more than 0.2-0.3 mm. This leaves the permeability coefficient as the major factor in determining whether the coating will provide the necessary protection.

The permeability coefficient is the combination of the solubility and diffusion coefficients because there are two major consecutive steps involved in the diffusion of a molecule through a solid film. First the migrating molecule must be absorbed into the surface of the coating. The rate of absorption is dependent on the affinity between the materials, i.e., water will be absorbed on a hydrophilic surface faster than on hydrophobic ones. In the second step, the absorbed molecule will proceed through the film moving from one "molecular hole" in its structure to another *(5)*. The availability and closeness of these molecular holes, and the facility of creating them will control the rate of this step. Crystallinity, crosslinking, sidechains, hydrogen bonding and/or any interaction which stabilizes the polymer chains in relation to one another will make the creation of such openings more difficult and it will result in the lowering of the diffusion coefficient.

The role of the natural skin and of an edible film, is not to separate hermetically the produce from the environment. The harvested produce is still subject to various aerobic and anaerobic respiratory processes and, therefore, the new coating must maintain appropriate gas transmission during storage. Too much atmospheric oxygen will result in rapid discoloration, while too little leads to physiological disorders. If the oxygen concentration falls below 1% inside the tissue, anaerobic respiration will occur, resulting in undesirable odors and flavors due to the accumulation of ethanol and acetaldehyde *(6)*.

The transmission characteristics of barriers formed by three major food component classes have been reported *(7)*. The general conclusion is that fats reduce water transmission, carbohydrate films reduce oxygen transmission and protein films provide mechanical stability. However, it should be pointed out that the characteristics of a film cast on the surface of an apple are not necessarily the same as those of the same film obtained when cast on an inert smooth surface.

Waxy coatings slow the migration of water because of their low polarity but the actual mechanism is not fully clarified. The diffusion of water vapor through fatty acid monolayers was reported to decrease logarithmically with the chain length *(8)*. Monoglyceride films exhibited the lowest water transmission at C_{18} but water migration increased sharply when the fatty acid chain was increased to 20 carbon atoms. Similar increase was noted with the change from stearic to oleic acid, i.e., saturated to unsaturated fatty acid *(9)*. Another surprising maximum barrier to water was observed in chitosan salts with C_{12} fatty acids, i.e., lauric acid *(10)*.

Polysaccharides, such as starch, cellulose, alginate, pectin and their derivatives have been studied for possible use as edible films, but they have not been applied commercially. Most of the polysaccharides are hydrophilic and, therefore, they do not prevent moisture loss. Polysaccharides that form aqueous gels have relatively reduced water transport properties. Carrageenan, a sulphated polysaccharide D-galactose derivative, was found to form a three-dimensional double-helix structured gel which acted as "sacrificing agent", i.e., the water from the gel will be lost first before the coated sample desiccate *(11, 12)*. Alginates and pectins formed gels when reacted with calcium ions to crosslink the polymer chains through the free carboxylic groups *(13, 14)*. Crosslinking aligns the polymer chains and facilitates the formation of hydrogen bonding between neighboring chains which strengthens the film. Calcium also contributes to decreased browning and reduced carbon dioxide and ethylene evolution *(15)*.

Carbohydrates, in spite of their limited resistance to water migration, have good potential for edible film formation. An inverse relation between water and oxygen permeability was reported for certain dry carbohydrate films *(16)*. The permeability of oxygen through high amylose starch films was found to be practically zero even with a plasticizer which is known to increase gas permeability *(17)*.

Casein, gelatin and albumin films exhibited low water resistance, only zein showed transmission properties in a desirable range. Wheat gluten films have also been suggested *(18)*. Desirable transmission properties were observed when a film was prepared from the mixture of gelatin and gum arabic, a natural polysaccharide *(19)*.

Casein films modified by crosslinking with tannic and lactic acids had improved water resistance *(20)*. Sodium hypochlorite was found to cause coupling of the aromatic rings of phenylalanine and tyrosine in casein *(21)*. Enzymatic crosslinking of casein and soy protein was reported using peroxidase/H_20_2 systems *(22)*. A casein film with strong mechanical strength, but unknown transmission characteristics, was obtained from casein solution with transglutaminase *(23)*. Casein treated with acyl esters of N-hydroxysuccinimide yielded a fatty acid acylated casein with hydrophobic characteristics *(24)*. Various acrylic monomers were grafted to casein and other proteins using water soluble redox systems *(25, 26)* and these functional groups were shown to be highly beneficial for crosslinking and obtaining gels *(27)*. Such strengthening of the casein structure can develop a film with good transmission characteristics, however the crosslinking and crosslinking agents may affect edibility, flavor and safety of the product.

Composite edible films

Composite barriers or films were evaluated as a means of obtaining the full range of desired functional properties in a single barrier. These were applied as an aqueous emulsion containing casein (0-10%), alginic acid (0-1%) and Myvacet 5-07 (0-15%).

Myvacet 5-07 is a commercial fatty material approved for food uses, a monoglyceride which is 48-51% acetylated. It was found that all three components were required since the absence of any of the three components caused significant increases in the rate of water loss. This was completely unexpected in view of the individual properties of proteins and carbohydrates *(7)*.

The weight loss of an uncoated piece of apple was 80% after seventy-two hours of exposure to air at room temperature. The absence of the lipid from the combination resulted in almost the same weight loss as shown in Figure 1. Combinations with lower than 15% of lipid were not suitable for effective protection against dehydration.

When the alginic acid was left out of the combination the weight loss was accelerated, while the addition of 0.5-1.0% of alginic acid produced very effective water barriers as shown in Figure 2. The effectiveness was unexpected, since alginic acid is hydrophilic and the moisture losses should be less in its absence. The relatively small amount of alginic acid needed, suggested a surface effect. This assumption became even more plausible when the casein was left out of the composition and the moisture barrier of alginic acid-lipid coating was negligible (Figure 3.) A protein would not be expected to have this influence on water transmission.

The data shown in Figures 2 and 3 can lead to a rational explanation of the protection provided by the casein-alginic acid-lipid coatings. The cut surface is hydrophilic and the lipid does not form a coating on it, while the alginic acid can quite easily interact with the surface. Since the lipid and the alginic acid do not mix well, a bridge between the two components is needed to create a uniform protective coating. Since, casein is an emulsifier its presence assures their effective coupling. One of the best coatings which provided for up to three days of protection against loss of moisture and discoloration, was created from the aqueous emulsion of 10% casein, 1% alginic acid and 15% of an acetylated monoglyceride at 55-60 °C *(28)*.

While the alginic acid/casein/lipid coating provides considerable protection against moisture loss, there are several drawbacks in the use of emulsion coating. (1) Emulsion coatings remain wet, creating handling problems with the coated products. (2) Emulsion coating may act as a sacrificial layer rather than a true barrier to moisture. (3) Emulsion stability is difficult to control. (4) The viscosity of emulsified products varies with the components used, and the actual amount of emulsion applied also changes. All these make it difficult to relate the effectiveness of an emulsion coating to the nature of its components. Such variations represent a handicap to commercial application.

A mixture of carbohydrate and fat without the need of emulsification should develop film uniformity with better barrier properties. In another series of experiments, it was shown that some chitosan-fatty acid salts form a dry film with high water barrier properties; however, chitosan is not approved for food uses in the United States. In order to minimize the uncertainties caused by the variation in the quality of emulsion, a bilayer coating was developed where the apple pieces were coated first with microcrystalline cellulose and then with a acetylated monoglyceride as described in one of our earlier publications *(29)*.

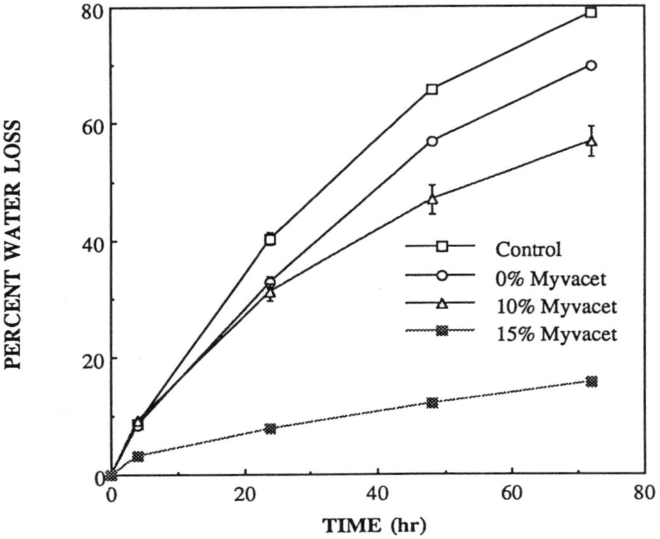

Figure 1. Weight loss of emulsion coated apple pieces as a function of lipid concentration in the emulsion at 23°C.

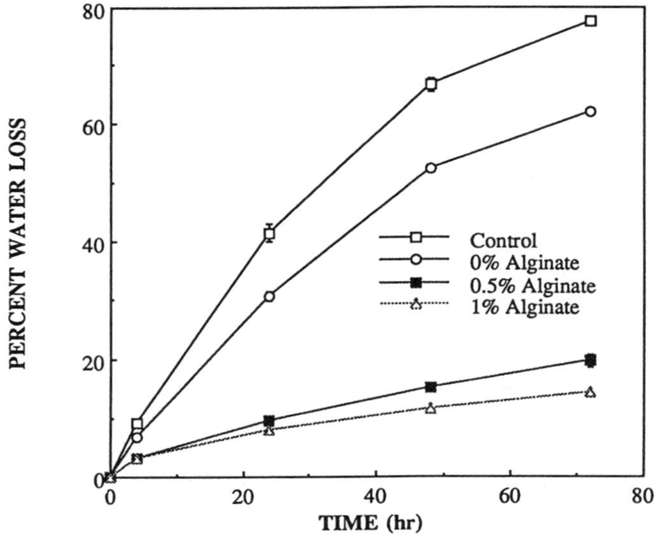

Figure 2. Weight loss of emulsion coated apple pieces as a function of alginate concentration in the emulsion at 23°C.

Effect of relative humidity

Initial experiments with both emulsion and bilayer type of coatings were carried out with single apple pieces supported in such way that all surfaces were exposed to an infinitely large volume of air. However, in practical application the water loss by processed pieces would be reduced because of increased total resistance to water transport and decreased surface area. The resistance to water transport will include surface resistance from apple to headspace through the coating and headspace to surrounding environment through the container wall. The surface area for transport would be reduced by virtue of the packing of the pieces in the container and the packing of the containers themselves. Assuming also a certain degree of ventilation, these factors would determine the headspace composition at equilibrum.

Since partial vapor pressure of water and relative humidity at equilibrum plays an important role in the rate of water loss, the effect of various relative humidity levels (36-100%) was studied. Bilayer coated apple pieces with all surfaces exposed were kept in a controlled humidity chamber at room temperature up to 15 days. As shown in Figures 4. a-e., the results were as expected. At 36%, 44%, 75% and 85% relative humidity the elapsed time to 15% weight loss was 6, 7, 10 and 12 days respectively. At 100% relative humidity, the weight loss was less than 2% even after 15 days, but mold formation was observed, which can be increasingly expected at higher humidities. In order to eliminate the need of any additive for the prevention of mold formation, a balance has to be found between the safe relative humidity level and the desired shelf-life. Furthermore, the need to maintain a lower than fully saturated atmosphere will require a more sophisticated packaging strategy with air space surrounding each container.

As a rule of thumb, mold formation can be expected around 80% RH or higher. As shown in Figure 5, the weight loss after 10 days is still high at 75% RH, therefore, additional factors need to be investigated for longer shelf-life. Since the rate of weight loss is influenced by the vapor pressure of the water, the storage temperature was expected to have an influence on it. When the apple pieces were kept at 4°C the weight loss curve at 44%RH was nearly identical with one obtained at room temperature and 86% RH. (Figure 6.) Therefore it appears that the combination of low temperature with relative humidity values below the danger level for mold formation can provide considerable protection.

Conclusion

The water losses of lightly processed fruits and vegetables can be considerably decreased by coating them with a combination of edible components, selected from proteins, carbohydrates and lipids. Preliminary experiments indicated that the coatings also provide protection against enzymatic browning, but further study is needed on this subject. Using lower temperatures and moderate relative humidities in the containers used for transportation the protection can be economically and commercially feasible. More research is needed to identify the best components and packaging materials.

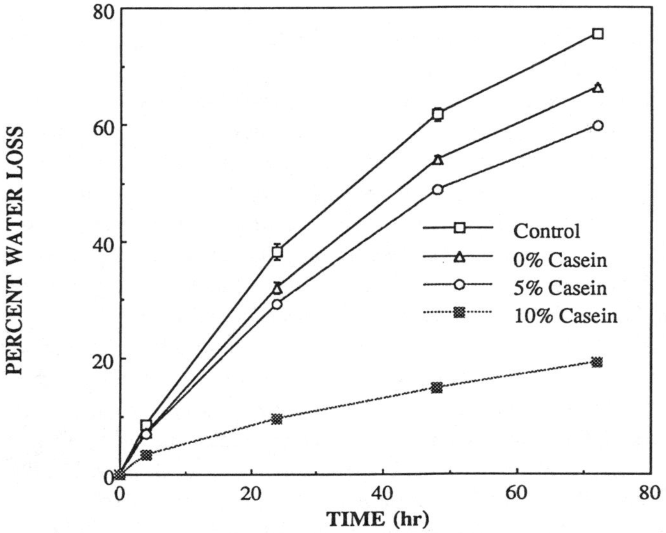

Figure 3. Weight loss of emulsion coated apple pieces as a function of casein concentration in the emulsion at 23°C.

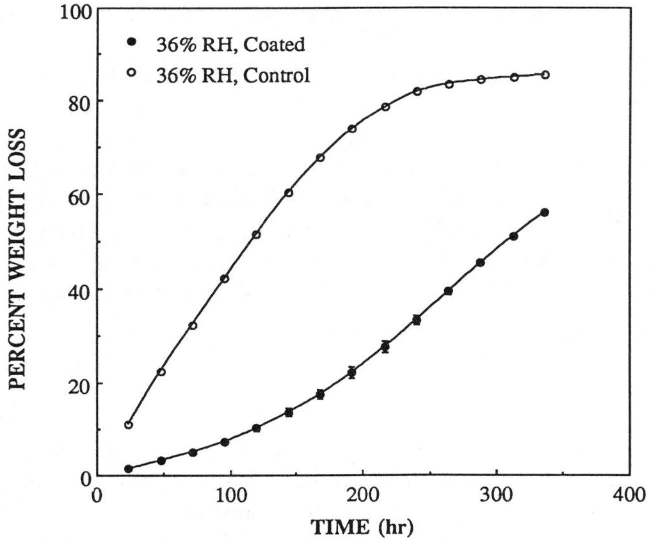

Figure 4.a. Weight loss of bilayer coated apple pieces at 23°C and 36% relative humidity.

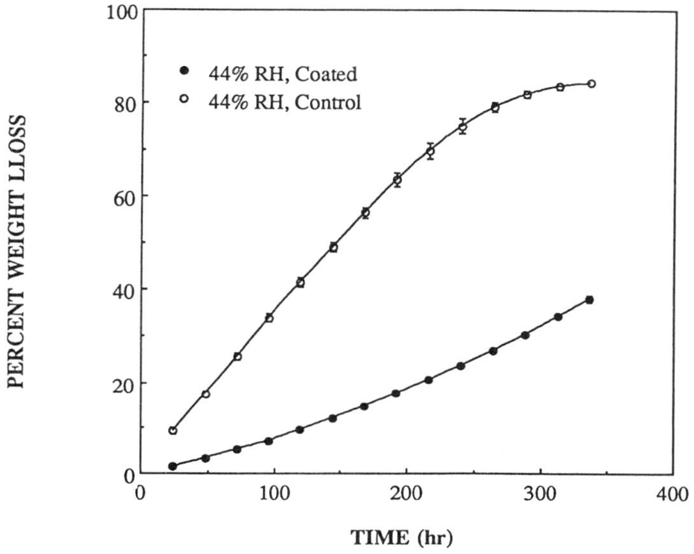

Figure 4.b. Weight loss of bilayer coated apple pieces at 23°C and 44% relative humidity.

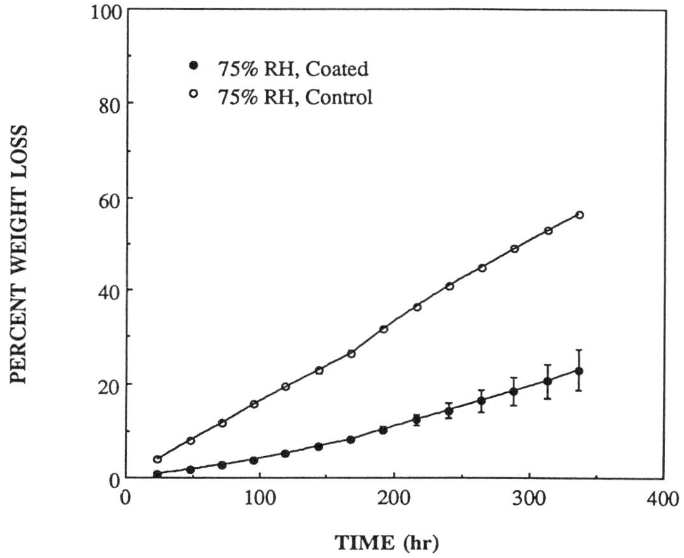

Figure 4.c. Weight loss of bilayer coated apple pieces at 23°C and 75% relative humidity.

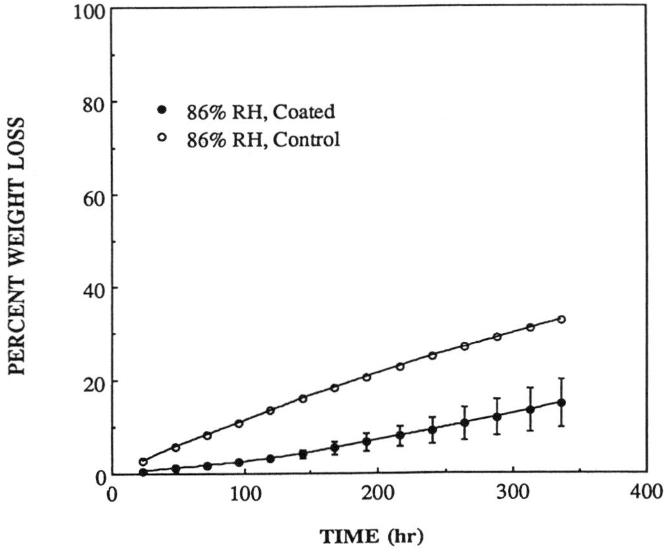

Figure 4.d. Weight loss of bilayer coated apple pieces at 23°C and 86% relative humidity.

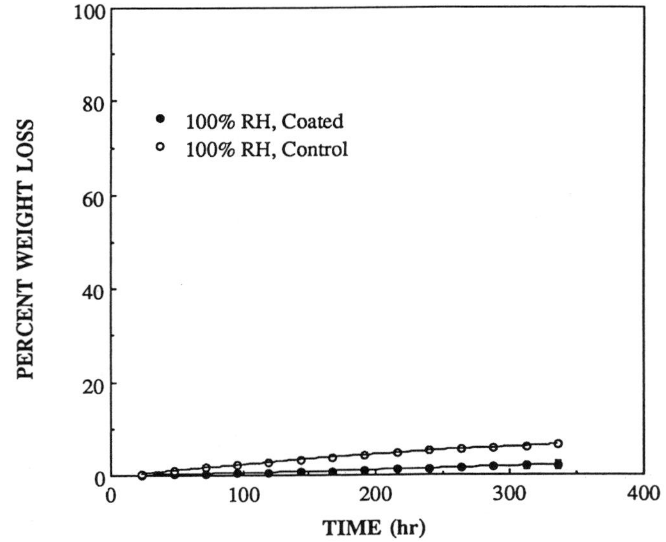

Figure 4.e. Weight loss of bilayer coated apple pieces at 23°C and 100% relative humidity.

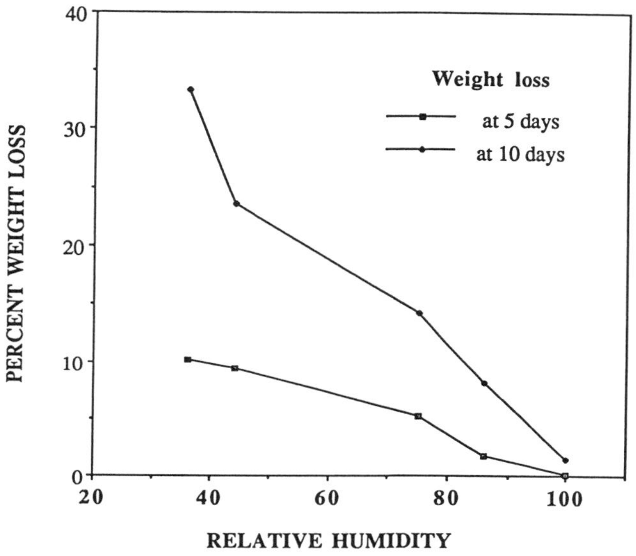

Figure 5. Weight loss of bilayer coated apple pieces after 5 and 10 days as a function of relative humidity.

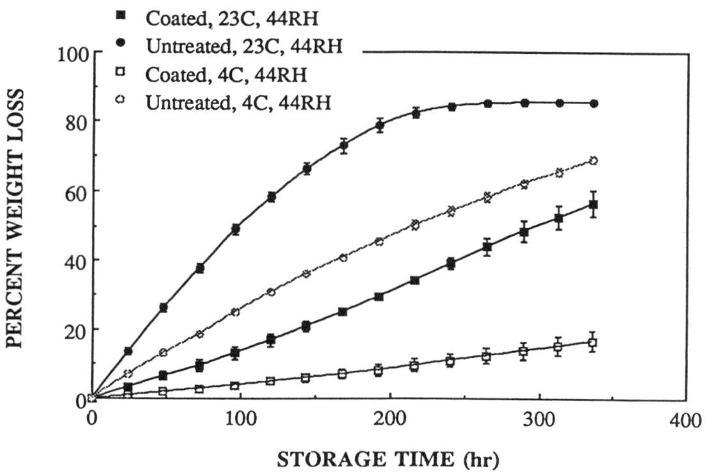

Figure 6. Weight loss of bilayer coated apple pieces at 4°C & 23°C and 44% relative humidity.

Literature Cited

1. Hardenburg, R.E. Agricultural Research Service Bulletin 51-15, United States Department of Agriculture, Washington, D.C., 1967.
2. Kaplan, H.J. In Fresh citrus fruits; Wardowski, W.F.; Nagy, S.; Grierson, W.,Eds.; AVI, Westport, CT, 1986, p.379.
3. Cosler, H.B. U.S. Patent No. 2,791,509, 1957.
4. DeLong, C.F.; Shepherd, T.H. U.S. Patent No. 3,669,691, 1972
5. Frisch, H.L.; Stern, S.A. CRC Critical Reviews in Solid State and Materials Sciences, 1981, II(2), 123-187.
6. Kader, A.A. Outlook **1986**, 13(2), 9.
7. Kesler, J.J.; Fennema, O.R. Food Technology **1985**, 40.(12), 47.
8. LaMer, V.K.; Healy, T.W.; Aylmore, L.A.G. J. Colloid Sci. **1964**, 19, 673.
9. Roth, T.; Loncin, M. In Properties of water in foods in relation to quality and stability; Simatos, D.; Multon, J.L., Eds.; Martinus Nijhoff Publishing, Dordrecht, The Netherlands, 1985, p.331.
10. Wong, D.S.W.; Gastineau, F.A.; Gregorski, K.S.; Tillin, S.J.; Pavlath, A.E. J. Agr. Food Chem. **1992**, 540.
11. Glicksman, M. In Food hydrocolloids; Glicksman, M., Ed.; CRC Press, Boca Raton, FL 1982, Vol. 1; p 47.
12. Glicksman, M. In Food hydrocolloids, Glicksman, M., Ed.; CRC Press, Boca Raton, FL 1983, Vol. 2; p 73.
13. Allen, L.; Nelson, A.I.; Steinberg, M.P.; McGill, J.N. Food Technol. **1963,** 17, 1437.
14. Morris, V.J. In Functional properties of food macromolecules; Mitchell, J.R.; Cedward, D.A., Eds.; Elsevier Applied Science Publishers, London, England, 1986, p. 121.
15. Poovaiah, B.W. Food Technol. **1986**, 40(5), 86.
16. Banker, G.S.; Gore, A.Y.; Swarbrick, J. J. Pharm. Pharmac. **1966**, 18, 457.
17 Mark, A.M.; Roth, W.B.; Mehltretter, C.L.; Rist, C.E. Food Technol. **1966**, 20, 75.
18. Anker, C.A.; Foster, G.A.; Leader, M.A. U.S. Patent No. 3,653,925, 1972.
19. Deasy, P.B. Microencapsulation and related drug processes. Marcel Dekker, Inc., New York, 1984.
20. Guilbert, S. In Food packaging and preservation : Theory and practice; M. Mathlouti, M., Ed.; Elsevier Applied Science Publishers, London, England., 1986, p 371.
21 Matoba, T.; Shiono, T.; Kito, M. J. Food Sci., **1985**, 50, 1738.
22. Matheis, G.; Whitaker, J.R. J. Protein Chem., **1984**, 3(1) 35.
23. Motoki, M.; Nio, N. Japan Kokai Tokyo Koho, JP 61,152,247, 1986.
24. Nippon Shinyaku Co. Ltd.Japan Kokai Tokyo Koho JP 59,155,396, 1984.
25. Mohan, D.; Radhakrishnan, G.; Nagabhushanam, T. J. Appl. Polymer Science, **1980**, 25. 1799.

26. Mohan, D.; Radhakrishnan, G.; Rajadurai, S.; Nagabhushanan, T; Joseph, K.T. J. Appl. Polymer Science, **1984**, 329
27. Pavlath, A.E. Textile Res. J. **1974**, 44. 658.
28. Pavlath, A.E.; Wong, D.S.W.; Kumosinski, T.F. CHEMTECH **1993**, 36.
29. Wong, D.S.W.; Tillin, S.J.; Hudson, J.S.; Pavlath A.E. J. Agr. Food Chem. **1994**, 2278.

Chapter 9

Biodegradable Polymers from Agricultural Products

John M. Krochta and Cathérine L. C. De Mulder-Johnston

Department of Food Science and Technology, University of California, Davis, CA 95616

Biological recycling of biodegradable polymers is an important option for reducing municipal solid waste. Legislation, including prohibitions on disposal of non-biodegradable waste at sea, is making development of biodegradable products more urgent. Biodegradable polymers based on cellulose, starch, protein, microbial polyesters and polylactic acid each have properties which suggest possible applications. Mechanical properties of synthetic polymers are generally easier to match with biodegradable polymers than are moisture and oxygen barrier properties. However, strategies exist for utilizing or enhancing barrier properties.

Decreasing the use of non-renewable resources and reducing the generation of solid waste have become high priorities for this environmentally-conscious era and, thus, legal requirements for a growing number of municipalities. The general public considers metal and glass to be recyclable, and paper packaging to be both biodegradable and recyclable. However, plastic is generally considered to be neither biodegradable nor recyclable by the public. Thus, plastic products have been the subject of intense scrutiny. This concern has focused largely on packaging, but carries over to the area of disposable food service, medical, agricultural and other plastic disposable items.

Biodegradable polymers have been a subject of interest for many years because of their potential to reduce non-biodegradable synthetic plastic waste. The objectives of this paper are to: 1) place biodegradable polymer products in the context of the main approaches to reducing municipal solid waste (MSW), 2) discuss issues surrounding the concept of biodegradable polymers, 3) summarize information on materials available for biodegradable polymer products, and 4) draw conclusions on status of usefulness of biodegradable polymers and future trends.

0097–6156/96/0647–0120$15.25/0

Approaches to Reducing Municipal Solid Waste

Approaches to reducing MSW can be placed into four categories: 1) source reduction, 2) product reuse, 3) recycling (including biodegradation), and 4) energy recovery. The technique of "life-cycle analysis" has emerged as an approach to material choices based on energy use in product material extraction/refining, manufacturing, conversion, operations, transport, collection after use and disposal (*1*).

Source Reduction. Source reduction has been pursued by the food industry for years as a way to reduce packaging cost. Through improved package design, the 1970 90-g high density polyethylene (HDPE) milk jug has been reduced in weight to today's 65-g version (*2*). Yogurt pots have been reduced in weight from 12-g polystyrene (PS) to 7-g polypropylene (PP), polyethylene (PE) bread bags have been reduced from 40μ to 25μ thick, and cereal box liners have been downsized by 12% (*1*). These reductions parallel reductions in non-plastic packaging, where aluminum can weight has dropped by 25% since 1972, and 16-oz glass container weight has dropped 30% since 1980 (*3*). The overall result is that, while the generation of MSW has increased from 3.1 to 3.5 lb/person/day since 1970, generation of packaging waste has decreased from 1.05 to 0.95 lb/person/day (*3*). Efforts at source reduction will surely continue, for both economic and environmental reasons.

Reuse. Although still significant in certain parts of the world (e.g., glass milk bottles in the U.K.), package reuse is limited by the trend towards centralized production facilities in the U.S. The energy and environmental gain in package reuse can easily be overwhelmed by the cost of collection and transportation back to the food plant.

Recycling. Material recycling can take the form of biological, mechanical and chemical processes.

Biological Recycling. Development of biodegradable polymers to replace conventional synthetic plastic products provides opportunities for reducing waste through biological recycling to the biosystem. Use of biodegradable polymers is also seen as an approach for slowing the introduction of fossil-fuel-derived carbon dioxide into the atmosphere (*4*). Carbon dioxide obtained from biodegradation processes is considered "recycled", whereas CO_2 from synthetic fossil-fuel-derived polymers is "new" and will add to the greenhouse effect.

Biodegradation involves enzymatic and chemical degradation by living organisms (usually bacteria and fungi) (*5*). Enzymatic degradation of polymers occurs mostly by hydrolysis and oxidation (*6*). The biodegradation process includes two essential steps: 1) depolymerization or chain cleavage, and 2) mineralization. The depolymerization step usually occurs outside the microorganism through the use of extracellular enzymes. Mineralization is the conversion of polymer into biomass, minerals, water, CO_2, CH_4, and N_2. The mineralization step usually occurs intracellularly.

Biodegradation requires three key elements: 1) a vulnerable substrate, 2) appropriate organisms, and 3) a well-tuned environment (*6*). The first requirement for biodegradation is that the polymer must contain chemical bonds that are susceptible to enzymatic hydrolysis or oxidation. Proteins, e.g., contain peptide bonds which can easily be hydrolyzed enzymatically. Factors that negatively affect enzymatic degradation are branching, hydrophobicity, high molecular weight, and crystallinity (*6*). Proteins and polysaccharides are good substrates for enzymatic attack, because they are hydrophilic, and they are usually not very branched or crystalline. The second requirement for biodegradation is the presence of microorganisms that are capable of synthesizing the enzymes required to depolymerize and mineralize the polymer. These two steps may or may not be carried out by the same microorganism. Proteins and polysaccharides are easily biodegraded since many microorganisms produce the enzymes required to metabolize these compounds. The third (and unfortunately often forgotten) requirement for biodegradation is a well-tuned environment in which the desired microorganisms thrive. This includes factors like appropriate temperature range, moisture level, presence and concentration of salts, oxygen level (aerobic versus anaerobic), pH, redox potential, etc. If any of these elements are outside the desired range, the entire biodegradation process may come to a halt (*6*). Sewage, marine water, landfills, compost, and soil all provide very different environments, and polymers that are degradable in one environment may not necessarily degrade in another environment (e.g., newspapers do not biodegrade in a landfill, even though cellulose itself is perfectly biodegradable when enough moisture is provided).

A number of tests have been developed to assess biodegradability (*7-10*). These methods include enzyme assays, plate tests, biological oxygen demands (BOD), respirometry, simulated laboratory scale accelerated systems, and exposure to natural environments. In general, more than one method is needed to fully assess the biodegradability of a given polymer (*6*).

Some municipalities are now providing curbside pickup for large-scale composting which could be used for biodegradable polymer products. Equipment for small-scale domestic composting is also available. In some cases, biodegradable products can be handled by domestic disposals and, subsequently, municipal waste treatment plants. Opportunities exist for biodegradable polymer products when recovery of conventional synthetic plastics for mechanical and chemical recycling or energy recovery is impractical (*7*). Such products include: 1) trash bags for kitchen, restaurant and yard waste; 2) disposable food service utensils, plates, cups and packaging; 3) agricultural mulching films; 4) diapers and other sanitary products; 5) medical clothing, disposables and trash bags; and 6) fishing nets and related gear (*11*).

Mechanical Recycling. Material reuse through mechanical recycling (collecting, cleaning, and forming into new products) is the dominant recycling technique in place now and has been receiving major attention as a means of reducing MSW. Packaging at approximately one-third of the waste stream contributes more recycling than the remainder of the waste (*3*). Generally, only single-component plastic packages (e.g., polyethylene terephthalate (PET) soda bottles, HDPE milk bottles) are recyclable. Recycled PET now finds use as ski jacket insulation, carpeting and egg cartons (*2*). Mechanical recycling along with chemical recycling (below) may be the best options in cases where the unique properties of synthetic polymers are required and cannot be matched by biodegradable polymers.

Chemical Recycling. Mechanically recycled packaging material is presently unacceptable for direct food contact. However, depolymerization back to monomers and then repolymerization has been approved by FDA for food use in the case of PET, with similar possibilities emerging for LDPE, HDPE and PP (*1*). Multicomponent, multilayer plastic packages (e.g., coextruded films) are hardest to recycle, either chemically or mechanically. There is growing pressure to replace multilayer plastic packages with simple designs more easily recycled.

Energy Recovery. Finally, although energy recovery from waste incineration is an important disposal technique in Europe and Japan, environmental issues have limited its application in the U.S. Generation of hydrogen chloride and possibly dioxins from polyvinyl chloride (PVC) and polyvinylidene chloride (PVDC) is the main concern, although modern incineration techniques should be able to deal with these compounds (*1*). Life-cycle analysis may find that energy recovery through incineration is the best environmental option for some plastic products. Energy recovery is an alternative option for biodegradable polymers, which have the additional advantage of being based on renewable resources.

Biodegradable Polymers

Development of biodegradable polymers has been the subject of intense interest and investigation in recent years (*12-15*). One of the primary goals in the development of synthetic polymers has been stability, and conventional synthetic polymers are clearly resistant to microbial attack (*6, 16*). Therefore, achievement of biodegradable polymers as an approach to reducing waste requires use of new polymers degradable by common microorganisms under the proper conditions (*16*). The goal of this approach is to directly replace non-biodegradable synthetic polymers with biodegradable materials.

Early predictions on replacement of synthetic plastic with biodegradable materials were overly optimistic, with as high as 15% replacement by 1992 once predicted (*17*). Biodegradable polymers made up 5 million pounds (approx. 0.08% of the total polymer resin market) in 1992, and are predicted to reach 8.4 million pounds in 1995/96 (*18*). The EPA believes there are many unanswered questions concerning the impact of biodegradable polymers and has opposed legislation mandating increased use (*4*). Several companies have undertaken research on biodegradable plastics, with nonfood packaging, personal and health care, and other disposables considered the key markets for biodegradable polymers for the nearer term (*18, 19*). One of the main challenges for biodegradable polymers is the low costs of conventional synthetic polymer resins, which recently have been in the $0.50-0.75 per pound range (*20*). Recent per pound prices for finished films from conventional synthetic polymers range from approximately $1.00 for low density polyethylene (LDPE) to $1.80 for PP to $2.00 for PS to $3.00 for PET (*21*).

The effort to develop biodegradable polymers for packaging for all uses including food has been especially urgent for the Navy, whose ships along with non-military vessels are now legally prohibited from dumping persistent waste at sea (*22*). However, there is also much debate as to whether biodegradable packaging can significantly reduce MSW.

For biodegradable materials to replace non-biodegradable synthetics, the essential mechanical, optical and/or barrier properties for the intended application will have to be matched. Use of conventional synthetic plastic processing technology will probably be necessary in order to ensure economic viability (*6*). In addition, the biodegradable characteristics must not limit product storage conditions, result in insect infestation, or produce safety problems in food applications (*22*). Thus, the challenge for biodegradable polymer products is controlled lifetime: performing intended functions, remaining stable during storage and use, and then biodegrading at the intended time and conditions (*6*).

Biodegradability is an often misunderstood option which would require major infrastructure changes. Since landfills are designed to prevent biodegradation, special composting systems are required to promote biodegradation (*4*). Successful biodegradable polymer materials could be degraded in backyard composting, or municipal composting systems would have to be introduced. However, two issues would have to be considered. Easy, efficient sorting of biodegradable from non-biodegradable packaging would be essential, lest both mechanical/chemical recycling and biodegradation be confounded (*23*). Secondly, biodegradation is slow and produces significant methane, with greenhouse-effect implications. In comparison, incineration is fast and can capture important energy from natural (biodegradable) and synthetic polymers alike (*24*).

Materials Available for Biodegradable Products

A number of naturally-occurring polysaccharides, proteins and polyesters, as well as synthetic biodegradable polymers are being considered for biodegradable products (*6, 11, 25, 26*). In addition, various polysaccharide and protein materials being considered for edible coatings and films also have potential as biodegradable packaging films (*27, 28*). Starch-based polymers, as well as polyhydroxybutyrate/valerate (PHBV) copolymers and polylactic acid (PLA) polymers, are believed in particular to have considerable promise as biodegradable materials (*4, 23*).

Following is a summary of the main materials which have been proposed and investigated for biodegradable products. Information on physical properties of specific materials is presented in the accompanying tables.

Cellulose-based Materials

Cellophane. Cellulose is a readily-available, fully-biodegradable, water-insoluble plant material. Cellophane, which is regenerated cellulose, is produced by extruding a viscous alkaline colloidal dispersion of cellulose xanthate into an acid/salt bath. After being treated in a plasticizer (usually glycerol) bath and then drying, the result is a transparent film with excellent clarity and sparkle (*29*). These characteristics led to its replacement of paper, waxed paper and glassine in many packaging applications when it was introduced. Cellophane dominated the transparent packaging film market until the introduction of synthetic thermoplastic packaging films in the 1950s. Since peaking in the mid-1960s at about 750 million pounds per year in the U.S., cellophane sales have dropped by 90% because of replacement by synthetic plastic films, including PP (*30*).

Cellophane tends to get brittle under dry conditions, so plasticizers like glycerol are used to improve flexibility. In addition, changes in relative humidity can cause shrinkage. Cellophane film by itself has high water vapor permeability (WVP) (*31, 32*), compared to synthetic films such as LDPE and HDPE (*33*), PP (*34*), oriented polypropylene (OPP) (*35*) and even ethylene-vinyl alcohol copolymer (EVOH) (*36*). In addition, cellophane is not a thermoplastic and is not heat-sealable. Because of its rather poor barrier properties and non-sealability, cellophane is usually coated with nitrocellulose-wax (NC-W) blend or PVDC and often additionally laminated (e.g., with metalized polyester) (*37*). NC-W- and PVDC-coated cellophane have WVPs comparable to HDPE (Table I). The choice of the specific type of cellophane to be used depends largely on the barrier properties required for a specific application (*29, 32*).

Because of its hydrogen-bonding capability, cellophane has low oxygen permeability (OP) at low relative humidity, comparable to EVOH and PVDC films (*38*). However, its oxygen-barrier properties are greatly dimished at intermediate to high relative humidity (Table II) (*32*). Cellophane OP is reduced significantly at high relative humidities by coating with NC-W or, especially, with PVDC (Table II). Cellophane is also a reasonable barrier for fats, oils and flavors (*39*).

Cellophane makes a fairly strong package, with tensile strength (TS) greater than LDPE, HDPE and PP, and elongation (E) comparable to OPP film (*34, 35, 40*) (Table III). Cellophane can be strengthened by lamination with OPP (*37*). Cellophane film tears easily when notched, which makes it an excellent material for easy-opening applications with tear tape (*39*). It has excellent printability characteristics and is one of the most machineable packaging materials. Cellophane-based film can be used for packaging a large variety of food products (*37*). In particular, NC-W-coated cellophane is useful as a wrap for baked goods, fresh and processed meats, fresh produce and candies. PVDC-coated cellophane is used in packaging applications where the highest moisture barrier and flavor protection are needed, including cookies, snacks, nuts, dried fruit and coffee. Depending on the type of cellophane, cost for finished film is approximately $2.00-3.50 per pound (*11, 15*).

Uncoated and NC-W-coated cellophane have been found to be totally biodegradable (*41, 42*). Uncoated films disintegrated in 10 to 14 days, whereas two-side coated films took 28 to 84 days to disintegrate. Time for total degradation of uncoated cellophane was 28-60 days, whereas coated films took 80-120 days. While PVDC coating disintegrated to inert dust, the NC-W coating biodegraded totally in six months or longer (*42*).

Cellulose Acetate. Cellulose acetate is a thermoplastic material produced by mixing cotton fibers with glacial acetic acid and acetic anhydride, with sulfuric acid used as catalyst (*39, 43*). Films can be obtained by either extrusion or solvent casting from acetone.

The WVP of cellulose acetate film is high, similar to uncoated cellophane (Table I). The OP of cellulose acetate appears to be an order of magnitude higher than cellophane, at comparable test conditions (Table II). Although moisture and gas barrier properties are poor, cellulose acetate film is considered excellent for fresh produce and pastries because it breathes and does not fog up (*44*). Like cellophane, cellulose acetate film has good resistance to oils and greases (*39, 43*). Cellulose acetate resin cost is $1.60-2.10 per pound (*15*) and finished film price is approximately $4.00 per pound (*21*).

Cellulose acetate films are crystal clear and tough, possessing good tensile and impact strengths (*39, 43*). TS and E are similar to cellophane (Table III). Applications which take advantage of cellulose acetate's

TABLE I. Water Vapor Permeabilities (WVP) of Biodegradable Polymer Films Compared to Conventional Synthetic Polymer Films

Reference	Material	Test Conditions*	Permeability g·mm/m²·d·kPa
(*31*)	Cellophane	25°C, 100/0%RH	17.6
(*32*)	Cellophane (~15% Plasticizer)	38°C, 90/0%RH	7.3
(*32*)	Cellophane, NC-W-coated	38°C, 90/0%RH	0.033
(*32*)	Cellophane, PVDC-coated	38°C, 90/0%RH	0.030
(*39*)	Cellulose Acetate (10% Plasticizer)	35°C, 90/0%RH	10.3
(*39*)	Cellulose Propionate (10% Plasticizer)	35°C, 90/0%RH	10.3
(*39*)	Ethyl Cellulose (10% Plasticizer)	35°C, 90/0%RH	0.7
(*39*)	Methyl Cellulose	35°C, 90/0%RH	4.8
(*54*)	HPMC	27°C, 0/85%RH	9.1
(*54*)	HPMC:SA (1.25:1)	27°C, 0/85%RH	0.026
(*57*)	HPMC:PEG (9:1)	25°C, 85/0%RH	6.5
(*57*)	SA:PA:HPMC:PEG	25°C, 85/0%RH	0.048
(*59*)	BW/SA:PA:MC:HPMC:PEG	25°C, 97/0%RH	0.058
(*31*)	Amylose	25°C, 100/0%RH	31.6
(*31*)	Amylose, NC-W-coated	25°C, 100/0%RH	0.36
(*53*)	Zein:Gly (4.9:1)	21°C, 85/0%RH	9.6
(*53*)	Gluten:Gly (3.1:1)	21°C, 85/0%RH	53
(*93*)	Gluten:Gly (5.0:1)	30°C, 100/0%RH	5.1
(*99*)	Gluten:BW:Gly (5:1.5:1)	30°C, 100/0%RH	3.0
(*94*)	SPI:Gly (4.0:1)	28°C, 0/78%RH**	39
(*95*)	WPI:Gly (4:1)	25°C, 0/77%RH**	70
(*98*)	WPI:BW:Sor (3.5:1.8:1)	25°C, 0/98%RH**	5.3
(*96*)	Sodium Caseinate	25°C, 0/81%RH**	36.7
(*96*)	Calcium Caseinate:BW (1.7:1)	25°C, 0/97%RH**	3.6
(*33*)	LDPE	38°C, 90/0%RH	0.079
(*33*)	HDPE	38°C, 90/0%RH	0.020
(*34*)	PP	38°C, 90/0%RH	0.047
(*35*)	OPP	38°C, 90/0%RH	0.017
(*36*)	EVOH (68% VOH)	38°C, 90/0%RH	0.25

*RHs are those on top and bottom sides of film (top/bottom); ** Corrected RH shown.
HPMC = Hydroxypropylmethyl cellulose, MC = Methyl Cellulose, SA = Stearic Acid, PA = Palmitic Acid, BW = Beeswax, PEG = Polyethylene Glycol, NC-W = Nitrocellulose-Wax, SPI = Soy Protein Isolate, WPI = Whey Protein Isolate, Gly = Glycerol, Sor = Sorbitol, PVDC = Polyvinylidene Chloride, LDPE = Low Density Polyethylene, HDPE = High Density Polyethylene, PP = Polypropylene, OPP = Oriented Polypropylene, EVOH = Ethylene-Vinyl Alcohol Copolymer, VOH = Vinyl Alcohol

TABLE II. Oxygen Permeabilities (OP) of Biodegradable Polymer Films Compared to Conventional Synthetic Polymer Films

Reference	Material	Test Conditions	Permeability $cm^3mm/m^2 \cdot d \cdot kPa$
(*38*)	Cellophane (~15% Plasticizer)	23°C, 0%RH	0.7
(*32*)	Cellophane (~15% Plasticizer)	23°C, 50%RH	16
(*32*)	Cellophane (~15% Plasticizer)	23°C, 95%RH	252
(*32*)	Cellophane, NC-W-coated	23°C, 50%RH	10
(*32*)	Cellophane, NC-W-coated	23°C, 95%RH	68
(*32*)	Cellophane, PVDC-coated	23°C, 95%RH	1.9
(*39*)	Cellulose Acetate (10% Plasticizer)	25°C, 50%RH	136
(*39*)	Cellulose Propionate (10% Plasticizer)	25°C, 50%RH	233
(*39*)	Ethyl Cellulose (10% Plasticizer)	25°C, 50%RH	7780
(*55*)	Methyl Cellulose	24°C, 50%RH	97
(*55*)	Hydroxypropylmethyl Cellulose	24°C, 50%RH	272
(*68*)	Amylomaize Starch (AS)	25°C, <100%RH	<65
(*69*)	Hydroxypropylated-AS	25°C, <78%RH	~0
(*92*)	Collagen	RT, 0%RH	< 0.04-0.5*
(*92*)	Collagen	RT, 63%RH	23.3
(*92*)	Collagen	RT, 93%RH	890
(*100*)	Zein:PEG+Gly (2.6:1)	25°C, 0%RH	38.7-90.3
(*101*)	Gluten:Gly (2.5:1)	25°C, 0%RH	6.1
(*102*)	SPI:Gly (2.4:1)	25°C, 0%RH	6.1
(*103*)	WPI:Gly (2.3:1)	23°C, 50%RH	76.1
(*103*)	WPI:Sor (2.3:1)	23°C, 50%RH	4.3
(*38*)	LDPE	23°C, 50%RH	1870
(*38*)	HDPE	23°C, 50%RH	427
(*34*)	PP	23°C, 50%RH	933
(*35*)	OPP	23°C, 50%RH	622
(*39*)	Polyester	23°C, 50%RH	15.6
(*38*)	EVOH (70% VOH)	23°C, 0%RH	0.1
(*38*)	EVOH (70% VOH)	23°C, 95%RH	12
(*38*)	PVDC-based films	23°C, 50%RH	0.4-5.1

*Based on values for PVDC-based films (*37*).
RT = Room Temperature

TABLE III. Mechanical Properties of Biodegradable Polymers Compared to Conventional Synthetic Polymers

Reference	Material	Tensile Strength (TS) MPa	Elongation (E) %
(*32*)	Cellophane (~15% Plasticizer)	55-124	16-60
(*32*)	Cellophane, NC-W-coated	55-124	16-60
(*32*)	Cellophane, PVDC-coated	55-124	22-60
(*44*)	Cellulose Acetate (16% Plasticizer)	69	20-45
(*39*)	Cellulose Propionate (10% Plasticizer)		80
(*39*)	Ethyl Cellulose (10% Plasticizer)		30
(*55*)	Methyl Cellulose	62	10
(*55*)	Hydroxypropylmethyl Cellulose	69	10
(*70*)	Starch	49	7
(*70*)	Amylose 70	23	
(*70*)	Amylose:Gly (3:1)	31	47
(*69*)	Amylomaize Starch (AS)	76	5
(*69*)	Hydroxypropylated AS	56	14
(*80*)	PVOH	21-50	112-600
(*10*)	Starch:PVOH (Novon M1801)	38	100
(*78*)	Starch:PVOH:Gly (6:6:1)	29	168
(*10*)	Starch:PVOH (Novon M4900)	16-22	414-450
(*91*)	[a]Collagen:Cellulose:Gly (3.4:0.8:1)	3.3-10.8	25-50
(*91*)	[b]Collagen:Cellulose:Sor+Gly (5.3:0.4:1)	6.2-9.4	38-57
(*100*)	Zein:PEG+Gly (5.9:1)	22-28	6-9
(*100*)	Zein:PEG+Gly (3.6:1)	7-16	43-198
(*100*)	Zein:PEG+Gly (2.6:1)	3-7	173-213
(*104*)	Gluten:Gly (2.5:1)	2.6	276
(*94*)	SPI:Gly (4.0:1)	13	17
(*105*)	SPI:Gly (1.7:1)	4.5	86
(*103*)	WPI:Gly (2.3:1)	14	31
(*107*)	Zein	38	4.2
(*107*)	Crosslinked Starch:Zein (5:1)	29	2.6
(*107*)	SPI	31	2.5
(*108*)	SPI:Starch (3:2)	5-15	
(*115*)	PHB/V	30-40	
(*113*)	PHB/V	20-31	8-42
(*119*)	PLA	48-101	6-15
(*119*)	PLA (17% lactide)	16	288
(*40*)	LDPE	9-17	500
(*40*)	HDPE	17-35	300
(*40*)	PP	42	300
(*40*)	OPP	165	50-75
(*40*)	Polyester	175	70-100
(*43, 81*)	PS	35-55	1

[a]Wet process; [b]Dry process
PHB/V = Polyhydroxybutyrate/valerate Copolymer, PLA = Polylactic Acid, PS = Polystyrene

mechanical properties include tool handles, combs and spectacle frames (*45*).

Chemical substitution of cellulose generally reduces biodegradability (*46*). Investigation of cellulose acetate biodegradability provided evidence of mineralization (*6, 47*). The rate of biodegradation can be increased by substituting plasticizers commonly used with specific esters and other low molecular weight components to a level of >30% (*45*). The resulting material retains thermoplastic properties.

Cellulose Propionate. Cellulose propionate is a thermoplastic material produced by treating cellulose with propionic acid and propionic anhydride, with sulfuric acid used as a catalyst. Films made from this material generally have properties similar to cellulose acetate (Tables I-III). However, cellulose propionate films reportedly possess twice the impact strength, absorb less moisture and swell less compared to cellulose acetate films (*43*). Cellulose propionate may be resistant to biodegradation (*6*).

Cellulose Ethers. Cellulose may be converted to a number of ethers, including two which are thermoplastic: hydroxypropyl cellulose (HPC) (*48*) and ethyl cellulose (EC) (*39*). EC possesses better deep-draw properties than any other thermoplastic material (*39*). EC films have WVP an order of magnitude lower than uncoated cellophane, but are still rather poor moisture barriers (Table I). EC films have much higher OP than other cellulose-based films (Table II), but they have been reported to be good oil and fat barriers for products like butter and olive oil (*39*). EC films appear to possess mechanical properties similar to those of cellophane (Table III), and they have excellent impact strength at low temperatures (*39*).

Methyl cellulose (MC) and hydroxypropylmethyl cellulose (HPMC) are not thermoplastic, but films can be manufactured by casting out of solvent on a stainless steel belt (*39*). MC and HPMC solubility in water is an advantage for unit packaging of dry products which are destined for addition to water. A number of studies have investigated the effects of MC molecular weight, film-forming solution composition, plasticizers, drying conditions and film testing conditions on film properties (*49-53*). MC and HPMC appear to have properties similar to cellulose acetate (*39, 54, 55*) (Tables I-III). Composite films made by combining fatty acids, beeswax and/or paraffin with MC and HPMC to form bilayer films have been studied extensively (*54, 56-66*), and have had WVPs as low as LDPE (Table I). MC and HPMC also have good resistance to oils and greases (*39, 55, 67*).

Starch-based Materials

Starches are low-cost, fully-biodegradable polymers which can be used to form biodegradable products. Aqueous solutions of amylose (which makes up ~25% of starch, with amylopectin the remainder) give better films upon drying than native starch solutions. Amylose films are poor moisture barriers, but can be coated with NC-W blend to yield low WVP comparable to NC-W-coated cellophane at the same test conditions of the study (Table I) (*31*).

Amylomaize starch (AS) with 71% amylose content and hydroxypropylated AS films have excellent oxygen barrier properties (Table II) (*68, 69*). Amylose films have mechanical properties similar to cellophane films (*70*). However, starch, AS and hydroxypropylated AS have mechanical properties inferior to amylose (Table III).

Starch exhibits thermoplastic behavior when a plasticizer such as water is added (*11, 71-74*). Amylose and hyroxypropylated AS can also be extruded (*75, 76*). The influence of water on the stability of starch-based materials limits their usefulness. This is perhaps the greatest problem with all biologically-based polymers, because of their inherent hydrophilic nature.

Starch and polyvinylalcohol (PVOH) have been blended to yield fully biodegradable, thermoplastic materials (*26, 72, 77-79*). PVOH is available over a broad range of properties (*80*). This provides opportunity for obtaining a broad range of starch-PVOH blend properties (*10*) (Table III). PVOH content, degree of hydrolysis and molecular weight have significant effect on mechanical properties and water resistance of sheets formed from injection molded high amylose starch-PVOH blends (*79*). Thus, some starch-PVOH blends have potential for replacement of PE films in applications where moisture barrier is not important; whereas other starch-PVOH blends may have replacement possibilities for polystyrene (PS) applications (e.g, disposable food service plates, cups, utensils), where the strength of PS is matched (*43, 81*) (Table III).

Several companies have attempted to commercialize extrudable blends containing high percentages of starch with synthetic biodegradable polymers (*4, 6, 22*). Novamont North America, Inc. developed a thermoplastic resin blend containing 60% or more corn starch with the balance a proprietary low-molecular weight biodegradable material (*22*). NOVON Products, formerly a division of Warner Lambert Co., successfully blended high percentages of starch with other completely biodegradable components in a patented process (*4*). Both the Novamont and NOVON resin blends have prices in the $1.50-3.00 per pound range (*4*). NOVON Products, which was the leading U.S. producer of starch-based biodegradable polymers, closed its plant in November 1993 (*19*), but now exists as NOVON International, a subsidiary of Churchill Technology, Inc.

The U. S. Army Natick RD&E Center Food Engineering Directorate evaluated starch-based biodegradable products, such as utensils and trash bags manufactured from resin supplied by Novamont North America, Inc. (*22*). Starch-based spoons met all specifications, except for rigidity. Forks and knives molded from the starch-based resin had similar rigidity problems. Nonetheless, consumers (aircraft carrier-based sailors) indicated that the starch-based spoons were comparable to PS spoons. Starch-based trash bags met requirements for impact resistance, tear resistance, load capacity and seam continuity, but mechanical properties were generally inferior to PE bags. When the utensils and trash bags were stored at 38°C/90% RH, fungal growth was evident after 2 weeks (*22*). Lower temperatures and RHs did not produce fungal growth.

Michigan Biotechnology Institute (MBI) entered into a joint venture with Japan Corn Starch to commercialize injection-molded products such as disposable utensils made from thermoplastic, modified starch-based AMYPOL resin developed by MBI (*11*). In addition, National Starch and Chemical Company and NOVON International each developed starch-based, expanded loose-fill packaging material as a replacement for expanded PS (*11*). Starch has been combined in a number of ways with non-biodegradable polymers in an attempt to improve the film properties of the former and increase the degradability of the latter (*6, 15, 23, 26, 82-85*). Starch-PE mixtures can be processed via co-extrusion, injection molding, or film blowing for the production of films or bottles (*82*). Generally, the starch component of these films can be metabolized by certain amylolytic bacteria. However, the PE component does not biodegrade; rather, it disintegrates (*86-89*). A number of approaches are possible to enhance biodegradation of the synthetic component of starch-synthetic polymer blends (*26, 85*).

Protein Materials

Proteins are natural polymers which are fully biodegradable like cellulose and starch. Proteins do not have the regular repeating units of other polymers, due to the different side groups on the amino acid residues which make up the protein chain. Nonetheless, considerable effort has gone into exploring and developing protein-based films (*90*).

Collagen. The most successful films based on a protein are collagen films used commercially as edible casings and wraps for meat products. Collagen is a fibrous, structural protein in animal tissue, with common repeating unit: glycine, proline and hydroxyproline (*6*). Unfortunately, collagen is not thermoplastic; thus, collagen film must be made by extruding a viscous colloidal acidic dispersion into a neutralizing bath, washing and then drying (*91*).

No data on the WVP of collagen films has been reported in the scientific literature. The OP is extremely low at 0% RH, increasing rapidly with RH (Table II) (*92*). In this regard, the OP characteristics of collagen film are similar to cellophane. The mechanical properties of collagen film depend on composition (e.g., plasticizer content) and whether longitudinal or transverse sections are tested (*91*). While the E of typical collagen film is similar to cellophane, the TS is considerably lower (Table III). Lowering the plasticizer content would increase film strength, but with the cost of decreased E. Achievement of mechanical properties similar to PS is thus possible. However, collagen film mechanical properties are inferior to LDPE and HDPE (Table III).

Zein, Gluten, Soy Protein, Whey Protein and Casein. The other proteins which have been studied for film formation include corn zein (CZ), wheat gluten (WG), soy protein isolate (SPI) and whey protein isolate (WPI), which are globular in nature, and casein (CS), which has an open, extended structure. Films from these proteins are generally formed out of ethanolic solution (CZ and WG) or aqueous solution (CZ latex, SPI, WPI and CS). However, some or all of these proteins may exhibit thermoplastic behavior under the right conditions.

All of these proteins have fairly large WVPs (Table I) (*53, 93-96*). Composite films which include wax have considerably lower WVP (*96-99*). However, protein film coating (or lamination) analogous to NC-W-coated cellophane to achieve lower WVP has not been studied; and protein composite films have not attained the low WVP of cellulose ether based composites.

The OPs of films made from CZ, WG and SPI are quite low at 0% RH (Table II) (*100-102*). As with collagen film, RH has a large effect on OP for films from these materials. The OPs at 50% RH indicate that WPI films may be better barriers than films from other globular proteins (*103*). Nonetheless, at low to intermediate RHs, protein films have OPs significantly lower than the PEs and PPs, and comparable to polyester.

As with permeability properties, mechanical properties are strongly affected by plasticizer content (and RH). On the basis of limited data, it appears that plasticized films made from the globular proteins have a combination of TS and E similar to collagen film, but inferior to cellophane (*94, 100, 103-105*). Among the globular proteins, CZ and WG films made from ethanolic solutions appear to possess superior mechanical properties compared to SPI and WPI films made from aqueous solution (Table III). The limited mechanical property data for protein films makes it difficult to foresee applications, but edible food coatings or paper coatings are possibilities (*28, 106*).

Soy protein isolate and corn zein have been shown to have properties resembling thermoset plastics and can be compression molded into shapes which have mechanical properties similar to PS (Table II) (*107-110*). Formaldehyde cross-linked starch-zein gave molded shapes with significantly greater water resistance than starch, zein or soy protein alone (*107, 111*). Thus, crosslinked starch-zein has potential for molding into utensils and containers usually fabricated from PS (*107*). Mixtures of soy protein and native corn starch have thermoplastic properties which are suitable for extrusion and injection molding (Table III) (*108*).

Microbial Polyesters

Other naturally-occurring polymers being considered for biodegradable products are the microbial polyesters (polyhydroxyalkanoates) produced by nutrient-limited fermentation of sugar feedstock. The copolymer poly(3-hydroxybutyrate)-co-(3-hydroxyvalerate) (PHB/V) is thermoplastic and fully biodegradable (*6, 11, 26, 112*).

PHB/V is reported to possess good oxygen, moisture and aroma barrier properties (*113*). Good moisture barrier properties are consistent with the fact that PHB/V is relatively more hydrophobic than polysaccharides and proteins. Furthermore, while the oxygen barrier may not be as good as that for polysaccharides and proteins at low RH, it should not be as sensitive to increasing RH.

Inclusion of HV in PHB results in flexibility, tensile strength and melting point to the resulting polymer ranging from PP-like (low HV) to PE-like (high HV) (*114*). PHB is stiff and brittle, but HV content improves flexibility and toughness (*115*). Thus, low-HV films have high TS and low E, while high-HV films have low TS and high E (Table III). The TS of typical PHB/V film overlaps that of PE and is somewhat less than that of PP and PS. E is greater than that for PS, but smaller than for PP and PE.

Targeted products for fabrication from PHB/V include bottles, sheets, films and coatings (*113*). PHB/V has been commercialized as Biopol™, and a biodegradable Biopol™ bottle was introduced in 1990 for a biodegradable shampoo (*116*). Other uses being considered include beverage bottles, coated paperboard milk cartons, motor oil containers, disposable razors, kitchen films and diapers (*116*). The main limitations are processability and cost of PHB/V (*26*). Current price for PHB/V resin is $8-10 per pound (*4*), but projected price is lower (*11, 113*).

Polylactic Acid

Polylactic Acid (PLA) is not a natural polymer, but it is based on lactic acid which can be produced by fermentation of simple sugars. Advances in recent years have produced more economical PLA polymers of sufficient molecular weight to possess useful mechanical properties (*11*). PLA is being commercialized by three companies in the U.S.: Cargill (Minneapolis, MN), Ecochem (Newport, DE) and Chronopol (Golden, CO), and Argonne National Labs has licensed technology to the Japanese firm Kyowa Hokko. Product targets include food service containers and utensils, yard waste and grocery bags, and agricultural films. PLA is thermoplastic, insoluble in water, and biodegradable in compost or seawater with microorganisms present (*117*).

PLA has mechanical properties similar to PS (Table III) (*118, 119*). However, PLA mechanical properties can be modified by varying molecular weight and crystallinity, copolymerization and other modifications, with resulting properties which also mimic PE, PP or PVC (*117, 120*). By plasticizing PLA with its lactide monomer, mechanical properties are modified to more closely resemble those of PE (Table III) (*119*). Cost of PLA at $1-3 per pound is similar to that for starch-PVOH blends (*4, 11*).

Other Biodegradable Materials

Films can be produced from a number of other renewable biodegradable materials by casting from solution. Materials include carrageenan, alginate, pectin, chitosan and pullulan (microbial polysaccharide). Limited property data is available. Although films from these materials are poor moisture barriers, they generally appear to have potential as oxygen and oil/fat barriers (*6, 121-129*).

Summary and Conclusions

Based on their properties, biodegradable polymers appear to have a place in the on-going effort to conserve non-renewable resources and reduce waste. However, applications may involve some trade-offs in performance and/or cost. The following specific conclusions are possible:

1) By their very nature, biodegradable polymers are generally not good moisture barriers (i.e., they have high WVPs). In the case of cellophane, coating with a nitrocellulose-wax (NC-W) blend provides a moisture barrier comparable to LDPE without sacrificing biodegradability. Excellent moisture-barrier films have also been produced experimentally by coating amylose with nitrocellulose-wax blend and from cellulose ethers in combination with fatty acids and wax. Utilization of biodegradable films could increase significantly with successful production of a low-WVP, coated, biodegradable film utilizing conventional extrusion technology.

2) Biodegradable polymer films are naturally good oxygen barriers (i.e., they have low OP) at low RH. However, their OPs increase exponentially as RH increases. Coating biodegradable polymer films with a moisture barrier layer would help retain oxygen barrier characteristics at medium to high RH. Another approach would be to use double bagging. This would involve protecting oxygen-sensitive products with a biodegradable polymer film bag acting as an oxygen barrier, and then inserting this bagged product in a synthetic polymer film bag with low WVP (e.g., LDPE) to protect the product and biodegradable polymer films bag from moisture. This approach would replace multilayer synthetic polymer film packages, which are inherently difficult to recycle, with a

single-component synthetic polymer film package which is more easily recycled.

3) Cellulose-based films have been studied and used most extensively among biodegradable polymers. In spite of not being thermoplastic, cellophane has a significant presence commercially, with use of coating and lamination to improve properties. However, only uncoated and NC-W-coated cellophane are easily biodegraded. Cellulose acetate is thermoplastic and biodegradable, but has not been coated or laminated to improve barrier and mechanical properties. Successful coating of cellulose acetate or other thermoplastic cellulose derivatives with a biodegradable hydrophobic material at low cost could expand applications in packaging and other areas.

4) Starch and PVOH can be combined to form thermoplastic blends which can be extruded to form films with mechanical properties similar to LDPE. Such films can be used to fabricate compostible bags for yard, kitchen and restaurant waste, or be used as biodegradable agricultural mulch. Different starch-PVOH blends can be used to mold biodegradable food service items (dishes, cups, utensils, etc.) and other biodegradable products with properties which resemble PS.

5) Various proteins can be formed into films which generally have high WVPs similar to cellulose- and starch-based films. Protein-based films analogous to NC-W-coated cellophane and cellulose ether bilayer composite films with low WVP have not been achieved. However, protein-based films appear to be excellent oxygen barriers at low RH. The relatively poor mechanical properties of plasticized protein flexible films limit their use for packaging to relatively small pouches/bags, food casings/wraps, edible food film coatings and paper coatings. Unplasticized proteins, especially when crosslinked with starch, have properties similar to PS.

6) PHB/V copolymer is a thermoplastic material which has adjustable properties dependent on the monomer ratio. PHB/V appears to have greater moisture resistance and lower WVP than the other biodegradable polymers. These properties allow production of bottles and other containers useful for high-moisture products. PHB/V is a versatile polymer which will find increasing use if the cost of production decreases.

7) PLA is a versatile thermoplastic material whose properties can be modified to either resemble LDPE or PS, with the same potential uses and costs mentioned above for starch-PVOH blends.

Legend of Abbreviations

AS	=	amylomaize starch
BW	=	beeswax
Cal	=	calcium
CS	=	casein
CZ	=	corn zein
E	=	% elongation
EC	=	ethyl cellulose
EVOH	=	ethylene-vinylalcohol copolymer
Gly	=	glycerol
HDPE	=	high density polyethylene
HPC	=	hydroxypropyl cellulose
HPMC	=	hydroxypropyl methyl cellulose
HV	=	hydroxyvalerate
LDPE	=	low density polyethylene

MC	=	methyl cellulose
MSW	=	municipal solid waste
NC-W	=	nitrocellulose-wax
OP	=	oxygen permeability
OPP	=	oriented polypropylene
PA	=	palmitic acid
PE	=	polyethylene
PEG	=	poly(ethylene glycol)
PET	=	poly(ethylene terephthalate)
PHB/V	=	poly(3-hydroxybutyrate)-co-(3-hydroxyvalerate)
PLA	=	poly(lactic acid)
Plas	=	plasticizer
PP	=	polypropylene
PS	=	polystyrene
PVDC	=	poly(vinylidene chloride)
PVOH	=	poly(vinyl alcohol)
RH	=	relative humidity
RT	=	room temperature
SA	=	stearic acid
Sor	=	sorbitol
SPI	=	soy protein isolate
TS	=	tensile strength
WG	=	wheat gluten
WPI	=	whey protein isolate
WVP	=	water vapor permeability

Literature Cited

1. Brown, D. *Trends Food Sci. & Technol.* **1993,** *4*(9), pp 294-300.
2. Fleming, R. A. *Chem. Tech.* **1992,** *22*(6), pp 333-335.
3. Marsh, K. In *Activities Report and Minutes of Work Groups & Sub-Work Groups of the R&D Associates;* Research and Development Associates for Military Food and Packaging Systems, Inc.: Boston, MA, 1994.
4. Beach, E. D.; Price, J. M. In *Industrial Uses of Agricultural Materials-Situation and Outlook Report IUS-1;* Glaser, L.; Gajewski, G., Eds.; Commodity Economics Division, Economic Research Service, USDA: Washington, D. C., 1993; pp 41-48.
5. Albertsson, A.-C.; Karlsson, S. In *Chemistry and Technology of Biodegradable Polymers;* Griffin, G. J. L., Ed.; Blackie Academic & Professional: London, 1994; pp 6-17.
6. Kaplan, D. L.; Mayer, J. M.; Ball, D.; McCassie, J.; Allen, A. L.; Stenhouse, P. In *Biodegradable Polymers and Packaging;* Ching, C.; Kaplan, D.; Thomas, E., Eds.; Technomic Publishing Co., Inc.: Lancaster, PA, 1993; pp 1-42.
7. Swift, G. *Acc. Chem. Res.* **1993,** *26*(3).
8. Narayan, R. In *Biodegradable Polymers '94;* Executive Conference Management, Plymouth, MI: Ann Arbon, MI, 1994.
9. Seal, K. J. In *Chemistry and Technology of Biodegradable Polymers;* Griffin, G. J. L., Ed.; Chapman & Hall, Inc.: New York, 1994; pp 116-134.
10. Chapman, G. M. In *Recycle '95 - Environmental Technologies;* Davos, Switzerland, 1995.
11. Narayan, R. In *Polymers from Agricultural Coproducts;* Fishman, M. L.; Friedman, R. B.; Huang, S. J., Eds.; American Chemical Society: Washington, D.C., 1994; pp 2-28.

12. *Biodegradable Polymers and Packaging*; Ching, C.; Kaplan, D. L.; Thomas, E. L., Eds.; Technomic Publishing Co., Inc.: Lancaster, PA, 1993.
13. *Chemistry and Technology of Biodegradable Polymers*; Griffin, G. J. L., Ed.; Blackie Academic & Professional: New York, 1994.
14. *Polymers from Agricultural Coproducts*; Fishman, M. L.; Friedman, R. B.; Huang, S. J., Eds.; ACS Symposium Series 575, American Chemical Society: Washington, D.C., 1994.
15. Beach, E. D.; Ahmed, I.; Glaser, L. K.; Price, J. M. In *Advances in Solar Energy: An Annual Review of Research and Development*; American Solar Energy Society: Boulder, CO, 1994; pp 383-458.
16. Rodriguez, F. *Chem. Tech.* **1971,** *1*(7), pp 409-415.
17. *Hydrocarbon Proc.* **1989,** *68*(3), pp 15.
18. Beach, D.; Ahmed, I. In *Industrial Uses of Agricultural Materials-Situation and Outlook Report IUS-1;* Glaser, L.; Gajewski, G., Eds.; Commodity Economics Division, Economic Research Service, USDA: Washington, D. C., 1993; pp 12-15.
19. Beach, D.; Ahmed, I. In *Industrial Uses of Agricultural Materials-Situation and Outlook Report IUS-2;* Glaser, L.; Beach, D., Eds.; Commodity Economics Division, Economic Research Service, USDA: Washington, D. C., 1993; pp 10-12.
20. *Modern Plastics.* **1995,** *72*(4), pp 32.
21. Grimsley, M. L., *Personal communication.* 1995, Hutchison-Miller Sales Co.
22. Ross, J. M. In *Activities Report and Minutes of Work Groups & Sub-Work Groups of the R&D Associates;* Research and Development Associates for Military Food and Packaging Systems, Inc.: Boston, MA, 1993.
23. Evans, J. D.; Sikdar, S. K. *Chem. Tech.* **1990,** *20*(1), pp 38-42.
24. Rowatt, R. J. *Chem. Tech.* **1993,** *23*(1), pp 56-60.
25. Griffin, G. J. L. In *Chemistry and Technology of Biodegradable Polymers;* Griffin, G. J. L., Ed.; Blackie Academic & Professional: New York, 1994; pp 1-6.
26. Chapman, G. M. In *Polymers from Agricultural Coproducts;* Fishman, M. L.; Friedman, R. B.; Huang, S. J., Eds.; American Chemical Society: Washington, D.C., 1994; pp 29-49.
27. Nisperos-Carriedo, M. O. In *Edible Coatings and Films to Improve Food Quality;* Krochta, J. M.; Baldwin, E. A.; Nisperos-Carriedo, M. O., Eds.; Technomic Publishing Co., Inc.: Lancaster, PA, 1994; pp 305-336.
28. Gennadios, A.; McHugh, T. H.; Weller, C. L.; Krochta, J. M. In *Edible Coatings and Films to Improve Food Quality;* Krochta, J. M.; Baldwin, E. A.; Nisperos-Carriedo, M., Eds.; Technomic Publishing Co., Inc.: Lancaster, PA, 1994; pp 201-277.
29. Hanlon, J. F. In *Handbook of Package Engineering*; McGraw-Hill Book Co.: New York, 1971; pp 3.9-3.16.
30. Jenkins, W. A.; Harrington, J. P. In *Packaging Foods with Plastics*; Technomic Publishing Co., Inc.: Lancaster, PA, 1991; pp 35-36.
31. Rankin, J. C.; Wolff, I. A.; Davis, H. A.; Rist, C. E. *I&EC.* **1958,** *3*(1), pp 120-123.
32. Taylor, C. C. In *The Wiley Encyclopedia of Packaging Technology;* Bakker, M., Ed.; John Wiley & Sons: New York, 1986; pp 159-163.
33. Smith, S. A. In *The Wiley Encyclopedia of Packaging Technology;* Bakker, M., Ed.; John Wiley & Sons: New York, 1986; pp 514-523.
34. Miglaw, I.; Pirog, E. In *The Wiley Encyclopedia of Packaging Technology;* Bakker, M., Ed.; John Wiley & Sons, Inc.: New York, 1986; pp 315-317.

35. Hasenauer, R. J. In *The Wiley Encyclopedia of Packaging Technology;* Bakker, M., Ed.; John Wiley & Sons, Inc.: New York, 1986; pp 320-325.
36. Foster, R. In *The Wiley Encyclopedia of Packaging Technology;* Bakker, M., Ed.; John Wiley & Sons: New York, 1986; pp 270-275.
37. *Film Types*; Flexel, Inc.: Atlanta, GA, 1993.
38. Salame, M. In *The Wiley Encyclopedia of Packaging Technology;* Bakker, M., Ed.; John Wiley & Sons: New York, 1986; pp 48-54.
39. Hanlon, J. F. In *Handbook of Package Engineering*; Technomic Publishing Co., Inc.: Lancaster, PA, 1992; pp 3.1-3.59.
40. Briston, J. H. In *The Wiley Encyclopedia of Packaging Technology;* Bakker, M., Ed.; John Wiley & Sons: New York, 1986; pp 329-335.
41. Banerjee, S. K.; Koch, W. J.; Wielicki, E. A. *Package Eng.* **1970**, pp 57-60.
42. Whitehouse, A. B. *Pollution Eng.* **1990**, *22*(5), pp 71-72.
43. Hanlon, J. F. In *Handbook of Package Engineering*; Technomic Publishing Co., Inc.: Lancaster, PA, 1992; pp 8.1-8.84.
44. *Clarifoil*; Courtaulds Chemicals: Derby, United Kingdom, 1992.
45. Ach, A. *J. Macromol. Sci.-Pure Appl. Chem.* **1993**, *A30*(9 & 10), pp 733-740.
46. Rossall, B. *Int. Biodeterior. Bull.* **1974**, *10*(4), pp 95-103.
47. Buchanan, C. M.; Gardner, R. M.; Komarek, R. J. *J. Appl. Polym. Sci.* **1993**, *47*, pp 1709-1719.
48. Ganz, A. J. *Food Prod. Dev.* **1969**, *3*(6), pp 65-74.
49. Donhowe, I. G.; Fennema, O. *J. Food Process. Preserv.* **1993**, *17*, pp 231-246.
50. Donhowe, I. G.; Fennema, O. *J. Food Process. Preserv.* **1993**, *17*, pp 247-257.
51. Park, H. J.; Weller, C. L.; Vergano, P. J.; Testin, R. F. *J. Food Sci.* **1993**, *58*(6), pp 1361-1370.
52. Debeaufort, F.; Voilley, A.; Meares, P. *J. Membrane Sci.* **1994**, *91*, pp 125-133.
53. Park, H. J.; Chinnan, M. S. *J. Food Eng.* **1995**, *25*, pp 497-507.
54. Hagenmaier, R. D.; Shaw, P. E. *J. Agric. Food Chem.* **1990**, *38*(9).
55. *A Food Technologist's Guide to Methocel Premium Food Gums*; The Dow Chemical Co.: Midland, MI, 1990.
56. Rico-Pena, D. C.; Torres, J. A. *J. Food Process. Preserv.* **1990**, *13*, pp 125-133.
57. Kamper, S. L.; Fennema, O. *J. Food Sci.* **1984**, *49*, pp 1478-1481, 1485.
58. Kamper, S. L.; Fennema, O. *J. Food Sci.* **1984**, *49*, pp 1482-1485.
59. Kester, J. J.; Fennema, O. *J. Food Sci.* **1989**, *54*(6), pp 1383-1389.
60. Greener, I. K.; Fennema, O. *J. Food Sci.* **1989**, *54*(6), pp 1400-1406.
61. Martin-Polo, M.; Voilley, A.; Blond, G.; Colas, B.; Mesnier, M.; Floquet, N. *J. Agric. Food Chem.* **1992**, *40*(3), pp 413-418.
62. Martin-Polo, M.; Mauguin, C.; Voilley, A. *J. Agric. Food Chem.* **1992**, *40*(3), pp 407-412.
63. Koelsch, C. M.; Labuza, T. P. *Lebensm.-Wiss. u.-Technol.* **1992**, *25*(5), pp 404-411.
64. Debeaufort, F.; Martin-Polo, M.; Voilley, A. *J. Food Sci.* **1993**, *58*(2), pp 426-434.
65. Park, J. W.; Testin, R. F.; Park, H. J.; Vergano, P. J.; Weller, C. L. *J. Food Sci.* **1994**, *59*(4), pp 916-919.
66. Sapru, V.; Labuza, T. P. *J. Food Process. Preserv.* **1994**, *18*(5), pp 359-368.
67. Nelson, K. L.; Fennema, O. R. *J. Food Sci.* **1991**, *56*(2), pp 504-509.
68. Mark, A. M.; Roth, W. B.; Mehltretter, C. L.; Rist, C. E. *Food Technol.* **1966**, *20*(1), pp 75-77.

69. Roth, W. B.; Mehltretter, C. L. *Food Technol.* **1967,** *21*(72), pp 72-74.
70. Wolff, I. A.; Davis, H. A.; Cluskey, J. E.; Gundrum, L. J.; Rist, C. E. *I&EC.* **1951,** *43*(4), pp 915-919.
71. Shogren, R. L.; Fanta, G. F.; Doane, W. M. *Starch.* **1993,** *45*(8), pp 276-280.
72. Loomis, G. L.; Hopkins, A. R.; George, E. R. In *Biodegradable Polymers and Packaging;* Ching, C.; Kaplan, D. L.; Thomas, E. L., Eds.; Technomic Publishing Co., Inc.: Lancaster, 1993; pp 43-51.
73. Griffin, G. J. L. In *Chemistry and Technology of Biodegradable Polymers;* Griffin, G. J. L., Ed.; Blackie Academic & Professional: New York, 1994; pp 135-150.
74. Willett, J. L.; Jasberg, B. K.; Swanson, C. L. In *Polymers from Agricultural Coproducts;* Fishman, M. L.; Friedman, R. B.; Huang, S. J., Eds.; American Chemical Society: Washington, D.C., 1994; pp 50-68.
75. Jokay, L.; Nelson, G. E.; Powell, E. L. *Food Technol.* **1967,** *21,* pp 1064-1066.
76. Morgan, B. H. *Food Prod. Dev.* **1971,** *5*(4), pp 75-77,108.
77. Westhoff, R. P.; Kwolek, W. F.; Otey, R. H. *Starch.* **1979,** *31*(5), pp 163-165.
78. Stenhouse, P. J.; Mayer, J. M.; Hepfinger, M. J.; Costa, E. A.; Dell, P. A.; Kaplan, D. L. In *Biodegradable Polymers and Packaging;* Ching, C.; Kaplan, D. L.; Thomas, E. L., Eds.; Technomic Publishing Co., Inc.: Lancaster, 1993; pp 151-158.
79. Park, E. H.; George, E. R.; Muldoon, M. A.; Flammino, A. *Polym. News.* **1994,** *19*(8), pp 230-238.
80. *Mono Sol*; Chris Craft Industrial Products: South Holland, IL.
81. Houston, J. S. In *The Wiley Encyclopedia of Packaging Technology;* Bakker, M., Ed.; John Wiley & Sons, Inc.: New York, 1986; pp 540-542.
82. Roper, H.; Koch, H. *Starch.* **1990,** *42*(4), pp 123-130.
83. Doane, W. M. *Starch.* **1992,** *44*(8), pp 293-295.
84. Pettijohn, R. M. *Chem. Tech.* **1992,** *22*(10), pp 627-629.
85. Griffin, G. J. L. In *Chemistry and Technology of Biodegradable Polymers;* Griffin, G. J. L., Ed.; Blackie Academic & Professional: New York, 1994; pp 18-47.
86. Cole, M. A. In *ACS Symposium Series 433*; American Chemical Society: Washington, D.C., 1990; pp 76-95.
87. Gould, J. M.; Gordon, S. H.; Dexter, L. B.; Swanson, C. L. In *ACS Symposium Series 433*; American Chemical Society: Washington, D.C., 1990; pp 65-75.
88. Barak, P.; Coquet, Y.; Halbach, T. R.; Molina, J. A. E. *J. Environ. Qual.* **1991,** *20,* pp 173-179.
89. Gilmore, D. F.; Antoun, S.; Lenz, R. W.; Goodwin, S.; Austin, R.; Fuller, R. C. *J. Ind. Microbiol.* **1992,** *10,* pp 199-206.
90. Krochta, J. M. In *Food Proteins and their Applications in Foods;* Damodaran, S.; Paaraf, A., Eds.; Marcel Dekker, Inc.: In press.
91. Hood, L. L. In *Advances in Meat Research;* Pearson, A. M.; Dutson, T. R.; Bailey, A. J., Eds.; Van Nostrand Reinhold Co.: New York, 1987; pp 109-129.
92. Lieberman, E. R.; Gilbert, S. G. *J. Polym. Sci.* **1973,** *Symp. No. 41,* pp 33-43.
93. Gontard, N.; Guilbert, S.; Cuq, J. L. *J. Food Sci.* **1992,** *57,* pp 190-195, 199.
94. Stuchell, Y. M.; Krochta, J. M. *J. Food Sci.* **1994,** *59*(6), pp 1332-1337.
95. McHugh, T. H.; Aujard, J. F.; Krochta, J. M. *J. Food Sci.* **1994,** *59*(2), pp 416-419, 423.
96. Avena-Bustillos, R. J.; Krochta, J. M. *J. Food Sci.* **1993,** *58*(4), pp 904-907.

97. McHugh, T. H.; Krochta, J. M. *JAOCS.* **1994,** *71*(3), pp 307-312.
98. McHugh, T. H.; Krochta, J. M. *J. Food Process. Preserv.* **1994,** *18,* pp 173-188.
99. Gontard, N.; Duchez, C.; Cuq, J. L.; Guilbert, S. *Int. J. Food Sci. Technol.* **1994,** *29,* pp 39-50.
100. Butler, B. L.; Vergano, P. J., *Degradation of edible film in storage, Paper No. 94-6551* ; ASAE: St. Joseph, MI, 1994.
101. Gennadios, A.; Weller, C. L.; Testin, R. F. *J. Food Sci.* **1993,** *58*(1), pp 212-214, 219.
102. Li, H.; Hanna, M. A.; Ghorpade, V., *Effects of chemical modifications on soy and wheat protein films, Paper No. 93-6529* ; ASAE: St. Joseph, MI, 1993.
103. McHugh, T. H.; Krochta, J. M. *J. Agric. Food Chem.* **1994,** *42*(4), pp 841-845.
104. Gennadios, A.; Weller, C. L.; Testin, R. F., *Modification of properties of edible wheat gluten films, Paper No. 90-6504* ; ASAE: St. Joseph, MI, 1990.
105. Brandenburg, A. H.; Weller, C. L.; Testin, R. F. *J. Food Sci.* **1993,** *58,* pp 1086-1089.
106. Trezza, T. A.; Vergano, P. J. *J. Food Sci.* **1994,** *59*(4), pp 912-915.
107. Jane, J.-L.; Lim, S.-T.; Paetau, I. In *Biodegradable Polymers and Packaging;* Ching, C.; Kaplan, D.; Thomas, E., Eds.; Technomic Publishing Co., Inc.: Lancaster, PA, 1993; pp 63-73.
108. Jane, J.-L. In *Activities Report and Minutes of Work Groups & Sub-Work Groups of the R&D Associates;* Research and Development Associates for Military Food and Packaging Systems, Inc.: Atlanta, Georgia, 1994.
109. Paetau, I.; Chen, C.-Z.; Jane, J.-L. *I&EC Res.* **1994,** *33*(7), pp 1821-1827.
110. Paetau, I.; Chen, C.-Z.; Jane, J. *J. Environ. Polym. Degrad.* **1994,** *2*(3), pp 211-217.
111. Lim, S.; Jane, J. *J. Environ. Polym. Degrad.* **1994,** *2*(2), pp 111-120.
112. Timmins, M. R.; Gilmore, D. F.; Fuller, R. C.; Lenz, R. W. In *Biodegradable Polymers and Packaging;* Ching, C.; Kaplan, D.; Thomas, E., Eds.; Technomic Publishing Co., Inc.: Lancaster, 1993; pp 119-130.
113. *Biopol - Nature's Plastic - Properties and Processing;* Zeneca Bio Products: Delaware, 1993.
114. Kemmish, D. In *Biodegradable Polymers and Packaging;* Ching, C.; Kaplan, D.; Thomas, E., Eds.; Technomic Publishing Co., Inc.: Lancaster, PA, 1993; pp 225-232.
115. Holmes, P. A. In *Developments in Crystalline Polymers;* Basset, D. C., Ed.; Elsevier: New York, 1988; pp 1-65.
116. Hocking, P. J.; Marchessault, R. H. In *Chemistry and Technology of Biodegradable Polymers;* Griffin, G. J. L., Ed.; Blackie Academic & Professional: New York, 1994; pp 48-96.
117. Naitov, M. H. *Plastics Technol.* **1995.**
118. *Eco-Pla Cast Sheet Properties;* Cargill: Minneapolis, MN, 1994.
119. Sinclair, R. G. In *Biodegradable Polymers '94;* ECM, Inc.: Ann Arbor, MI, 1994.
120. Gruber, P. R. In *Corn Utilization Conference V;* St. Louis, MO, 1994.
121. Conca, K. R.; Yang, T. C. S. In *Activities Report and Minutes of Work Groups & Sub-Work Groups of the R&D Associates;* Research and Development Associates for Military Food and Packaging Systems, Inc.: 1993; pp 41-53.

122. Cottrell, I. W.; Kovacs, P. In *Handbook of Water-soluble Gums and Resins;* Davidson, R. L., Ed.; McGraw-Hill Book Co.: New York, 1980; pp 1-43.
123. Davies, D. H.; Elson, C. M.; Hayes, E. R. In *Chitin and Chitosan: Sources, Chemistry, Biochemistry, Physical Properties, and Applications;* Skjåk-Bræk, G.; Anthonsen, T.; Sandford, P., Eds.; Elsevier Applied Science: New York, 1989; pp 467-472.
124. Hosokawa, J.; Nishiyama, M.; Yoshihara, K.; Kubo, T. *Ind. Eng. Chem. Res.* **1990,** *29,* pp 800-805.
125. Hosokawa, J.; Nishiyama, M.; Yoshihara, K.; Kubo, T.; Terabe, A. *Ind. Eng. Chem. Res.* **1991,** *30,* pp 788-792.
126. Kester, J. J.; Fennema, O. R. *J. Food Sci.* **1986,** *40*(12), pp 47-59.
127. Sandford, P. A. In *Chitin and Chitosan: Sources, Chemistry, Biochemistry, Physical Properties, and Applications;* Skjåk-Bræk; Anthonsen, T.; Sandford, P., Eds.; Elsevier Applied Science: New York, 1989; pp 51-69.
128. Schultz, T. H.; Miers, J. C.; Owens, H. S.; Maclay, W. D. *J. Phys. Colloid Chem.* **1949,** *53,* pp 1320-1330.
129. Yuen, S. *Process Biochem.* **1974,** *9*(9), pp 7-9, 22.

Chapter 10

Advances in Alternative Natural Rubber Production

Katrina Cornish and Deborah J. Siler

Western Regional Research Center, Agricultural Research Service, U.S. Department of Agriculture, 800 Buchanan Street, Albany, CA 94710

Natural rubber (*cis*-1,4-polyisoprene) is a plant-derived raw material essential to commercial, transportation, medical and defense needs. Virtually all natural rubber is obtained from the Brazilian rubber tree (*Hevea brasiliensis*), a species restricted to the tropics. As developing countries move away from rubber farming toward higher valued agriculture, declining production, coupled with increasing global demand, are causing natural rubber shortages and higher prices. In this chapter, we discuss research results underlying promising approaches to the genetic engineering of temperate-zone annual plants into crops capable of commercial yields of natural rubber. Also, we report the biotechnological development of hypoallergenic latex to address markets unsuitable for *H. brasiliensis* latex and of sufficient value to permit the immediate commercialization of a major domestic guayule crop.

Natural rubber is a vital raw material used in enormous quantities by commercial, medical, transportation and defense industries. The United States and similar temperate regions are wholly dependent upon imports from developing countries. At present, all commercial natural rubber comes from a single plant species, the Brazilian rubber tree (*Hevea brasiliensis* Muell. Arg.), largely from plantation-grown clonal material. Continuity of this source is endangered by many factors, including diminishing acreage, increasing global demand, changing political climes, and crop disease. The recent widespread occurrence of life-threatening "latex allergy" to *H. brasiliensis* rubber products makes development of an alternative source of natural rubber imperative. A domestic natural rubber crop will guarantee sufficient natural rubber to supply existing United States strategic demands as well as the new, major, health-related, hypoallergenic rubber market. A sustainable, domestic rubber commercial crop also will enhance rural economics as an alternative and profitable crop.

In any development of a new crop, it must have an existing market and also be profitable for the grower and the buyer of the crop. Approaches to achieving commercial competitiveness may be grouped in two categories, (i) increasing the yield of the agricultural commodity in question and (ii) adding value to the agricultural commodity. Our research program involves both approaches and employs multiple biotechnological strategies to develop new crops for domestic production of natural rubber in the temperate climate of the United States.

Increased Yield.

Recombinant DNA techniques have great promise in the development of domestic rubber-producing crops. Genetic manipulation of a biochemical process requires an understanding of the biochemistry of rubber biosynthesis and of the identity and *in vivo* function of essential enzymes and structural proteins, and their genes. With this understanding we could isolate, insert and express selected genes in suitable plant systems and then optimize these transformed systems to produce viable yields of high quality rubber.

Natural Rubber Biosynthesis. Natural rubber (*cis*-1,4-polyisoprene) is produced in more than 2,000 diverse plant species, and at least two fungal genera (*1,2*). Rubber is contained in cytoplasmic rubber particles. Rubber molecular weight varies considerably, and high quality rubber is related to high molecular weight. *H. brasiliensis* and *Parthenium argentatum* Gray (guayule) both accumulate high quality rubber with a mean molecular weight of about 1,500,000 Da (*3*). In contrast, *Ficus elastica* Roxb. (Indian rubber tree) produces mostly poor quality rubber with a molecular weight of about 1,000 Da (*3*). Rubber yield varies greatly from species to species. This wide range of rubber-producing plants may be exploited when addressing different aspects of the biotechnological development of a new rubber crop. For example, a highly developed rubber-biosynthetic plant, such as *H. brasiliensis* which forms high quality rubber as a tapable latex within a specialized laticiferous network (*4*), provides an excellent system for biochemical studies. *P. argentatum*, however, produces rubber in individual parenchyma cells (*2*), and may provide the simplest source of key genes since it lacks genes specific to rubber-producing laticifers.

Natural rubber is made almost entirely of isoprene units derived from isopentenyl pyrophosphate (IPP) in a side-branch of the ubiquitous isoprenoid pathway (Figure 1). The polymerization of the rubber polymer is catalyzed by the enzyme rubber transferase (EC 2.5.1.20). Rubber transferase, a *cis*-prenyl transferase, requires divalent cations (such as Mg^{2+} or Mn^{2+}) for activity, but rubber transferase, IPP and Mg^{2+} together will not result in rubber biosynthesis. Another substrate, an allylic pyrophosphate, is needed to initiate the polymerization process (*5-7*). Hence, rubber molecule formation requires three distinct biochemical steps: (i) initiation, which requires an allylic pyrophosphate molecule (synthesis of allylic pyrophosphates is catalyzed by soluble *trans*-prenyl transferases); (ii) elongation, the rubber transferase-catalyzed *cis*-1,4-polymerization of isoprene units from IPP; and (iii) termination, the release of the polymer from the rubber transferase. The first two steps are detailed in Figure 2. The plant prenyl transferases are comparable to animal enzyme systems in which *trans*-prenyl transferases are soluble enzymes, whereas *cis*-prenyl transferases are membrane-bound (*8,9*).

The rubber polymerization reaction must take place at the particle surface because (i) the substrates from which rubber is made are hydrophilic and available in the cytoplasm and (ii) the hydrophobic rubber molecules produced are

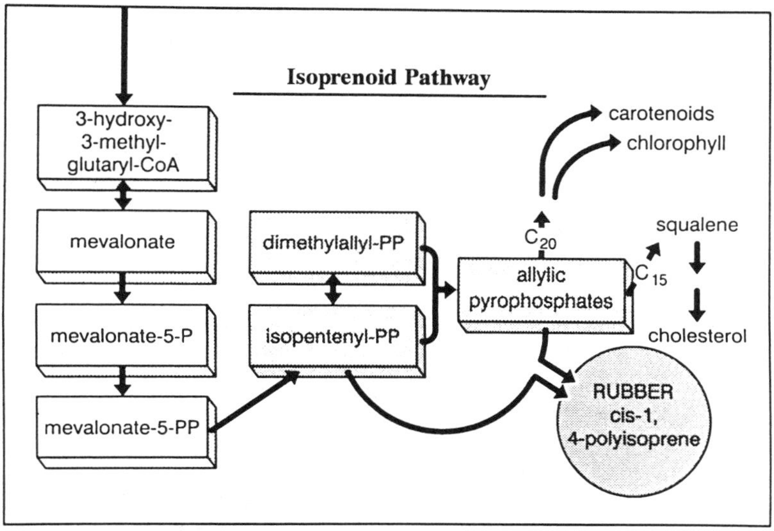

Figure 1. The isoprenoid pathway, illustrating the alternative branch to natural rubber biosynthesis.

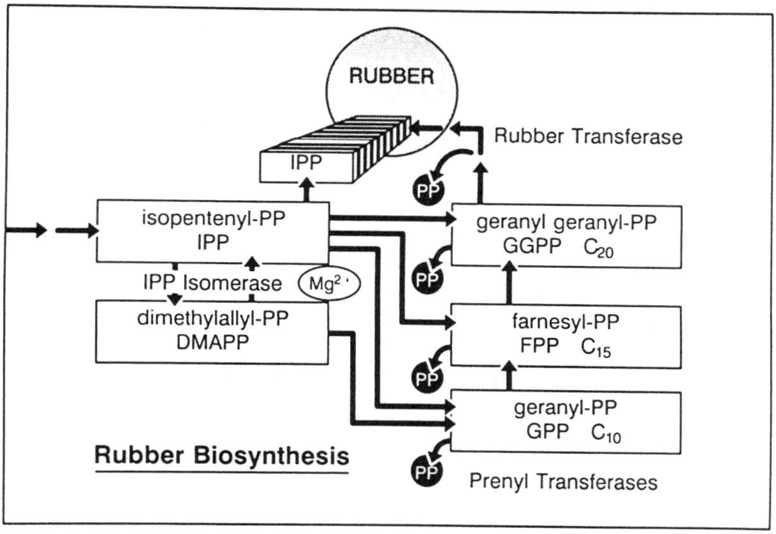

Figure 2. Natural rubber biosynthesis from isopentenyl pyrophosphate.

packaged inside rubber particles. Particle-bound rubber transferases have been demonstrated in the rubber-producing species *H. brasiliensis* (*1,5,9,10*), *P. argentatum* (*6,7*) and *F. elastica* (*11*). Kinetic studies give linear Eadie-Hofstee plots (V against V/[S]) for IPP and for FPP (*6*) indicating the presence of a single particle-bound rubber-biosynthesizing enzyme.

Our genetic engineering program involves three approaches leading from our research; namely, (i) increasing the endogenous levels of rubber transferase, (ii) increasing the endogenous levels of specific substrates, and (iii) extending rubber-production duration in *P. argentatum*. Our rationale and progress in these three areas are described below.

Rubber transferase. The identification and isolation of rubber transferase is a prerequisite to any attempt to enhance yield through the over-production of the enzyme *in vivo*. However, rubber transferase identification has proved problematical. Enzymatically-active rubber particles contain many proteins, particularly in *H. brasiliensis*. The *H. brasiliensis* rubber transferase was reported to be the result of an association between a small membrane-bound protein and a soluble FPP-synthase (*12,13*), but this theory has been contradicted (*9*). Also, the classical approach of solubilizing enzyme activity as a prelude to identification of a purified enzyme has not yet been effective. One report of *P. argentatum* protein solubilization implicated a 52 kD protein as the rubber transferase (*14*). However, subsequent cDNA cloning and sequencing, as well as enzyme analysis, has proved this protein to be a cytochrome P450, not a rubber transferase (*15*). An immunoinhibition approach has led to the identification of the rubber transferase in *F. elastica* within a 376 kD protein (*16,17*). This identification is of limited use since *F. elastica* produces predominantly poor quality, low molecular weight rubber. Furthermore, we do not yet know how polymer length is regulated. A series of cross-specific immunological approaches has begun to characterize rubber transferase proteins in high molecular weight rubber-producing species (*16,18*). When this is accomplished, a definitive identification of the *P. argentatum* rubber transferase may be possible.

Substrate levels. As described earlier, rubber biosynthesis requires two distinct substrates, IPP and allylic pyrophosphate, which could be rate-limiting in any rubber-producing species. Since the allylic pyrophosphates are used by many other enzymes in the isoprenoid pathway (Fig. 1), it seems likely that raising their endogenous concentrations could enhance the production of rubber. We performed detailed kinetic analyses of rubber molecule initiation and polymerization by enzymatically-active rubber particles purified from *P. argentatum* (*19*) so that we can predict the likely effect *in vivo* of altering endogenous substrate levels. We compared the effects of four different allylic pyrophosphate initiators on the rate of [^{14}C]IPP incorporation into rubber. The rate of rubber biosynthesis increased with the chain length of the initiator up to C_{15} (Figure 3, taken from (*19*)). Similar results were observed in *F. elastica* (*20*) and *H. brasiliensis* (*5,20*). Also, we compared the rates of rubber molecule initiation and polymerization using [^{14}C]IPP and the allylic pyrophosphates [^{3}H]FPP or [^{3}H]GPP (Figure 4, taken from (*19*)). Under non-limiting substrate concentrations, new rubber molecules were initiated at a constant rate *in vitro* by both initiators, which implies that rubber molecule termination and release from the rubber transferase also proceeded at a constant rate *in vitro*. The differences in rubber biosynthesis for the different initiators (Fig. 3) are reflected in the differences in rubber molecule initiation (Fig. 4). Our results indicated that initiation regulates the overall rate of rubber biosynthesis (*19*) and so we have designed strategies to enhance *in vivo* rubber yield by genetic manipulation of the rubber molecule initiation system. Furthermore, our results suggest that increasing the endogenous levels of the longer allylic pyrophosphates will increase the overall rate of rubber biosynthesis. However, we may need also to manipulate the endogenous levels of IPP, if this substrate should prove rate-limiting in the transformants. The isoprenoid pathway is a popular field of study and various genes encoding allylic pyrophosphate synthases, including FPP (*21*) and GGPP (*22*), have already been identified and cloned.

***Parthenium argentatum*.** This species has received much more attention as a candidate for domestic rubber-production than any other. Various approaches have been, and will be, attempted to generate yields commercially competitive with imported *H. brasiliensis* rubber.

Plant Breeding. The classical approach of plant breeding has been pursued in support of domestic rubber production by *P. argentatum* (guayule). As has been described in detail (*23*), rubber yields were between 220 and 560 kg/ha during the 1950's; and by the Second Guayule Regional Variety Trials (1985-1988), annual yields had been increased to between 600 and 900 kg/ha (*24*), and over 1,100 kg/ha in breeding plots (*25*). However, the complex genetics of guayule (a highly variable, facultative apomictic plant with varying ploidy levels) does place limits on the yield gains obtainable through plant breeding.

Agronomy. Rubber yield also can be affected by cultivation practices, which have been extensively reviewed for *P. argentatum* (*24,26*). First harvest by pollarding at three years, followed by regrowth and reharvest at two year intervals maximizes the profitability of the current best lines. However, growing *P. argentatum* in areas climatically unsuitable for its environmentally-induced rubber biosynthetic pathway (to be discussed) will not give acceptable rubber yields.

Production period. In the perennial *P.argentatum*, rubber is accumulated predominately in the winter months, and the biochemical pathway may be induced by low temperature (*26,27*). The seasonal dependence of rubber biosynthesis was investigated in detail by assaying rubber transferase activity *in vitro* in rubber

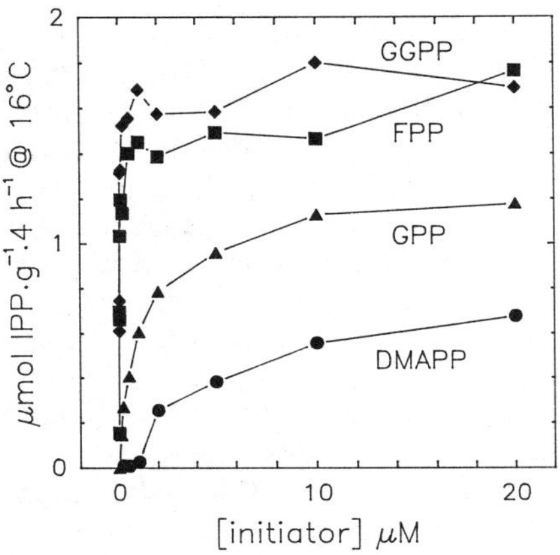

Figure 3. Allylic diphosphate concentration dependence of IPP-incorporation into rubber by isolated, enzymatically-active rubber particles from *P. argentatum*. Incorporation rates were determined in the presence of 1 mM [^{14}C]IPP (from *19*).

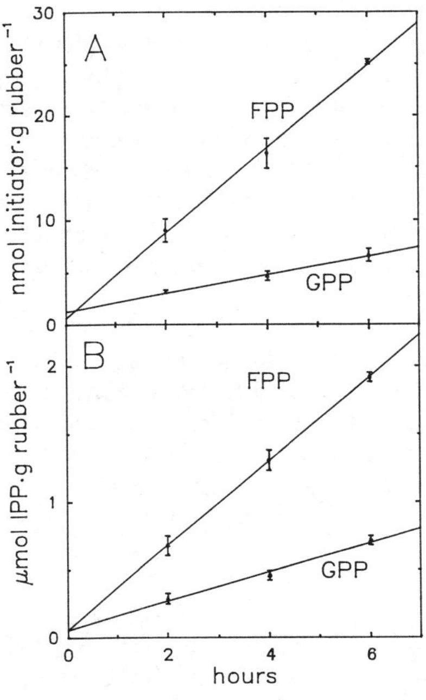

Figure 4. Time course of active incorporation of (A) [^{3}H]GPP and [^{3}H]FPP, and of (B) [^{14}C]IPP in the presence of either GPP or FPP, into rubber by isolated, enzymatically-active rubber particles from *P. argentatum*. Incorporation rates were determined in the presence of 5 mM IPP and 20 μM GPP or FPP. Each value is the mean of 3 $\pm$ s.e.(from *19*).

particles isolated, over the course of a year from mature, field-grown plants. It was observed that small fluctuations in activity occurred during the summer months, and a 100-fold increase in activity took place in less than one week during late October (*28*, Figure 5). This environmentally-induced rubber transferase activity was observed in two different guayule lines (11591 and 593), and in young, medium and old bark tissues (*28*). The smaller fluctuations in activity observed throughout the non-winter months indicate that guayule continually synthesizes some new rubber transferase. However, an environmentally-induced, rubber transferase gene promoter is undoubtedly responsible for the huge and rapid increase in rubber transferase activity observed at the end of October (Figure 5).

Expression of the rubber transferase gene under a constitutive promoter is likely to effectively increase the level of rubber biosynthesis and extend the season of rubber production. This approach has considerable promise since even one month of additional rubber production would effectively increase rubber yield by about 25% (Fig. 5).

The variety of approaches discussed in this section on "**Increased Yield**" should lead to increased rubber yield in both annual plants and *P. argentatum.* However, research must continue to develop an understanding of the regulation of polymer length. Without this knowledge, it is entirely conceivable that a genetically-engineered plant would arise containing a high yield of poor quality, low molecular weight rubber, instead of the desired high molecular weight.

Added Value.

The value of an agricultural commodity may be increased through the discovery of a specific new use or market, or through an increased demand and/or decreased supply leading to lower competition. The global demand for natural rubber is on the increase (*29*), even though acreage has declined, resulting in substantial price rises. Also, latex rubber is more expensive than bulk (solid) rubber, which costs usually about 0.7x the latex price. Guayule has been extensively investigated with the hope of its commercialization as a true competitor in natural rubber markets currently occupied by *H. brasiliensis* bulk rubber (*4*). *P. argentatum* rubber has been used for fabricating automobile tires (*30*), but had not been used for dipped or extruded goods since their manufacture requires the raw rubber to be in latex form. Unlike laticiferous species, such as *H. brasiliensis* and *F. elastica,* guayule produces rubber in individual cells, located predominantly in the root and stem bark parenchyma. The traditional extraction methods used to produce bulk rubber from guayule (whether solvent or water based) have not led to competitively-priced product, even from the best bred lines. *P. argentatum*, as a source of more valuable latex rubber, has much greater commercial potential than bulk rubber, especially since "guayule latex" is hypoallergenic with respect to *H. brasiliensis* latex allergy.

In this "added value" section of this chapter, we review the feasibility of ameliorating and circumventing Type I latex allergy caused by the presence of *H. brasiliensis* proteins in latex products. Also, we report progress in hypoallergenic guayule latex development.

Latex and Latex Products. Natural rubber is used in over 40,000 different products, including more than 300 medical applications. High value dipped and extruded products are manufactured directly from the latex of *H. brasiliensis*. A life-threatening "latex allergy" has developed, in recent years, that is triggered by proteins present in latex. A hypersensitive individual must take care to avoid contact with rubber latex products, including gloves and condoms. Synthetic alternatives to *H. brasiliensis* latex products do not have the required range of physical properties that encompass resilience, strength, elasticity and viral impermeability. The unique properties of natural rubber, together with concerns for public safety make the need for an alternative, hypoallergenic source of natural rubber imperative. This special need creates an excellent opportunity for developing a new, hypoallergenic, natural rubber crop.

Latex allergy. Latex products had been safely used for many decades with only infrequent contact reactions (Type IV) occurring generally in response to chemical additives. However, life-threatening Type I allergy appeared in the 1980's, which is triggered by proteins present in *H. brasiliensis* latex. A hypersensitive individual must take care to avoid contact with all natural rubber products made from *H. brasiliensis* latex. Allergic reactions to *H. brasiliensis* rubber products include local urticaria, systemic urticaria, rhinitis, conjunctivitis, edema, bronchospasm, tachycardia, anaphylaxis and death (*31,32*). The Food and Drug Administration responded to this public health problem by issuing a medical alert concerning use of latex products (March 1991).

Causes of latex allergy. The surge of Type I latex allergy coincided with sudden world-wide increased demand for latex gloves in response to institution of universal precautions to prevent transmission of the blood-borne diseases HIV and Hepatitis B (*33*). New and inexperienced glove manufacturers entered the market, and short-cuts in manufacturing became common, in order to supply the increased demand (*33*). Over-production of these gloves led to a drop in price which, combined with the increasing demand, induced some other manufacturers to reduce costs through curtailed processing (*34*). Altered manufacturing processes included reduction and sometimes elimination of the leaching step normally used to wash the latex products (*33,34*). The washing process removes soluble latex components, as well as chemical additives, and underwashed products contain high levels of *H. brasiliensis* latex proteins.

The use of high protein latex products, especially of single-use latex gloves, as well as their extensive deployment throughout society, led to widespread development of Type I latex allergy. The problem was compounded by requirements for health-care workers to frequently change gloves. Proteins in gloves can bind to glove powder. When gloves are removed the powder becomes a source of air-borne, respirable, latex allergens (*35,36*).

The latex-hypersensitive population. The first reported incidents of latex allergy in the United States appeared in 1988, and increased to at least 500,000 by 1992, with multiple-surgery patients and health-care workers being the most severely affected. Estimates in 1994, based on analysis of donated blood from the general public, suggest that 17 million Americans, outside of the health care industry, are now affected (*37*). Many other countries around the world also have serious problems with latex allergy. The number of latex hypersensitive people continues to climb due both to increased usage of latex products and to the continued sale of high protein latex gloves and other products. Manufacturing methods known to reduce product protein levels are available, but it is important to realize that even the best processing practices currently available will not produce latex products that are protein-free.

Latex allergy amelioration. Since Type I latex allergy arose because of high protein latex products, the manufacture and use of low protein latex products should greatly diminish the incidence of new allergy cases. Various methods for removing protein from latex and latex products are available (*38*). However, future latex products will have to be shown to have low protein. Thus, the protein levels in a latex product must be accurately measured.

H. brasiliensis latex and latex products are difficult materials to assay for protein. Its latex is a complex and complete cytoplasm that also contains about 30% rubber in the form of microscopic particles. Latex proteins fall into several groups with some soluble (48%), some associated with organelles (26%), and some bound to the surface of rubber particles (26%) (*39*). Allergic patients can react to a large number of different latex protein allergens, which can originate from all of these protein groups. At least 57 different latex proteins, out of approximately 240, have been shown to be human allergens (*40*). They range in size from 2 to 100 kD (*41*). Nevertheless, soluble latex proteins and others not directly associated with rubber particles should be easier to remove from latex and/or latex products than rubber particle-bound proteins (*42*). Their removal would substantially lower the overall protein level, thus reducing the allergenicity of the final latex product for people who have not yet developed latex allergy.

Latex protein assays must be meaningful with respect to latex allergy and be of use to latex product manufacturers. Thus, the assay must effectively measure all classes of protein allergens in latex, or latex products, and eliminate interference from nonproteinaceous substances in the latex as well as additives such as ammonia. A suitable method for assaying protein levels in latex has been developed (*43*) and an ASTM panel has developed an assay for use with latex products (*44*).

Latex allergy circumvention. As discussed above, it is possible to produce low protein latex products that should moderate the incidence of new allergy cases. However, trace amounts of allergen can induce serious systemic reactions in a hypersensitive person. Thus, production of latex products safe for use by allergic individuals requires the elimination of all latex allergens.

Allergens that are not associated with the rubber particles can be removed from the latex, prior to product manufacture, by repeated purification cycles using centrifugation/flotation (*45*). Certain types of dipped products, such as some surgical gloves, electricians' gloves and condoms, are made from doubly-centrifuged latex. This is produced by centrifuging standard latex (30-35% dry rubber), diluting to 20-25% with water, and centrifuging again. This process can remove about 70% of the non-rubber substances (*38*). However, 25% of the proteins are bound to the rubber particles and even extensive rubber particle purification does not remove them - these proteins would persist in the final product. We tested the immunogenicity of various latex materials using antibodies, raised in rabbits, against proteins extracted from *H. brasiliensis* latex films. These *H. brasiliensis* latex protein antibodies recognized many of the proteins in complete *H. brasiliensis* latex (*45*). As expected, purified rubber particle preparations, from which the soluble latex proteins were removed, contained far fewer proteins than latex (*42,45*). Nevertheless, these highly purified *H. brasiliensis* rubber particle preparations still contained numerous proteins, many of which are immunogens (*45*). Also, a report that IgE antibodies from serum of latex-sensitized individuals recognized the 14.6 kD protein "rubber elongation factor" implicates this abundant rubber particle-bound protein as a major allergen in *H. brasiliensis* latex (*46*).

Thus, complete removal of both rubber particle-bound and soluble latex protein allergens is necessary to produce a *H. brasiliensis* latex material safe for use by a *H. brasiliensis* latex hypersensitive person. Removal of the rubber particle-bound proteins would necessitate drastic treatment *e.g.* with proteases and/or detergents, that would not only probably be prohibitively expensive, but also adversely affect the performance characteristics and quality of the resulting latex products.

Some manufacturers have introduced so-called "hypoallergenic" *H. brasiliensis* latex gloves into the market. Although some of these gloves are low protein, tests demonstrated that many contain substantial amounts of proteins that bind IgE antibodies from sera of *H. brasiliensis*-hypersensitive patients (*47*). Low protein hypoallergenic gloves would be unlikely to induce allergies in individuals not already allergic to *H. brasiliensis* latex proteins. However, these gloves would not be safe for use by *H. brasiliensis*-hypersensitive people.

Suggestions have been made to remove the allergens from *H. brasiliensis* latex by breeding or by using genetic engineering techniques (antisense technology) to eliminate the allergenic proteins. These approaches are biologically unsound. This is primarily because many different latex proteins have been shown to be allergens (see above). It is inconceivable to us that they could all be removed from the *H. brasiliensis* tree without dire effects on latex production. Also, many of these proteins undoubtedly provide essential functions in the plant and their elimination would be fatal to the plant. Furthermore, the genetic base of cultivated *H. brasiliensis* is extremely narrow, which would likely preclude a plant breeding approach. We confirmed that clonal selection is unlikely to eliminate latex antigens in experiments that demonstrated that *H. brasiliensis* latex protein antibodies recognized many proteins in samples from three different commercial lines of *H. brasiliensis* (*45*). The proteins recognized by the antibodies showed similar patterns in all three clones.

Nevertheless, different rubber-producing species may contain proteins different from those of *H. brasiliensis* and perhaps provide a source of hypoallergenic rubber (*42*). We examined two plant species, *F. elastica* and *P. argentatum*. Anti-*H. brasiliensis* latex protein antibodies did not recognize latex proteins from *P. argentatum* or *F. elastica* using ELISA (enzyme-linked immunosorbent assay). Proteins extracted from purified rubber particle preparations of *P. argentatum* and *F. elastica* showed no reaction even at concentrations at least 2 or 3 orders of magnitude above the *H. brasiliensis* rubber particle protein concentration that elicited a positive reaction (Fig. 6, (*45*)). Western blot analyses of the rubber particle proteins from *P. argentatum* or *F. elastica* with the *H. brasiliensis* antibodies confirmed the lack of cross-reactivity (*45*). These results establish that *H. brasiliensis* latex allergy can indeed be circumvented using rubber from other species.

Our results were confirmed in preliminary clinical trials on humans. In one trial, individuals with proven hypersensitivity to *H. brasiliensis* latex were skin prick tested with latex samples from different sources. All these subjects reacted strongly to *H. brasiliensis* latex (wheals and flare), whereas none responded to latices from *P. argentatum* and *F. elastica* (*48*). In a second trial, 56 "at risk" people (such as health care workers, spina bifida patients and multiple surgery cases), ages 16 months to 72 years, and three controls, were tested both by skin prick and by *in vitro* RAST (radioallergosorbent assay) assays (*49*). The RAST assay detects the presence of human IgE antibodies specific to *H. brasiliensis* latex proteins. Twenty five people reacted with *H. brasiliensis* materials in the skin prick test and 30 in the RAST. None reacted to *P. argentatum* or *F. elastica* latices, confirming the earlier clinical test results, and our experiments using the rabbit anti-*H. brasiliensis* latex protein antibodies.

The hypoallergenicity of *F. elastica* latex which, like *H. brasiliensis*, is a complete cytoplasm produced in laticifers, indicates that the *H. brasiliensis* latex allergens are species-specific. Thus, sources of hypoallergenic natural rubber are not restricted to non-laticiferous rubber-producing species such as *P. argentatum*.

The inferior quality of *F. elastica* rubber precludes this species as a possible commercial source. However, *P. argentatum* rubber latex should be suitable for the manufacture of high quality, hypoallergenic natural rubber products safe even for the *H. brasiliensis*-hypersensitive individual. Some other rubber-producing plant species also produce high molecular weight rubber, but currently none are suitable for commercial cultivation. In contrast, *P. argentatum* agronomics are well understood (*26*).

Thus, our diverse immunological tests have demonstrated that latex from *P. argentatum*, "guayule latex", is truly hypoallergenic and that guayule latex products could be used safely by *Hevea* latex-sensitive people. Of course, it is

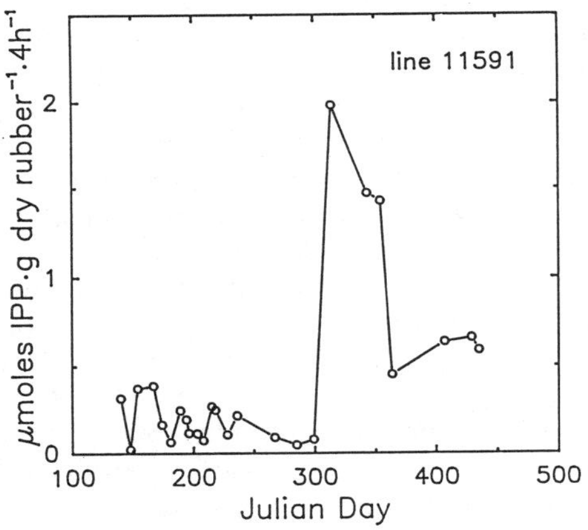

Figure 5. Seasonal dependence of IPP-incorporation rate into rubber by enzymatically-active rubber particles isolated from mature, field-grown, *P. argentatum* over the course a year. Incorporation rates were determined in the presence of 80 μM [^{14}C]IPP and 20 μM FPP. Each value is the mean of $\geq$3 (from *28*).

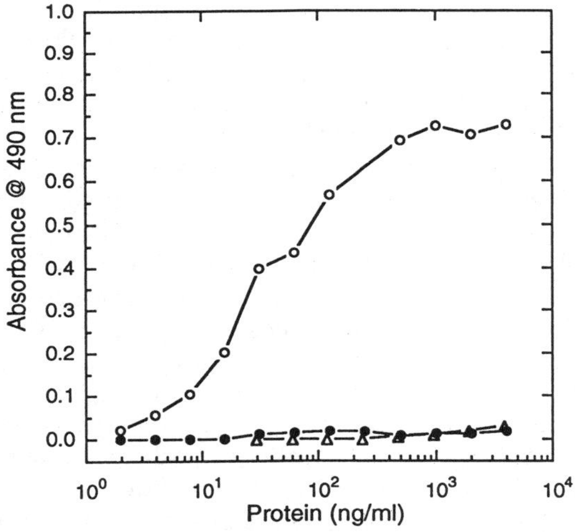

Figure 6. ELISA of purified latex samples from three plant species. The allergenicity of rubber particle proteins from *H. brasiliensis* (○-○), *F. elastica* (Δ-Δ), and *P. argentatum* (●-●) was quantified with respect to antibodies raised against proteins from *H. brasiliensis* latex films (Adapted from (*45*)).

essential that guayule latex products be maintained as low protein so that allergy problems don't redevelop later. The most difficult, and probably impossible, proteins to completely remove are those bound to the rubber particles themselves. *P. argentatum* rubber particles have less protein than *H. brasiliensis* (*20*) and, therefore, guayule latex could be produced with lower protein levels than will ever be possible with *H. brasiliensis* latex.

Production of hypoallergenic guayule latex. Natural rubber is packaged in cytoplasmic particles irrespective of their species-specific laticiferous or parenchymatous origin. If the rubber particles could be extracted intact from *P. argentatum* it should be possible to generate a latex. Rubber particle suspensions have been made in a variety of ways (*10,50,51*).

We have now developed large-scale processing methods to extract intact rubber particles and produce an artificially-produced purified guayule latex (*52*). These methods also should be suitable for the production of latex from other rubber producing species.

Non-*H. brasiliensis*-derived hypoallergenic latex presents a major new, high value, rapidly expanding market because hypersensitive people cannot safely use *H. brasiliensis* latex products. The U.S. market for latex gloves had a $2,800,000,000 retail value in 1992. Hypoallergenic natural rubber latex products would provide a sufficient premium to permit the immediate development of *P. argentatum* as a new crop even at its current rubber yield.

During the Spring of 1994, a production run was carried out in Arizona where 5,000 lbs of field-grown shrub were processed. We used a scaled-up processing system (*53*) to generate ammoniated guayule latex for allergy testing, performance testing and manufacturing trials. Several different trials have been completed on this material confirming that (i) guayule latex does not cross-react with antibodies raised against *H. brasiliensis* latex (*54*), (ii) guayule latex may be compounded using procedures similar to those for *H. brasiliensis* latex (*55*), and (iii) quality medical products can be manufactured. (*56*).

Conclusions.

Large scale production of guayule latex is feasible and the high quality of guayule natural rubber makes it suitable for hypoallergenic product manufacture. These products should be safe for use by the *H. brasiliensis*-hypersensitive population. Upon the generation of sufficient supplies, hypoallergenic guayule latex products first would become available to known hypersensitive patients, and the at-risk health-care profession, and then to the general population. Guayule latex provides the first natural rubber solution to the urgent global need for a reliable hypoallergenic source of natural rubber latex. Concerted and dedicated efforts should make safe guayule latex products a reality and the guayule crop a major feature of the farming landscape of the south-western United States. Continuing research also should lead to the introduction of genetically-engineered *P. argentatum* lines with enhanced rubber yield and extended growing range.

The domestic production of high quality natural rubber in annual plants remains a long-term goal of our research. We have developed several genetic engineering strategies that should enable us to achieve this goal and permit the production of natural rubber throughout the continental United States.

Literature Cited

1. Archer, B.L., Audley, B.G., Cockbain, E.G. and McSweeney, G.P., **1963,** The biosynthesis of rubber. *Biochem. J.* **89,** 565
2. Backhaus, R.A., **1986,** *Israel J. Bot.* 34, 283
3. Cornish, K., Siler, D.J., Grosjean, O.K., and Goodman, N., **1993,** *J. Nat. Rubber Res.* **8,** 275
4. d'Auzac J., Jacob, J-L. and Chrestin, H. (Eds), **1989,** Physiology of Rubber Tree Latex. *Physiology of Rubber Tree Latex.* CRC Press, Boca Raton, Florida, 470 pp.
5. Archer, B.L. and Audley, B.G., **1987,** New aspects of rubber biosynthesis. *Bot. J. Linn. Soc.* **94,** 181
6. Cornish, K. and Backhaus, R.A., **1990,** *Phytochemistry* **29,** 3809
7. Madhavan, S., Greenblatt, G.A., Foster, M.A. and Benedict, C.R., **1989,** *Plant Physiol.* **89,** 506
8. Ericcson, J., Runquist, M., Thelin, A., Andersson, M., Chojnacki, T. and Dallner, G., **1992,** *J. Biol. Chem.* **268,** 832
9. Cornish, K., **1993,** *Eur. J. Biochem.* **218,** 267
10. Berndt, J., **1963,** *U.S. Government Res. Rep.* AD-**601,** 729
11. Siler, D.J. and Cornish, K., **1993,** *Phytochemistry* **32,** 1097
12. Light, D.R. and Dennis, M.S., **1989,** *J. Biol. Chem.* **264,** 18589
13. Light, D.R., Lazarus, R.A. and Dennis, M.S., **1989,** *J. Biol. Chem.* **264,** 18598
14. Benedict, C.R., Madhavan, S., Greenblatt, G.A., Venkatachalam, K.V. and Foster, M.A., **1990,** *Plant Physiol.* **92,** 816
15. Pan, Z., Durst, F., Werck-Reichhart, D., Gardner, H.W., Camara, B., Cornish, K., and Backhaus, R.A., **1995,** *J. Biol. Chem.* **270,** 8487
16. Cornish, K., Siler, D.J., and Grosjean, O.K., **1994,** *Phytochem.* **35,** 1425
17. Siler D.J. and Cornish K. **1993,** *Phytochem.* **32,** 1097
18. Siler D.J. and Cornish K. **1994,** *Phytochem.* **36,** 623
19. Cornish, K. and Siler, D.J. **in press,** *J. Plant Physiol.*
20. Cornish, K. and Siler, D.J., unpublished data
21. Anderson, M.S., Yarger, J.G., Burck, C.L. and Poulter, C.D., **1989,** *J. Biol. Chem.* **264,** 19176
22. Kuntz, M., Römer, S., Suire, C., Hugueney, P., Weil, W.H., Schantz, R. and Camara, B., **1992,** *The Plant Journal* **2,** 25
23. Ray D.T., **1991,** *Guayule: A source of natural rubber,* In: New Crops J. Janick and JE Simon (Eds), pp 338
24. Thompson, A.E. and Ray, D.T., **1989,** *Plant Breeding Reviews* **6,** 93
25. Gathman, A.C, Ray,D.T. and Livingston, M., **1992,** *Industr. Crops and Products* **1,** 67

26. Whitworth J.W., and Whitehead, E.E. (Eds), **1991,** *Guayule Natural Rubber: A Technical Publication with Emphasis on Recent Findings.* Guayule Administrative Management Committee and USDA Cooperative State Research Service, Office of Arid Lands Studies, The University of Arizona, Tucson, 445 pp.
27. Ji, W., Benedict, C.R. and Foster, M.A., **1993.** *Plant Physiol.* **103,** 535-542
28. Cornish, K. and Backhaus, R.A., unpublished results
29. Kirschner, E., **1995,** *Chemical and Engineering News,* **August 14,** 11-16
30. Hammond, B.L. and Polhamus, L.G., **1965,** *USDA Tech.Bull.* **No. 1327,** pp.157
31. Slater, J.E., Mostello, L.A., Shaer, C., and Honsinger, R.W., **1990,** *Ann. Allergy* **65,** 411
32. Tomazic, V.J., Withrow, T.J., Fisher, B.R. and Dillard, S.F., **1992,** *Clinical Immunology and Immunopathology* **64,** 89
33. Bodycoat, I., **1993,** *In*: Latex Protein Allergy: The Present Position, Crain Communications Ltd., Rubber Consultants, Brickendonbury, Hertford, U.K., p.43
34. Russell-Fell, R., **1993,** *In*: Latex Protein Allergy: The Present Position, Crain Communications Ltd., Rubber Consultants, Brickendonbury, Hertford, U.K., p.3
35. Tarlo, S.M., Sussman, G., Contala, A., and Swanson, M.C., **1994,** *J. Allergy and Clin. Immunol.* **93,** 985
36. Tomazic, V.J., Shampaine, E.L., Lamanna, A., Withrow, T.J., Adkinson, N.F., and Hamilton, R.G., **1994,** *J. Allergy and Clin. Immunol.* **93,** 451
37. Ownby, D.R., Ownby, H.E., McCullough, J.A., and Shafer, A.W., **1994,** *J. Allergy and Clin. Immunol.* **93,** 282
38. Pailhories, G., **1993,** *Clinical Reviews in Allergy* **11,** 391
39. Kekwick, R.G.O., **1993,** *In:* Latex Protein Allergy: The Present Position. Crain Communications Ltd., Rubber Consultants, Brickendonbury, Hertford, U.K., p.21
40. Alenius, H., Kurup, V., Kelly, Y., Palosuo, T., Turijanmaa, K. and Fink, J., **1994,** *J. Laboratory and Clin. Med.* **123,** 712
41. Hamann, C.P., **1993,** *Am. J. Contact Dermatitis* **4,** 4
42. Siler, D.J. and Cornish, K., **1992,** International Latex Conference, Nov. 5-7, Baltimore, MD, p.46
43. Siler, D.J. and Cornish, K., **1995,** *Anal. Biochem.* **229,** 278
44. ASTM Designation: D 5712-95, **1995,** *Annual Book of ASTM Standards* **1402,** 1
45. Siler, D.J. and Cornish, K., **1994,** *Indust. Crops and Products* **2,** 307
46. Czuppon, A.B., Chen, Z., Rennert, S., Engelke, T., Meyer, H.E., Heber, M., and Baur, X., **1994,** *J. Allergy and Clin. Immunol.* **92,** 690
47. Yunginger, J.W., Jones, R.T., Fransway, A.F., Kelso, J.M., Warner, M.A., and Hunt, L.W., **1994,** *J. Allergy and Clin. Immunol.* **93,** 836

48. Carey, A.B., Cornish, K., Schrank, P.J., Ward, B., and Simon, R.A., **1994,** *Ann. Allergy, Asthma and Immunol.* **74,** 317
49. Ber, D.J., Munemasa, K.H., Gilumilion, R., Cornish, K., Klein, D.E., and Settipane, G.A., **1993,** *J. Allergy and Clin. Immunol.* **91,** 271
50. Madhavan, S., and Benedict, C.R., **1984,** *Plant Physiol.* **75,** 908
51. Jones, E.D., **1948,** *Industrial and Engineering Chemistry* **40,** 864
52. Cornish, K., **1993,** U.S. Patent pending S.N. 08/145,456.
53. Coates, W. and Cornish, K., unpublished data
54. Siler, D.J., Cornish, K. and Hamilton, R.G., unpublished data
55. Schloman, W.W. Jr., Wyzgoski, F., McIntyre, D., Cornish, K. and Siler, D.J., unpublished data
56. Bader, H.F. and, Cornish, K., unpublished data

Applications in Biotechnology

Chapter 11

Genetic Modification of Oilseed Crops To Produce Oils and Fats for Industrial Uses

Thomas A. McKeon, Jiann-Tsyh Lin, Marta Goodrich-Tanrikulu, and Allan Stafford

Western Regional Research Center, Agricultural Research Service, U.S. Department of Agriculture, 800 Buchanan Street, Albany, CA 94710

Vegetable oils represent an important component of chemical feedstock for industry. The advent of genetic engineering techniques provides the opportunity to "design" oils that contain specific fatty acid components. Such oils could have premium value depending on the ultimate use. Moreover, the reliance on plants for this chemical production provides a renewable resource that can serve as high value crops for farmers.

Oilseed crops hold great possibilities for producing chemicals. In addition to food uses, the vegetable oils produced by these crops have long served as a source of industrial chemicals for soaps, coatings, lubricants and plastics (*1*). The oil produced is derived from atmospheric CO_2 and the principal energy input is sunlight. The commodity price of common vegetable oils, ranging from \$0.15 to \$0.30/pound, reflects the relatively low cost of production. The oilseed, in addition to producing and storing the oil, maintains the capability for producing the same oil in its genetic coding. Recent success in engineering the composition of vegetable oil demonstrates considerable capability in controlling production of fatty acids for chemical uses. Thus, oilseeds are effective and renewable chemical factories.

Vegetable oils consist mainly of triacylglycerols, compounds that contain three fatty acids esterified to glycerol. Soybean, corn and cottonseed oils, the major vegetable oils produced in the U.S., are composed primarily of palmitate (hexadecanoate), oleate (cis-9-octadecenoate) and linoleate (cis,cis-9,12-octadecadienoate) (Table I). These oils are usually used in food products, but they also have important non-food applications. Some of the oils that are used primarily for industrial purposes world-wide are obtained from U.S. grown crops; these oils are composed of fatty acids that are

Table I

Major Fatty Acids in Selected Vegetable Oils[a] (Per Cent)

Source	16:0	18:0	18:1	18:2	18:3	Other
Soybean	11	4	22	53	7.5	20:0- 1 20:1- 1
Corn	13	2.5	30.5	52	1	
Canola	4	2	56	26	10	20:1- 2
Peanut	12.5	2.5	37	41	<1	>18:0- 5 >18:1- 2
Rape	3	1	16	14	10	20:1- 6, 22:1- 49
Linseed	6	3	17	14	60	
Palm kernel	9	2	14	2		<12:0- 3, 12:0- 49, 14:0- 16
Castor	1	1(di-OH)	3	4	tr	18:1(OH)-90

[a] Based on data compiled in "The Lipid Handbook" Second Edition, Gunstone, F.D., Harwood, J.L. and Padley, F.B., eds. Chapman & Hall, London, 1994. pp. 53-146.

polyunsaturated or have chain-lengths of 20 carbons or more. In addition to these oils, there are examples of seed oils that have commercial value as due to a high content of uncommon fatty acids. These are fatty acids that have structural modifications such as double bonds in uncommon positions, hydroxyl groups, epoxy rings, or triple bonds (*2-4*) (Figure 1). These modifications enhance the value of an oil or fatty acid for industrial use by imparting chemical reactivity that leads to new or improved applications, e.g. specialty polymers, lubricants or novel coatings, or that replaces a need currently met by expensive chemical processing of less exotic oils.

A problem associated with producing these uncommon oils is that the plant source is often not suitable for cultivation in temperate climates due to growth habitat. However, the plants do provide a source of genes coding for enzymes that will carry out specific fatty acid modifications. Genetic technology provides the means to transfer these genes into temperate oilseed crops, with consequent production of the desired fatty acid in the seed oil.

One new "designer" vegetable oil has come to market, with production at several million pounds per year, and several others are nearing production (*5*). These achievements are based on a body of biochemical research that provides a fundamental understanding of the biosynthesis of fatty acids and lipids in plants. In fact, incomplete understanding of the biochemistry that controls the amount of oil produced and the fatty acid composition of the oil are current limitations to enhancing oilseed value by genetic engineering. This article will describe key developments that have given rise to "designed" vegetable oils and will explain the principles underlying development of novel oils. It will also describe potential chemical uses for oils containing uncommon fatty acids and some of the challenges to be met in order to achieve control of vegetable oil production and composition.

Composition and uses of vegetable oils

Vegetable oils are generally thought of in terms of food uses, such as salad oils and cooking oils. However, non-food uses of fats and vegetable oils by industry represent one quarter of the market for fats and oils (*6*), and vegetable oils provide the source for 40% of the fat used by industry (*1, 6*). An important component of this use, the surfactant and detergent market, accounts for $9.6 billion in sales (*7*). Table I presents the composition of the common vegetable oils and Figure 1 the structure of fatty acids that are present in some or all oils. Most of the vegetable oils are composed of the same fatty acids, with unsaturates predominating. The saturates are palmitate and stearate (octadecanoate), with palmitate generally present in greater amount. Oleate, linoleate and α-linolenate (cis,cis,cis-9,12,15-octadecatrienoate) are the common unsaturated fatty acids and differences in their composition are often characteristic of the oil source. For some crops, there are varieties with very different compositions, such as high oleate versions of

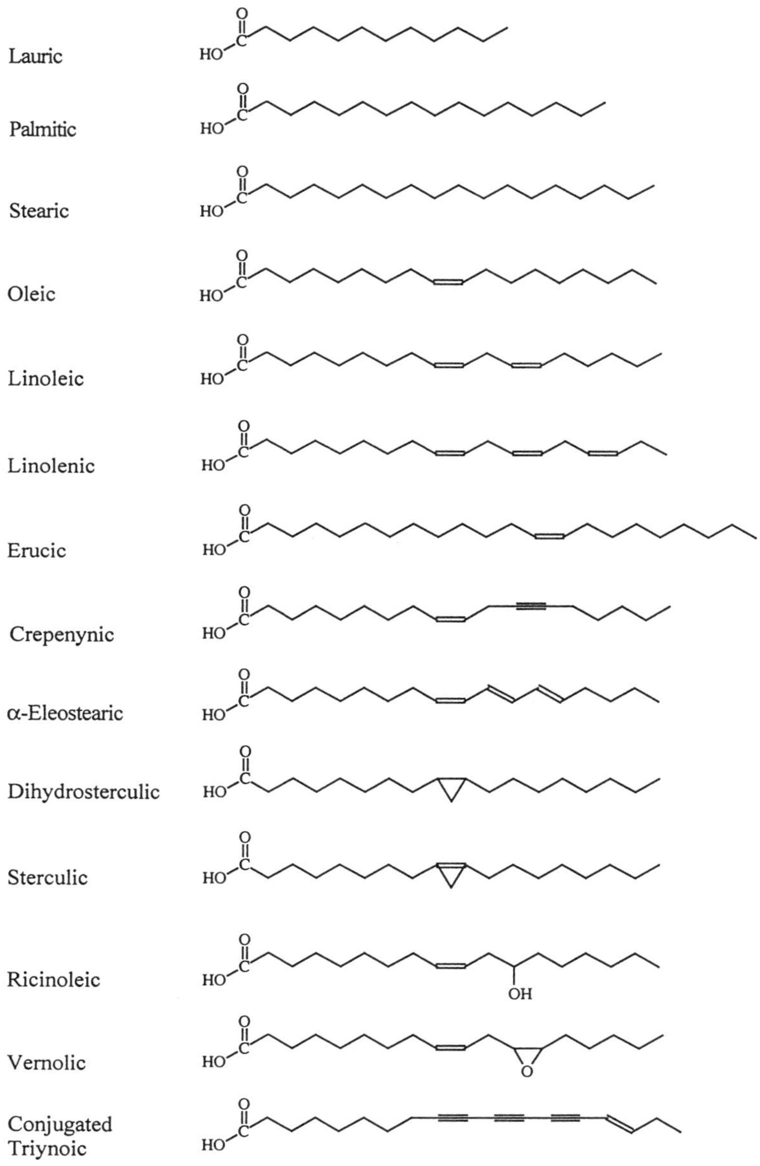

Figure 1. Structures of some fatty acids that have been found in seed oils.

peanut, safflower and sunflower oil, oilseeds that usually produce oils with higher linoleate content. The food oil from canola has a very different composition than that of rapeseed oil, although canola is the result of a breeding program that eliminated undesirable traits, including high erucate (cis-13-docosenoate), from rapeseed. There are also less dramatic differences in composition that arise from varietal and environmental conditions.

The composition of an oil inherently affects its food and non-food uses. Polyunsaturated fatty acids are readily air-oxidized; food oils high in polyunsaturates tend to oxidize and become rancid. However, this same oxidation process underlies the value of linseed oil (60% α-linolenate) as a drying oil and an ingredient in oil-based coatings. Oils designed in response to changing food needs can also meet certain industrial needs. Higher saturated fat levels result in a higher melting point and a more oxidatively stable oil for elevated temperature uses, such as frying. Animal tallow, tropical (coconut and palm kernel) oils and hydrogenated vegetable oils all meet this need. However, health concerns about cholesterol, cholesterogenic activity and dietary *trans*-fatty acids resulting from use of these oils continue to drive this market toward seeking improved alternatives. These alternatives may also serve industrial needs, such as improved lubricant stability at higher temperatures. High oleate oils are considered desirable in terms of health benefits, but they are also of industrial interest in providing a feedstock of oleate for production of plastics, surfactants and lubricants.

Plant oils that are used primarily for industrial purposes generally have unusual compositions and properties. These oils include castor oil, which is the only commercial source of ricinoleate (12-hydroxy-oleate). The hydroxyl group greatly increases the viscosity of ricinoleate, making it suitable for greases, and also provides a second functionality to the molecule that allows polymer formation. Another industrial oil is palm kernel oil, which is high in laurate (dodecanoate), a fatty acid that is used in many surfactants. Table I illustrates the composition of some common vegetable oils, including some with primarily food uses and others with industrial uses.

Fatty acid biosynthesis in plants

Plants have two physically separated pathways that are both involved in the biosynthesis of fatty acids. These fatty acids are produced for membrane lipids during growth and development, and for storage in the seed oil (Figure 2). Initial reactions of fatty acid biosynthesis are localized in the chloroplast or the plastid (the equivalent organelle in non-photosynthetic tissue such as seed). The plastid/chloroplast reactions are distinctive in their use of dissociable acyl-carrier protein (ACP) to hold the growing acyl chain (*8-11*). Condensation of acetyl-CoA with malonyl-ACP and successive condensations with malonyl-ACP to add 2-carbon units are carried out on a dissociable fatty acid synthase complex. When the chain is 16 carbons long

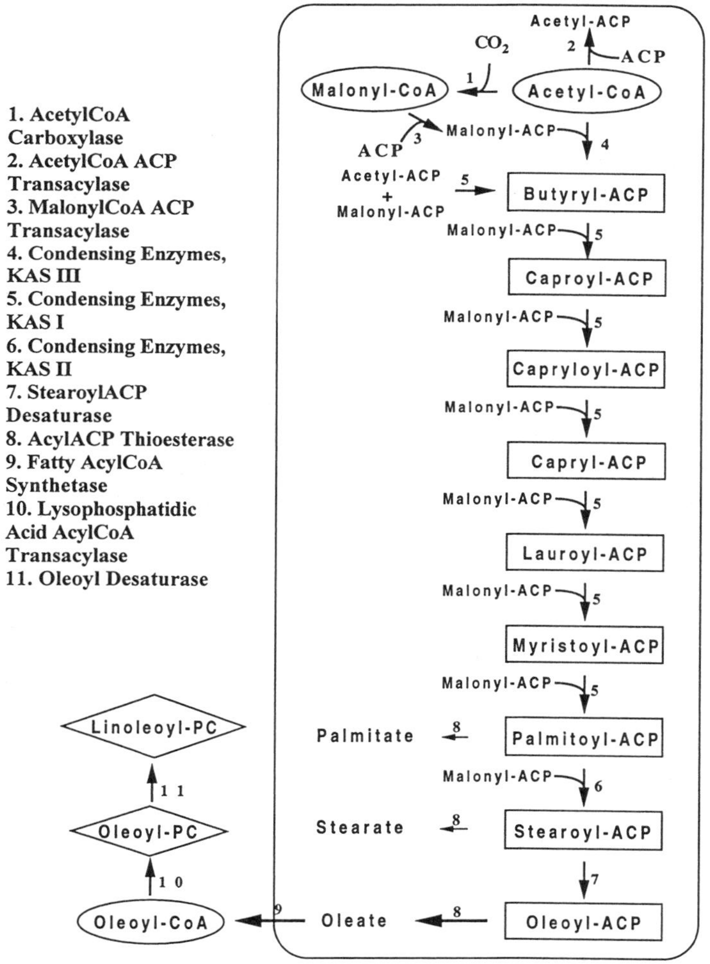

Figure 2. Fatty acid biosynthesis in plants.

(palmitoyl-ACP) it is elongated to stearoyl-ACP by a specific condensing enzyme, KAS II, and then desaturated to oleoyl-ACP by the highly specific stearoyl-ACP desaturase. Free oleate is released by an acyl-ACP thioesterase that has a strong preference for hydrolyzing oleoyl-ACP. This pathway (the ACP track) leads primarily to production of oleate by the plastid (*8, 12*). Prokaryote fatty acid biosynthesis (bacteria) also uses a dissociable ACP and fatty acid synthase, therefore, the plastid pathway is often referred to as the prokaryotic pathway.

Free oleate is converted to oleoyl-Coenzyme A thioester (oleoyl-CoA) in the plastid envelope and is incorporated into complex lipids, e.g. phospholipids, in the endoplasmic reticulum (the microsomal pathway of fatty acid biosynthesis). Further reactions, including additional desaturations, occur in the endoplasmic reticulum. There had been great debate about the primary substrate of the microsomal pathway, based on the ready utilization of oleoyl-CoA as the source of labelled fatty acid. However, the oleate from oleoyl-CoA is rapidly incorporated into phospholipid and it is now accepted that desaturation of oleate occurs using phosphatidylcholine as the carrier for the substrate fatty acid (*9, 10*). Thus acyl-CoA is the substrate used to incorporate the acyl chain into complex lipid. These transacylase reactions are of key importance to seed oil production and membrane synthesis for normal growth and development. Desaturation reactions in the chloroplast, occurring on different complex lipids (the mono- and di-galactosyl diacylglycerols), are principally important for the development and maintenance of chloroplast membranes.

For most plants, oleate serves as the central substrate for fatty acid production in seed oils (*8*). This fact has made it possible to manipulate fatty acid composition of vegetable oils by altering expression of single genes that encode the enzyme for a selected step in the biochemical pathway leading to or from oleate production (*13-15*).

Genetic Manipulation of Oil Composition

The reactions described above are those responsible for the fatty acid composition of a "normal" vegetable oil. For those vegetable oils with uncommon fatty acids, such as laurate or ricinoleate, the activity of a single enzyme can often account for the altered composition. This single-gene change approach forms the foundation for most of the current genetic engineering success in modifying oil composition.

Thus, progress in developing new oils has depended on the ability to identify enzyme activities that can carry out the reaction of interest and to clone the corresponding cDNA (the DNA copy of the mRNA produced from genes). The crop plant is then genetically transformed to express the desired activity in the oilseed during maturation, in some cases interacting with endogenous enzymes to effect the change. For example, plants that produce oils containing high levels of laurate contain an acyl-ACP thioesterase that is specific for shorter acyl chain length (Fig. 3a), leading to accumulation of laurate (*16, 17*). Additional products of interest include very long chain fatty acids (Fig. 3b), hydroxy fatty acids (Fig. 3c), and uncommon monounsaturated fatty acids (Fig. 3d) (*18*). The reactions depicted in Fig. 3b and 3d are examples of biosynthetic reactions requiring interaction with endogenous enzymes.

Techniques for plant transformation and for obtaining seed-specific expression have become accessible over the past decade or more, and there are continual improvements in this technology (*19*). A key limitation in designing new oils has been the availability of genetic material encoding enzymes that would carry out the desired reaction during oilseed development (*19, 20*).

Approaches for Selecting Genes. Techniques that are used to obtain the genetic material include "reverse genetics" that uses the isolated protein to develop probes for the corresponding cDNA, DNA probes that are derived from related cDNA or genes ("homologous probing"), genetic complementation that "cures" mutations, gene inactivation with specific DNA markers ("gene tagging") that facilitate retrieval of the inactive gene, as well as detection of expressed products in transgenic organisms.

Reverse Genetics. The "reverse genetics" method uses the purified enzyme to obtain antibodies that can detect genetic clones expressing the enzyme. Alternatively, the enzyme's amino acid sequence provides a DNA sequence that is used to detect clones by hybridization, that is, binding to DNA that carries the complementary base-sequence. Antibodies have been used for detecting clones of the stearoyl-ACP desaturase from castor (*21*). Protein sequence information has provided deduced DNA sequence for isolating clones of stearoyl-ACP desaturase from safflower (*22*), a lauroyl-ACP thioesterase from bay laurel (*17*) and a medium chain lysophosphatatidic acid acyl transferase (LPAAT) from coconut (*23*). The cDNA for each of these enzymes has been used to develop transgenic plants that produce vegetable oils with altered fatty acid composition.

Several alternative methods to clone cDNA for enzymes that carry out useful fatty acid transformations have been developed and applied, since some enzymes are not readily purified.

Homology Cloning. A number of plants in the Umbelliferae family, e.g. parsley and fennel, produce petroselinic acid (cis-6-octadecenoic) as a major component of the seed oil. This monounsaturate has commercial value because it is readily oxidized to lauric acid and adipic acid, a precursor to Nylon 6. It was thought that the desaturase that made petroselinate would be similar to the stearoyl-ACP desaturase, since both types of desaturase produce a monounsaturated 18-carbon fatty acid. Based on this presumed

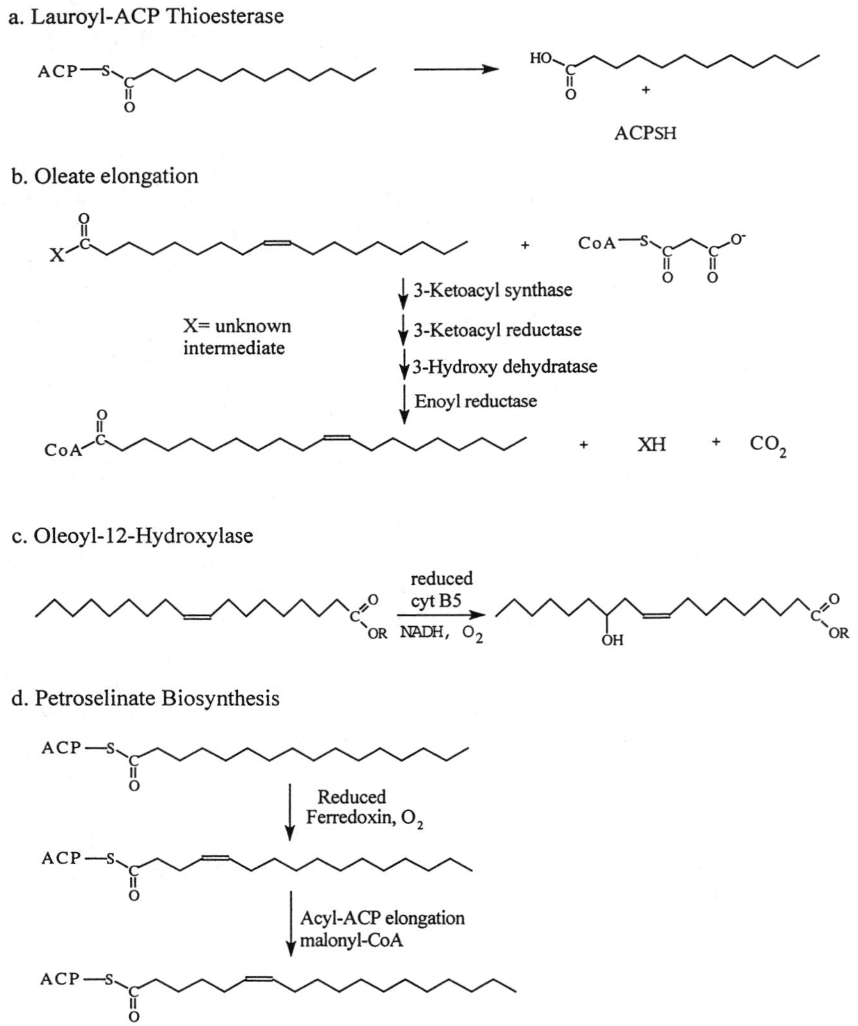

Figure 3. a-d. Reactions that produce uncommon fatty acids in plants.

homology, the genetic material for the enzyme that produces petroselinate (18:1Δ^6) was cloned, using antibody to stearoyl-ACP desaturase to screen a coriander seed cDNA library (*24*). The cloned cDNA was expressed in a transgenic plant, and shown to be a palmitoyl-ACP Δ^4-desaturase (Figure 3d). A number of fatty acid desaturases have been cloned from plants on the basis of homology to previously cloned desaturases (*13, 25, 26*).

Complementation. Genetic complementation uses externally added non-mutant DNA to complement or "cure" mutants that are defective in synthesizing a product of interest. The most notable use of this technique in elucidating fatty acid biosynthesis provided the first isolation of a membrane-bound desaturase from plants. An *Arabidopsis* mutant carrying the defective gene *fad3* was deficient in the biosynthesis of linolenate and accumulated linoleate in non-plastid membranes. The defective gene mapped near two specific chromosome markers. Fragments of DNA containing these markers were found to hybridize to specific pieces of *Arabidopsis* DNA that are maintained as yeast artificial chromosomes (YAC), a means of storing foreign DNA in yeast that allows it to replicate like the yeast chromosomes. These pieces of *Arabidopsis* DNA were used, in turn, to screen a *Brassica* seed cDNA library to find a cDNA that hybridized to the YAC. The *Arabidopsis* mutant was transformed with this cDNA and produced high levels of linolenate, demonstrating the mutation had been "cured" (*27*).

DNA Tagging. The process of gene inactivation (also known as transposon tagging) is the reverse of genetic complementation. The tag, T-DNA, is derived from *Agrobacterium tumefaciens*; this specific piece of DNA can insert into numerous sites on the chromosomes of a plant that has been genetically transformed with the T-DNA. If the site of insertion is in an active gene, the T-DNA will disrupt the gene, causing loss of the gene-coded enzyme activity. Samples from numerous T-DNA transformants of *Arabidopsis* were screened for fatty acid content, and a mutant with lower levels of linolenate was isolated. Since the disrupted gene has the piece of T-DNA inserted in it, a T-DNA probe hybridized to the T-DNA in the disrupted gene. The presence of T-DNA in the gene allowed detection of the gene, providing a direct means to isolate genomic DNA corresponding to the linoleoyl desaturase disrupted by the T-DNA (*28*). Other gene-disrupting DNA pieces, such as translocating elements, can be used analogously (*29*).

Gene Expression. An additional approach for isolation of genes is the expression of the gene in an alternate organism. This technique has proven effective for a number of genes, and it is a standard way of confirming the function of a cloned gene. The bacterium *Escherichia coli* is commonly used, as is the yeast *Saccharomyces cerevisiae*. For plant expression, *Nicotiana* transformation is often used. Expression in *Nicotiana* has recently been used to identify an oleoyl-12-hydroxylase from castor, by demonstrating the production of low levels of the product, ricinoleate (*30*). Our research group is currently examining transgenic expression in the fungus *Neurospora* as a potentially useful way to evaluate activity of plant genes that modify fatty acids. *Neurospora* has many similarities in the microsomal pathways of fatty acid and lipid biosynthesis to those of plants. We have found expression at the mRNA level of the plant microsomal linoleoyl desaturase cDNA, FAD3. We have also developed a collection of *Neurospora* mutants that are defective in fatty acid desaturation (*31, 32*). These could serve as indicators for expressing genes or cDNA that cure the mutation, some of which appear to be in the desaturation process while others appear to involve lipid biosynthesis. *Neurospora* as a model system should thus provide a convenient way to identify or confirm the identity of cloned genes.

Novel vegetable oils

Practical Considerations. With the appropriate cDNA cloned, the next step is to express it in the desired crop. For practical purposes, the companies involved in applying biotechnology to oilseed improvement work with a limited number of crops (Table II). This limitation is primarily related to the economic importance of the crops. In the U.S., soybean, cotton and corn are major sources of vegetable oil. Canola is an important oilseed in Canada, and although a minor crop in the U.S., its cultivation is expanding rapidly due to the popularity of the high monounsaturate oil for food uses. Flax is being bred and engineered for use as a cooking oil in Australia because it is one of the important oilseeds in Australian agriculture.

The economics of genetically engineering a new oil crop is a significant limitation to applying the technology broadly. The inherent cost and risk in undertaking a major research effort to develop the appropriate approach is a significant commercial hurdle. Identification of high value end uses for the new crop is, therefore, a basic need for industrial interest in the genetic engineering approach. Other practical considerations include finding growers for the new crop, protecting ownership or licensing rights for the crop, and maintaining the desired properties of the oil through production and processing.

Transgenic Commercial Crops. Transgenic plants producing altered oils are already in commercial production. As shown in Table II, the principal crops modified for altered oil production are canola and soybean. As technology advances, cotton, corn and flax oils will be developed. Eventually, sunflower, peanut, safflower and other oilseeds are likely to be engineered.

As described in the section on fatty acid biosynthesis, production of a high laurate canola is based on transgenic expression of lauroyl-ACP

Table II

Genetic Modifications of Oilseeds for Commercial Uses [b]

Organization	Oilseed Crop	Modification
Calgene	Canola*	laurate
	Canola	high stearate
	Canola	medium chain FA
	Cotton*	bromoxynil tolerance
	Cotton	insect resistance (Bt)
DuPont	Soybean	high oleic, low saturate
	Soybean	high stearate, low polyunsaturate
	Soybean	high lysine
	Corn	high oil, high oleic
InterMountain Canola	Canola	high oleic
	Canola	high oleic, low saturate
	Canola	high stearate, low polyunsaturate
AgrEvo Canada Inc.	Canola*	herbicide tolerance
	Soybean	herbicide tolerance
Monsanto Company	Canola	glyphosate tolerance
	Cotton	insect resistance (Bt)
	Soybean	glyphosate tolerance
Univ. of Saskatchewan	Flax*	sulfonylurea tolerance

[b]From: Anon. (1995) Transgenic oilseed harvests to begin in May. INFORM 6: 152-157.

* In commercial use, 1995; others are targeted for introduction from 1996 to 1999.

thioesterase during seed development. The oil is now commercially produced and is being used by the detergent industry to replace high-laurate palm and coconut oils (*5*). Canola oil containing high levels of medium chain fatty acids is similarly produced, using transgenic expression of a *Cuphea* sp. medium chain (C8-10) acyl-ACP thioesterase (*5, 33*). In addition to meeting chemical needs in the detergent industry (*34*), triacylglycerols based on the medium-chain fatty acids have applications in the snack and pharmaceutical industries as rapidly metabolized, high energy food components (*5*).

Anti-sense DNA technology is used to block expression of the stearoyl-ACP desaturase, resulting in high stearate canola. In the anti-sense process, the plant is transformed to express an mRNA complementary (anti-sense) to the mRNA (sense) for the desaturase. The result is diminished sense mRNA translation, and therefore limited desaturase production, causing accumulation of stearate by the seed. Canola oils containing high levels of stearate (up to 40%) have been produced and are solid at room temperature (*13*). This type of fat could provide a source of oil that is stable to high-temperature and oxidation. Anti-sense technology has been used to block desaturation of oleate to linoleate, leading to high oleate oil and to block desaturation of linoleate to linolenate, providing an oil with low linolenate content that is less susceptible to oxidation (*18, 29, 36*).

An alternate way to block gene expression in plants is co-suppression, in which the genomic insertion of a second copy of the gene in the sense direction interferes with the expression of both the original and the inserted copy. The mechanism for co-suppression is currently unclear but the gene that acts as the suppressor behaves as a simply inherited dominant trait (*37*). There are breeding advantages to using co-suppression, and it has been effective in the development of high oleate, high stearate, and low linolenate soybean oils (*36*).

Other Developments. At this time, other transgenic oilseeds including flax, cotton and corn have been modified for herbicide resistance and are near commercial use (*5*); flax has also been modified to block stearoyl desaturase, producing a flax oil more suited to cooking. Other, non-commercial transgenics with altered oil composition have also been developed (e.g. *24, 28, 30*). These include the plants *Nicotiana* and *Arabidopsis* that are readily manipulated genetically and are useful for research purposes. While these plants have little direct use as sources of oil, they do provide the means for evaluating both the expression of cDNA's for enzymes that can alter fatty acid biosynthesis and the effect of such expression on fatty acid composition. These model oilseed systems underlie the technology that allows the development of new vegetable oils.

Unusual fatty acids and their uses

The plant kingdom contains numerous examples of plants that produce seed oils containing unusual fatty acids (see Fig. 1) with such variations as double bonds in uncommon positions, epoxy functions, cyclopropyl rings or triple bonds (*3, 4, 38*). While most of the plants that produce these oils are not cultivated due to unsuitable growth habit, their oils are intrinsically valuable. The presence of specific functional groups on a fatty acid endows it with value for industrial uses in several ways:

Direct Uses. As a result of chemical reactivity or altered physical properties, some unusual fatty acids or oils are useful without additional modification. An epoxy fatty acid can serve directly as a plasticizer or as a replacement for epoxy compounds now used in paints. Due to the mid-chain hydroxyl group in ricinoleate, castor oil is very viscous, providing the basis for its use in lubricating oils and greases.

Chemical Modifications. Reactions arising from the chemical reactivity of the fatty acid provide new sources of material for existing applications. Oxidation (e.g. by ozonolysis) of a monounsaturate cleaves the double bond, forming a dibasic and a monobasic acid. Petroselinate ($18{:}1\Delta^6$) would thus provide a source of adipic acid (for Nylon 6) and lauric acid (for surfactants) (*18*). The presence of two reactive functional groups, e.g., the carboxyl and hydroxyl on ricinoleate, imparts the ability to form interchain crosslinks, leading to formation of polymers or, by sequential reaction, composite materials. The hydrogenation of dihydrosterculate, a C-19 cyclopropyl fatty acid, results in a mixture of products, 85% of which are fatty acids having an 18-carbon chain length with a methyl or methylene substituent branching off from the middle of the chain (*39*). Such fatty acids have low freezing points and would be useful as additives for low temperature lubricants.

Uses and Potential Uses. The fatty acids found in common vegetable oils have important industrial uses, especially palmitate for surfactants, oleate as a precursor for plastics and polyunsaturates for drying oils, and after epoxidation, as plasticizers (*1*). Table III lists several additional fatty acids that currently have important industrial uses or have potential uses. Several of these are derived from cultivated plants, but most of the oils are imported, including castor, coconut, rapeseed and tung oil. U.S. grown flax meets over half of the U.S. need (*6*). The need for epoxy fatty acids is currently met by chemical conversion of soybean oil (*18*). The medium chain (C8/C10) have potential uses in surfactants, lubricants and high value food products, and will soon be available in canola (*5*). The uncommon monounsaturated fatty acids have potential uses as feedstock for polymer production, but are not yet

Table III

Industrially Useful Fatty Acids from Plants

Type	Example	Plant Source	Uses
Poly-unsaturate	α-Linolenic	Flax (Linseed Oil)	Drying Oil
Hydroxy	Ricinoleate	Castor	Greases, Coatings
Epoxy	Vernolate	Vernonia sp.	Plasticizer
Medium Chain	Laurate (C12)	Coconut	Surfactants
Medium Chain	Caprylate (C8)	Cuphea	Surfactants, Pharmaceuticals
Very Long Chain	Erucate (22:1)	Rapeseed	Slip/Anti-Block Agent for Films
Cyclo-propenoid	Malvalate	Malva sp.	Low-Temp. Lubricant
Triple Bond	Crepenynic	Crepis sp.	Specialty Chemical
Conjugated Double Bonds	α-Eleostearate	Aleurites fordii (Tung oil)	Drying Oil

available from suitable crops. There is no current demand for cyclopropenoid and alkynoic fatty acids, but each of these functional groups can polymerize via several reaction mechanisms and are also susceptible to addition reactions. The ability to react these fatty acids with other monomers by free radical combination or esterification provides the capability of selectively building a polymeric material one step at a time. Such fatty acids would be excellent substrates for building composite plastics with novel properties. A recent example is the use of diacetylenes to form supramolecular structures with an ordered array of functional groups (*40*). These materials have potential applications in electronics and nanotechnology.

There is also a potential market for selected fatty acids as precursors of chemicals with high-value end uses, such as pharmaceuticals. These uses are exemplified by the conversion of sterols from the Mexican yam into steroidal drugs. Plants, like any other organism, produce fatty acids that are specific in terms of chain length, geometric configuration and chirality. Biological activity relies on structural and chiral specificity and, generally, the specific structural requirements for biological activity in one organism correspond to the geometric and chiral structure of a potential precursor produced by another organism. This basic fact forms the foundation for the pharmaceutical industry. Fatty acids with multiple double or triple bonds, polysubstitution or unusual ring structures could provide suitable substrates for synthesis of pharmaceuticals or other biologically active compounds.

It is remarkable that plants can make fatty acids incorporating a range of substituents, e.g. three conjugated acetylene groups in fatty acids produced in some Santalaceae (*38*), which have only recently been duplicated in the laboratory. The value of these fatty acids will rely on the economy of having the plant carry out what could otherwise be an expensive or hazardous synthesis. One likely class of pharmaceuticals that could be synthesized from these fatty acids are the prostaglandins, which are made in vivo by the action of the cyclooxygenase on polyunsaturated fatty acids. Such complex fatty acids are produced by various plants, and a horticulturally suitable plant source should have a premium value for the grower.

Future Directions and Challenges

There is considerable success in the development of novel vegetable oils, but there remain some obstacles in the design and utilization of these oils. One problem is providing the genetic programming needed for optimal production of the desired oil. A second problem lies in broadening the range of fatty acids that can be introduced into vegetable oils. Finally, there is, currently, a limited scope of uses for novel oils, with "detergents, coatings and lubricants" the standard refrain for industrial uses of vegetable oils.

One clear challenge is maximizing the level of the desired fatty acid. In general, any oil destined for chemical feedstock will increase in value as the amount of the desired fatty acid reaches 90% of the total fatty acids. In the case of laurate canola oil, one limiting factor is the strong tendency in canola to have an unsaturated fatty acid in the sn2-position of the triacylglycerol (*41*). The solution to this problem was the insertion of the gene for lysophosphatidic acid acyl-transferase (LPAAT) from coconut (Fig. 4). This enzyme inserts a fatty acid in the sn2-position of phosphatidic acid, a phospholipid precursor of triacylglycerol. In most seeds, the LPAAT has a strong preference for inserting an unsaturated fatty acid, but the enzyme from coconut preferentially inserts laurate (*23*). Increasing production of other unusual fatty acids may require additional or alternative manipulation of lipid biosynthesis to enhance their incorporation into oil. The recent cloning of oleoyl-12-hydroxylase from castor illustrates the general problem of attaining high levels of a desired fatty acid. The ricinoleate produced in castor accounts for 90% of the fatty acid in oil, yet, in transgenic tobacco seed, it accumulates to only 0.1%. While one possible cause is the need for specific acyl transferases that insert ricinoleate in phospholipid and neutral lipid, other possibilities for such low accumulation are poor expression of the enzyme activity in tobacco seed or rapid metabolism of the ricinoleate as it is produced (*30*).

As more enzymes of a given type are cloned and expressed, it will be possible to identify sequences or structures that are involved in substrate recognition, chain length specificity and positional specificity for modification. Of enzymes under study, the primary desaturases are farthest along in terms of understanding the enzyme mechanism and in the availability of genetic stock that differs in positional and chain-length specificity (*42, 24*). The specificity of enzymes for fatty acid chain-length can be manipulated by site-selected mutagenesis (*43*). As more structural comparisons are made, it will be possible to engineer plants to introduce a double bond in a selected position on a fatty acid of desired chain-length, and to accumulate the fatty acid in vegetable oil. Since monounsaturates are readily converted to dicarboxylates, these "invented" oils could provide a resource for novel forms of Nylon and other polymers.

The capability to clone and express fatty acid-modifying enzymes in transgenic plants is advancing rapidly. It is likely that the ability to design and produce oils containing uncommon fatty acids will expand beyond current uses for such oils. With the constant development of new types materials, these materials could rapidly find their way into applications with the availability of suitable precursors. A recent example is the development of carbon chains that can act as "molecular wires" with possible applications in advanced electronic or optical devices (*44*). These carbon chains depend on the recently accomplished synthesis of conjugated polyacetylenes of defined chain-length. However, some plants make fatty acids with up to three conjugated triple bonds (*38*); a crop source of these fatty acids could replace the need for the complex synthesis, greatly reducing the cost of fabrication.

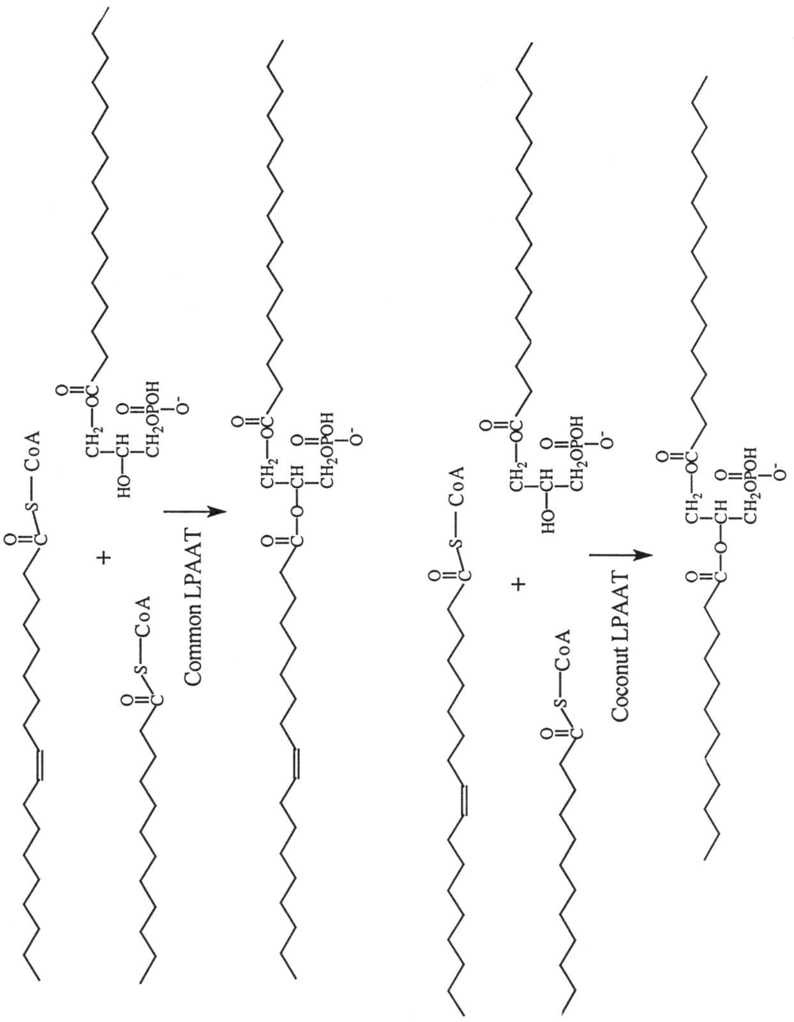

Figure 4. Lyso-phosphatidic acid acyl transferase (LPAAT) reaction.

The challenge in designing novel vegetable oils will eventually shift to the development of new uses for complex fatty acids produced in vegetable oils. As such oils become available, they will provide a source of chemicals for producing high value materials including composite polymers, organo-optics, organo-electronics and chiral pharmaceutical intermediates.

Literature Cited

1. Pryde, E.H. and Rothfus, J.A. (1989) Industrial and nonfood uses of vegetable oils. *in* Oil Crops of the World. Röbbeln, G., Downey, R.K. and Ashri, A., eds. McGraw-Hill, New York. pp. 87-117.
2. Princen, L.H. (1979) New crop developments for industrial oils. J. Am. Oil Chem. Soc. 56: 845-848.
3. Hirsinger (1989) New annual oil crops. *in* Oil Crops of the World. Röbbeln, G., Downey, R.K. and Ashri, A., eds. McGraw-Hill, New York. pp. 518-532.
4. van de Loo, F.J., Fox, B.G. and Somerville, C. (1993) Unusual fatty acids. *in* Lipid Metabolism in Plants. ed. T.S. Moore, CRC Press, Boca Raton. pp.
5. Anon. (1995) Transgenic oilseed harvests to begin in May. INFORM 6: 152-157.
6. Agricultural Statistics, 1994 (1995) Table 188. U.S. Government Printing Office, Washington, D.C.
7. Ainsworth, S.J. (1996) Soaps and Detergents. Chem. & Eng. News 74 (4) 32-54.
8. Stumpf, P.K., Kuhn, D.N., Murphy, D.J., Pollard, M.R., McKeon, T. A. and MacCarthy, J.J. (1980) Oleate - the central substrate in fatty acid biosynthesis. *in* Biogenesis and Function of Plant Lipids. Mazliak, P., Costes, C. and Douce, R., eds. Elsevier, North-Holland. pp. 3-10
9. Ohlrogge, J.B., Browse, J. and Somerville, C. (1991) The genetics of plant lipids. Biochim. Biophys. Acta 1082: 1-26.
10. Browse, J. and Somerville, C. (1991) Annu. Rev. Plant Physiol. Plant Mol. Biol. 42: 467.
11. Ohlrogge, J.B. and Browse, J. (1995) Lipid biosynthesis. The Plant Cell 7: 957-970.
12. McKeon, T.A. and Stumpf, P.K. (1982) Purification and characterization of the stearoyl-acyl carrier protein desaturase and the acyl-acyl carrier protein thioesterase from maturing seeds of safflower. J. Biol. Chem. 257: 12141-12147.
13. Knutzon, D.S., Thompson, G.A., Radke, S.E., Johnson, W.B., Knauf, V.C. and Kridl, J.C. (1992) Modification of *Brassica* seed oil by antisense expression of a stearoyl-acyl carrier protein desaturase gene. Proc. Nat. Acad. Sci. (USA) 89: 2624-2628.
14. Miquel, M. and Browse, J. (1995) Molecular biology of oilseed modification. INFORM 6: 108-111.

15. Topfer, R., Martini, N. and Schell, J. (1995) Modification of plant lipid synthesis. Science 268: 681-686.
16. Pollard, M.R., Anderson, L., Fan, C., Hawkins, D.H. and Davies, H.M. (1991) A specific acyl-ACP thioesterase implicated in medium-chain fatty acid production in immature cotyledons of *Umbellularia californica*. Arch. Biochem. Biophys. 284: 306-312.
17. Voelker, T.A., Worrell, A.C., Anderson, L., Bleibaum, J., Fan, C., Hawkins, D.J., Radke, S.E. and Davies, H.M. (1992) Fatty acid biosynthesis redirected to medium chains in transgenic oilseed plants. Science 257: 72-74.
18. Ohlrogge, J.B. (1994) Design of new plant products: engineering of fatty acid metabolism. Pl. Physiol. 104: 821-826.
19. Slabas, A.R., Simon, J.W. and Elborough, K.M. (1995) Information needed to create new oil crops. INFORM 6: 159-166.
20. Wolter, F.P. (1993) Altering plant lipids by genetic engineering. INFORM 4: 93-98.
21. Shanklin, J. and Somerville, C. (1991) Stearoyl-acyl-carrier-protein desaturase from higher plants is structurally unrelated to the animal and fungal homologs. Proc. Nat. Acad. Sci. (USA) 88: 2510-2514.
22. Thompson, G.A., Scherer, D.E., Foxall-Van Aken, S., Kenny, J.W., Young, H.L., Shintani, D.K., Kridl, J.C. and Knauf, V.C. (1991) Primary structures of the precursor and mature forms of stearoyl-acyl carrier protein desaturase from safflower embryos and requirement of ferredoxin for enzyme activity. Proc. Nat. Acad. Sci. (USA) 88: 2578-2582.
23. Knutzon, D.S., Lardizabal, K.D., Nelsen, J.S., Bleibaum, J.L., Davies, H.M. and Metz, J.G. (1995) Cloning of a coconut endosperm cDNA encoding a 1-acyl-*sn*-glycerol-3-phosphate acyltransferase that accepts medium-chain-length substrates. Pl. Physiol. 109: 999-1006.
24. Cahoon, E.B., Shanklin, J. and Ohlrogge, J.B. (1992) Expression of a coriander desaturase results in petroselinic acid production in transgenic tobacco. Proc. Nat. Acad. Sci. (USA) 89: 11184-11188.
25. Yadav, N.S., Wierzbicki, A., Aegerter, M., Caster, C.S., Perez-Grau, L., Kinney, A.J., Hitz, W.D., Booth, J.R. Jr., Schweiger, B., Stecca, K.L., Allen, S.M., Blackwell, M., Reiter, R.S., Carlson, T.J., Russell, S.H., Feldmann, K.A., Pierce, J. and Browse, J. (1993) Cloning of higher plant ω-3 fatty acid desaturases. Pl. Physiol. 103: 467-476.
26. Hitz, W.D., Carlson, T.J., Booth, J.R., Kinney, A.J., Stecca, K.L. and Yadav, N.S. (1994) Cloning of a higher-plant plastid ω-6 fatty acid desaturase cDNA and its expression in a cyanobacterium. Pl. Physiol. 105: 635-641.
27. Arondel, V., Lemieux, B., Hwang, I., Gibson, S., Goodman, H.G. and Somerville, C.R. (1992) Map-based cloning of a gene controlling omega-3 fatty acid desaturation in Arabidopsis. Science 258: 1353-1355.
28. Okuley, J., Lightner, J., Feldmann, K., Yadav, N., Lark, E. and Browse, J. (1994) *Arabidopsis* FAD2 gene encodes the enzyme that is essential for polyunsaturated lipid synthesis. Plant Cell 6: 147-158.

29. James, D.W., Lim, E., Keller, J., Plooy, I., Ralston, E. and Dooner, H.K. (1995) Directed tagging of the *Arabidopsis FATTY ACID ELONGATION1* (*FAE1*) gene with the maize transposon *Activator*. Plant Cell 7: 309-319.
30. van de Loo, F.J., Broun, P., Turner, S. and Somerville, C. (1995) An oleate 12-hydroxylase from *Ricinus communis* L. is a fatty acid desaturase homolog. Proc. Nat. Acad. Sci. (USA) 92: 6743-6747.
31. Tanrikulu, M.G., Stafford, A.E., Lin, J.T., Makapugay, M.M., Fuller, G. and McKeon, T.A. (1994) Fatty acid biosynthesis in novel ufa mutants of *Neurospora crassa*. Microbiology 40: 2683-2690.
32. Tanrikulu, M.G. Lin, J.T., Stafford, A.E., Makapugay, M.M., McKeon, T.A. and Fuller, G. (1995) *Neurospora crassa* mutants with altered synthesis of polyunsaturated fatty acids. Microbiology 141: 2307-2314.
33. Dehesh, K., Jones, A., Knutzon, D. and Voelker, T. (1996) Production of high levels of 8:0 and 10:0 fatty acids in transgenic canola by overexpression of *CH FatB2*, a thioesterase cDNA from *Cuphea hookeriana*. Plant J. in press.
34. Haumann, B.F. (1992) Lauric oils have their own niche. INFORM 3: 1080-1093.
35. Miquel, M.F. and Browse, J. (1994) High-oleate oilseeds fail to develop at low temperature. Pl. Physiol. 106: 421-427.
36. Fader, G.M., Kinney, A.J. and Hitz, W.D. (1995) Using biotechnology to reduce unwanted traits. INFORM 6: 167-169.
37. Niebel, F. C., Frendo, P., Van Montagu, M. and Cornelissen, M. (1995) Post-transcriptional cosuppression of ß-1,3-glucanase genes does not affect accumulation of transgene nuclear mRNA. The Plant Cell 7: 347-358.
38. Hitchcock, C. and Nichols, B.W. (1971) Plant Lipid Biochemistry. Academic Press, London and New York. 387 pp. ISBN 0-12-349650-0
39. Zarins, Z.M., White, J.L., Willich, R.K. and Feuge, R.O. (1982) Hydrogenation of cyclopropenoid fatty acids occuring in cottonseed oil. J. Am. Oil Chem. Soc. 59: 511-515.
40. Kane, J.J., Liao, R.F., Lauher, J.W. and Fowler, F.W. (1995) Preparation of layered diacetylenes as a demonstration of strategies for supramolecular synthesis. J. Am. Chem. Soc. 117: 12003-12004.
41. Oo, K.-C. and Huang, A.H.C. (1989) Lysophosphatidate acyltransferase activities in the microsomes from palm endosperm, maize scutellum, and rapeseed cotyledon of maturing seeds. Pl. Physiol. 91: 1288-1295
42. Fox, B.G., Shanklin, J., Somerville, C. and Munck, E. (1993) Stearoyl-acyl carrier protein Δ^9 desaturase from *Ricinus communis* is a diiron-oxo protein. Proc. Nat. Acad. Sci. (USA) 90: 2486-2490.
43. Yuan, L., Voelker, T.A. and Hawkins, D.J. (1995) Modification of the substrate specificity of an acyl-acylcarrier protein thioesterase by protein engineering. Proc. Nat. Acad. Sci. (USA) 92: 10639-10643.
44. Dagani, R. (1996) Capping carbon chains with metals gives 'molecular wires' and more. Chem. & Eng. News 74 (4) 22-24.

Chapter 12

Plants as Sources of Drugs

A. Douglas Kinghorn and Eun-Kyoung Seo

Program for Collaborative Research in the Pharmaceutical Sciences, Department of Medicinal Chemistry and Pharmacognosy, College of Pharmacy, University of Illinois, 833 South Wood Street, Chicago, IL 60612–7231

Plants have a historically important role as sources of prescription drugs in western medicine, and their active principles also serve as templates for synthetic drug optimization and provide intermediates that are used in the production of semi-synthetic drugs. Worldwide, hundreds of higher plants are cultivated for substances useful in medicine and pharmacy. Although the anticancer agent taxol (paclitaxel) is the first naturally occurring plant-derived drug to have gained the approval of the U.S. Food and Drug Administration for more than 25 years, analogs of other plant constituents such as artemisinin, camptothecin, and forskolin are currently under development for drug use in western medicine. Another recent development has occurred in Western Europe, where there is currently a well-developed phytomedicinal market, with extractives of many plants being used for therapeutic purposes. An underlying concern with the production of plant-derived drugs is the question of supply, which must be stable and reliable, and frequently necessitates cultivation rather than collection in the wild. Information on the cultivation, constituents, and therapeutic uses of two examples of important medicinal plants (*Panax ginseng*; Korean ginseng and *Ginkgo biloba*; Ginkgo) will be provided in this chapter.

For millenia, drugs from higher plants (gymnosperms and angiosperms) have been used to cure or alleviate human sickness, and, until relatively recently, were the major sources of medicines. The vast majority of plant drugs now used are classified as secondary metabolites of the producing organism, and, as such, they are derived from primary metabolites biosynthetically but have no apparent function in the primary metabolism of the plant. In the past two centuries, many plant secondary metabolites

0097–6156/96/0647–0179$15.00/0

have been purified and structurally characterized because of the medicinal properties of their species of origin, affording numerous compounds of medicinal and pharmaceutical importance (*1-6*). Many of the natural drugs presently used were obtained from plants used as toxins in their native habitats, such as physostigmine, *d*-tubocurarine, and certain cardiac glycosides (*7*). Other plant drugs have been obtained as a result of folkloric use for medicinal purposes, as exemplified by digitoxin, morphine, pilocarpine, and quinine (*7*). Plant secondary metabolites have two important additional roles, in being used as templates for the design and synthesis of completely novel drug entities, as well as serving as source material for the semi-synthesis of medicinal agents (*1-7*).

Importance of Plant-Derived Drugs in Modern Medicine

It has been estimated that 80% of the world's population who live in developing countries depend chiefly on traditional medical practices, inclusive of the use of medicinal higher plants for their primary health care needs (*1,4*). Some of these systems are very well developed, such as Chinese traditional medicine (*1,4*). Farnsworth and colleagues have determined that about 120 plant-derived substances from some 90 species are used as drugs in one or more countries, with 74% of these having been discovered as a result of laboratory studies conducted on the active principles of medicinal plants used in traditional medicine (*1,4*). Examples of well-established drugs of plant origin used in various countries throughout the world are shown in Table I, and it is to be noted that most of these are still produced commercially by cultivation, extraction, and purification, with only the more structurally simple compounds successfully synthesized for the marketplace. Large-scale production of plant secondary metabolites by tissue-culture techniques has attracted a great deal of interest, but thus far none of the plant drugs listed in Table I are produced commercially in this manner (*8*). In addition to pure drug entities of plant origin, extractives of medicinal plants containing partially purified secondary active principles are still widely used. Examples include standardized extractives from *Atropa belladonna* and *Datura metel* (with the solanaceous tropane alkaloids present used as mydriatics and anticholinergics), and from *Cephaelis acuminata* and *C. ipecacuanha* (having isoquinoline alkaloids that are commonly used as emetic agents in the treatment of domestic accidental poisonings), and from *Digitalis purpurea* (containing cardiotonic glycosides, which are still of major use in the treatment of congestive heart failure) (*1,3,4*). Drugs derived from higher plants, both in the pure and the crude extractive form, are contained in up to 25% of the prescriptions dispensed in community pharmacies in the United States, at an estimated value of some $8 billion in 1980 (*1,4*).

In addition to their value as drugs per se, plant secondary metabolites have a considerable history of use as lead compounds for the design of synthetic drugs, and perhaps the best examples are cocaine and morphine, which in turn have served as templates for the synthesis of novel local anesthetics and analgesics. Other examples are atropine, physostigmine, and quinine, which have been utilized as chemical models for the design and synthesis of anticholinergics, anticholinesterases, and antimalarials,

TABLE I. SOME PLANT-DERIVED DRUGS WITH CLINICAL USE[a]

Acetyldigoxin	Methyl salicylate[b]
Aescin	Morphine
Ajmalicine	Narcotine
Allantoin[b]	Nicotine
Andrographolide	Noscapine (narcotine)
Anisodamine	Ouabain
Anisodine	Papain
Arecoline	Papaverine[b]
Atropine	Physostigmine[b]
Benzyl benzoate[b]	Picrotoxin
Berberine	Pilocarpine
Borneol	Pinitol
Bromelain	Podophyllotoxin
Caffeine[b]	Protoveratrines A and B
Camphor	Pseudoephedrine[b]
Chymopapain	Pseudoephedrine, nor-[b]
Cocaine	Quinidine
Colchicine	Quinine
Curcumin	Rescinnamine
Danthron[b]	Reserpine
Deslanoside	Salicin
Deserpidine	Scillarins A and B
Digitoxin	Scopolamine
Digoxin	Sennosides A and B
L-DOPA[b]	Silymarin
Emetine	Sparteine
Ephedrine[b]	Strychnine
Etoposide[c]	Δ^9-Tetrahydrocannabinol[d]
Galanthamine	Theobromine[b]
Gitalin	Theophylline[b]
Hydrastine	Thymol
Hyoscyamine	*d*-Tubocurarine
Kawain[b]	Valepotriates
Khellin	Vinblastine
Lanatosides A, B, and C	Vincristine
α-Lobeline	Xanthotoxin
Menthol[b]	Yohimbine

[a] Information taken from Farnsworth *et al.*, 1985 (*1*) and Farnsworth and Soejarto, 1991 (*4*).

[b] Also produced by total synthesis; the remaining drugs in the table are produced by cultivation and extraction.

[c] Semi-synthetic.

[d] Used as a synthetic drug.

respectively (*1-7*). Plant constituents may also be employed as raw materials for the commercial production of drugs by partial synthesis, as exemplified by the production of hormones from diosgenin, hecogenin, and stigmasterol, for use in oral contraceptive formulations (*1-7*).

Only a small proportion (probably 10-15%) of the 250,000 or so higher plants that occur worldwide have so far been subjected to any form of scientific study concerning the chemical composition and biological properties of their secondary metabolites (*1,4,5,9*). This is especially true for plants of the tropical rainforests, where about one half or 125,000 of the world's flowering plants occur, and it has been estimated recently that only about one-eighth (48) of the potential drugs therein (about 375) have so far been discovered (*10*). As a result of the devastation of the tropical rainforests, which has been well-publicized, there is now a very great interest in preserving both the genetic resources and the indigenous knowledge associated with the use of medicinal plants in these regions (*9*). The issues of fostering scientific collaboration between scientists in western and developing countries, as well as providing appropriate compensation to persons in tropical countries where the greatest plant biodiversity occurs, have become increasingly important in the context of the drug discovery process from higher plants (*9*). The majority of the plant drugs presently on the market are alkaloid salts and glycosides, and hence reasonably water-soluble, a property which has facilitated their laboratory and clinical evaluation. However, most of the newer drugs, drug candidates, and lead compounds from plants have less than optimum water solubility properties, and some have proven to be problematic in terms of their stability or supply. New therapeutic agents of plant origin, which have either been recently introduced onto the market, or else have the promise of being introduced soon, are considered in the next section of this chapter.

New Medicinal Agents of Plant Origin

The semi-synthetic lignan, etoposide, derived from podophyllotoxin, a constituent of *Podophyllum peltatum* and *P. emodi* rhizomes, has been used for over 10 years as a cancer chemotherapeutic agent for the treatment of small-cell lung and testicular cancers (*11*). Very recently a second podophyllotoxin analog, teniposide (**1**) (Vumon®), has been introduced in the United States for use in combination chemotherapy to treat patients with refractory childhood acute lymphoblastic leukemia (*12*). Another recently introduced derivative, the tartrate salt of vinorelbine (**2**) (5′-noranhydrovinblastine, Navelbine®), a semi-synthetic derivative of the bisindole alkaloids, vinblastine and vincristine, which are in turn constituents of the leaves of *Catharanthus roseus*. Vinblastine and vincristine are well-established anticancer agents themselves, and have been used in therapy for over 30 years (*3,11*). Vinorelbine (**2**) was developed by a French pharmaceutical company (*6,13*), and is indicated for first-line therapy of non-small-cell lung cancer, either alone or in combination with cisplatin (*13*).

There is no question, however, that the plant-derived drug that has captured the attention of both the scientific community and the lay public to the greatest extent in recent years is the nitrogenous diterpenoid, paclitaxel (**3**; TAXOL®; previously known in the scientific literature as taxol). Paclitaxel was approved by the FDA for the treatment of refractory ovarian cancer in late 1992 and for refractory breast cancer in 1994. This is the first chemically unmodified plant secondary metabolite to have been introduced to the U.S. drug market in nearly 30 years (*5,14,15*). Paclitaxel was discovered more than 20 years ago as an antitumor constituent of the bark of *Taxus brevifolia* (Pacific or western yew), by M.E. Wall and M.C. Wani of Research Triangle Institute in North Carolina (*14,15*). Later Horwitz and co-workers demonstrated that the compound has a unique mechanism of action in being an antimitotic agent which stabilizes microtubules and prevents depolymerization (*14,15*). Clinical trials on paclitaxel began in the United States in 1984, and proved encouraging, so much so that the clinical demand created a supply crisis for paclitaxel when obtained by isolation and extraction from *T. brevifolia* bark (*14-16*). It was calculated that the amount of dried *T. brevifolia* bark required to produce 1 kg of paclitaxel is 15,000 pounds, and in 1991 over 1.6 million pounds of the bark of this species were harvested from Washington and Oregon for the production of paclitaxel (*16*). However, this large-scale collection of bark from naturally growing *T. brevifolia* specimens caused environmental concerns (*15*). Accordingly, various proposals have been put forth for the alternative production of large enough quantities of paclitaxel to both meet clinical demands and allay the environmental outcry. Several semisynthetic routes have been proposed for paclitaxel from diterpenoid building blocks (*14,15*), and the compound has actually been totally synthesized by the Holton and Nicolaou groups (*14,15*). Other methods of paclitaxel production are direct extraction and isolation from ornamental *Taxus* cultivars, plant cell culture, and fermentation of an endophytic fungus (a novel *Taxomyces* species) that biosynthesizes this compound *de novo* (*16*). However, because the molecule of paclitaxel (**3**) has 11 chiral centers, and the methods of total synthesis are complex and involve over 25 steps each, future production of this drug will involve partial synthesis from natural starting materials such as 10-deactylbaccatin III (**4**) and baccatin III (**5**), which are extractable in large quantities from renewable ornamental species such as *Taxus baccata* (European yew) and *T. wallichiana* (Himalayan yew) (*14-16*). Approval has been given by the FDA for the production of paclitaxel by semi-synthesis from diterpenoid precursors produced by *Taxus* species (*17*). To provide the paclitaxel diterpenoid synthetic intermediates, millions of seedlings of the cultivar *Taxus X media* "Hicksii" are being grown in nurseries in the United States and Canada each year (*18*). A semi-synthetic analog of paclitaxel that has been developed in France is docetaxel (Taxotere®), and this has shown good progress in clinical trials to date (*14,15*). According to a company profile of the manufacturer of docetaxel, this compound is produced from yew trimmings obtained from English country houses, French cemeteries, and in the Indian subcontinent (*19*).

Another group of plant-derived compounds with high potentiality for the treatment of cancer are analogs of the quinoline alkaloid, camptothecin (**7**), obtained from the Chinese tree, *Camptotheca acuminata*. The alkaloid was isolated and structurally characterized about 30 years ago by Wall and Wani, when it was found to exhibit antileukemic and antitumor activity (*20*). After an early disappointing clinical trial on the water-soluble sodium salt of camptothecin in the early 1970s in the United States, interest in camptothecin and its analogs increased substantially when these compounds were found to target specifically the enzyme DNA topoisomerase I (*21*).

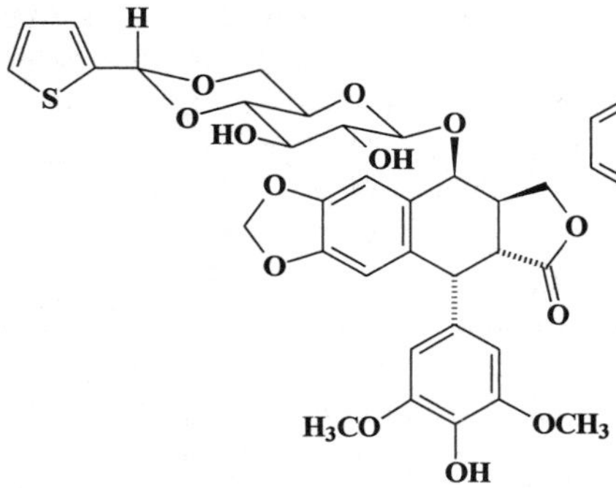

1

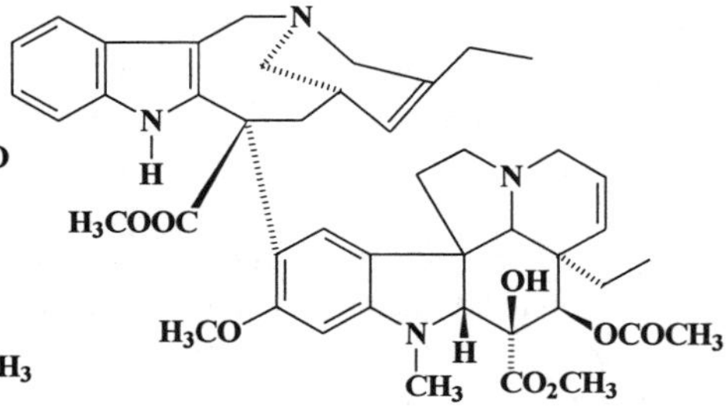

2

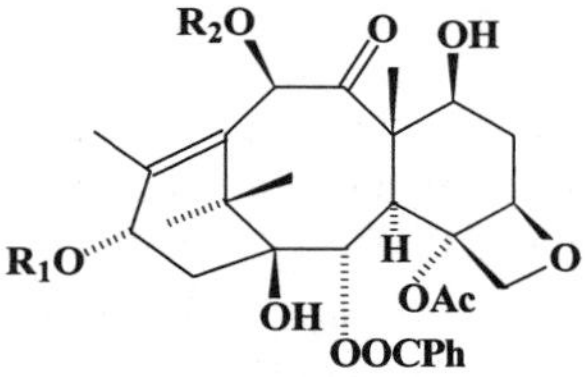

	R_1	R_2
3	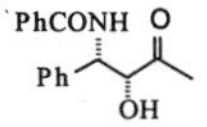	Ac
4	H	H
5	H	Ac
6	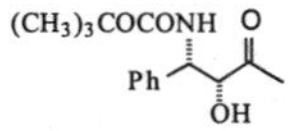	H

Ph = Phenyl

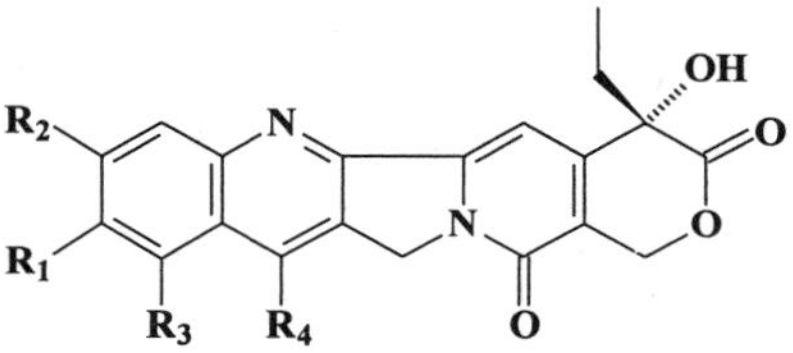

	R_1	R_2	R_3	R_4
7	H	H	H	H
8	H	H	NH_2	H
9	OH	H	$CH_2N(CH_3)_2$	H
10	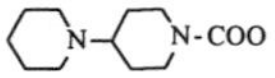	H	H	CH_2CH_3
11	$-OCH_2CH_2O-$		H	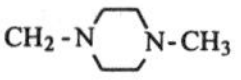

Following successful murine xenograft preclinical evaluation, at the present time camptothecin and four of its more water-soluble analogs are in clinical trials in several countries as anticancer agents, comprising 9-amino-20(*S*)-camptothecin (**8**), topotecan (**9**), irinotecan (**10**, CPT-11), and 7-[(4-methylpiperazino)methyl]-10,11-(ethylenedioxy)-20(*S*)-camptothecin (**11**), and several of the free base or salt forms of these compounds have shown significant activities in patients with refractory malignancies (*20-22*). Recently, irinotecan has been approved as an anticancer agent in Japan and France (M.C. Wani, Research Triangle Institute, Research Triangle Park, North Carolina, private communication). *C. acuminata* is native to several provinces including Szechwan in southern parts of the People's Republic of China, and requires frost-free conditions for growth (*20,21*). To meet the present demand, camptothecin is obtained from *C. acuminata* cultivated in India and the People's Republic of China (*16*). Camptothecin is extracted from *C. acuminata* cultivated at Ishigaku Island in southern Japan for the production of irinotecan after chemical modification, for use in Japan (*23*).

A group of plant-derived compounds based on the sesquiterpene lactone endoperoxide artemisinin (**12**; also known as qinghaosu), a constituent of the Chinese medicinal plant, *Artemisia annua*, offer good clinical prospects for the treatment of malaria (*24,25*). Artemisinin is itself poorly soluble in water and has only a short plasma half-life, but the sodium salt of the hemisuccinate ester (sodium artesunate, **13**) is readily water soluble. The latter compound is available in the People's Republic of China and several southeast Asian countries for the treatment of falciparum malaria. A more stable analog, also used as an antimalarial in the same countries as **13** is artemether (**14**), the methyl ether derivative of artemisinin (*24,25*). A major pharmaceutical company is planning to introduce artemether to the world market (*26*). Other derivatives of artemisinin with potential for clinical use are arteether (**15**) and sodium artelinate (**16**) (*24,25*). In terms of its production to meet the clinical demand, artemisinin is harvested from wild stands in the People's Republic of China, with varieties in Sichuan Province apparently affording yields of up to 0.5% (w/w). Plant growth improvement trials are also being undertaken in several countries, including Australia (Tasmania), India, the Netherlands, Saudi Arabia, Switzerland, the United States, and Yugoslavia (*27*).

Forskolin (**17**), a diterpenoid of the labdane type, is a consituent of the Indian plant, *Coleus forskohlii*, which has a history of use in ayurvedic medicine. This compound exhibits potent adenylate cyclase activating activity, and has been found to have hypotensive properties, with cardiotonic, platelet-aggregation inhibitory, and spasmolytic effects (*28,29*). While the biological activities of the parent compound have been known for some time, a number of forskolin analogs with potential for drug use have been developed in India (*29*). Of these, NKH 477 (**18**) is a derivative with an aminoacyl substituent that is water-soluble and is undergoing clinical trial in Japan, in the form of its hydrochloride salt, as a cardiotonic. HIL 568 (**19**) reduces the intraocular pressure of rabbits, and is being developed for the treatment of glaucoma (*29*).

Examples of Two Medicinal Plants [*Panax ginseng* (Korean Ginseng) and *Ginkgo biloba* (Ginkgo)] Used as European Phytomedicines

In addition to the above-mentioned promising pure compounds from plants with present drug use or potential for drug use, standardized extractives of plants with therapeutic efficacy are employed as in several countries in western Europe as "phytomedicines" (*30-32*). Phytomedicines, otherwise known as herbal remedies, are powdered vegetable drugs as well as various types of extracts made from powdered drugs. In Germany the Federal Health Agency (*Bundesgesundheitsamt*) has set up the so-called Commission E for Human Medicine, Section on Phytotherapy, to evaluate herbal drugs and produce monographs describing drug safety and efficacy, which are employed in the licensing of phytomedicines (*30,32*). Whereas in countries such as Germany, phytomedicines are supplied as prescription drugs, several of the same preparations are sold in the United States in health food stores without therapeutic claims (*30*). About 150 phytomedicines are listed as being beneficial by the European Scientific Cooperative on Phytotherapy (ESCOP) (*31*), and in a standard German textbook on herbal drugs recently translated into English, monographs of 181 of these drugs are provided (*32*).

Most of the European phytomedicines are produced for commercial purposes by cultivation, and to further emphasize such medicinal plants as being renewable resources obtained through agriculture, two examples have been selected, with both being of major market significance. The first of these is *Panax ginseng*, a plant which is native to eastern Asia, which has been used for thousands of years as a tonic and panacea (*32-35*). *P. ginseng* is used as a herbal tea drug in Germany as a tonic to combat feelings of lassitude and debilitation, a lack of energy and/or an inability to concentrate, and during convalescence (*32*). Ginseng is an adaptogen, and thereby enables the adaptation of the subject to external or internal disturbances (*32*). The experimental evidence for the clinical activity of ginseng is equivocal, although an immunostimulant activity has been demonstrated in laboratory animals, and it is apparent that the clinical benefits may not take place immmediately (*32)*. The active principles of *P. ginseng* are considered to be the ginsenosides, which are triterpene saponins, whose aglycones are of two major types. For example, ginsenosides Rb_1 (**20**), Rb_2 (**21**), and Rc (**22**) are based on the sapogenin, 20(*S*)-protopanaxadiol and ginsenosides Re (**23**), Rg_1 (**24**), and Rh_1 (**25**) are based on the sapogenin, 20(*S*)-protopanaxatriol (*32-34*). Many other ginsenosides have been structurally elucidated from *Panax* species, and it has been shown that non-triterpenoidal constituents of ginseng also exhibit biological activity (*32*). In Korean ginseng, ginsenosides Rb_1, Rc, and Rg_1 are the most abundant triterpene saponin constituents, but different ginsenoside profiles occur in different parts of the plant (*33*). Since the various ginsenosides have different biological activities, it is important that their concentration levels are standardized in commercially used *P. ginseng* preparations.

Panax ginseng is a perennial herb that grows best in shaded areas, and, although this is also cultivated for the drug market in Japan, the People's Republic of China, and Russia, much of the worldwide production is obtained from Korea, where modern farming and factory procedures are employed. Commercial Korean ginseng is now

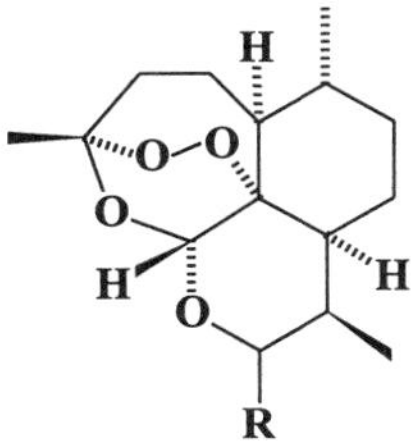

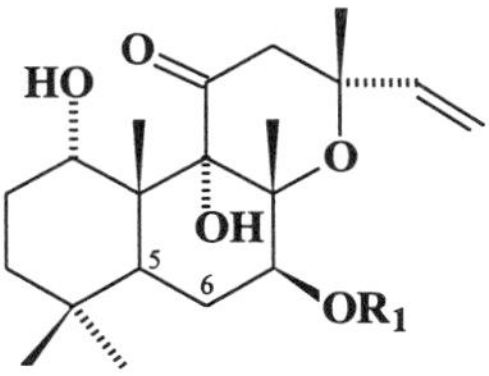

	R
12	=O
13	β-$OCOCH_2CH_2CO_2Na$
14	β-OCH_3
15	β-OCH_2CH_3
16	β-$OCH_2C_6H_4CO_2Na$

	R_1	Other
17	$COCH_3$	5α–H, 6β-OH
18	$COCH_3$	5α–H, 6β-$OCO(CH_2)_2N(CH_3)_2$
19	$CONHCH_3$	$\Delta^{5,6}$

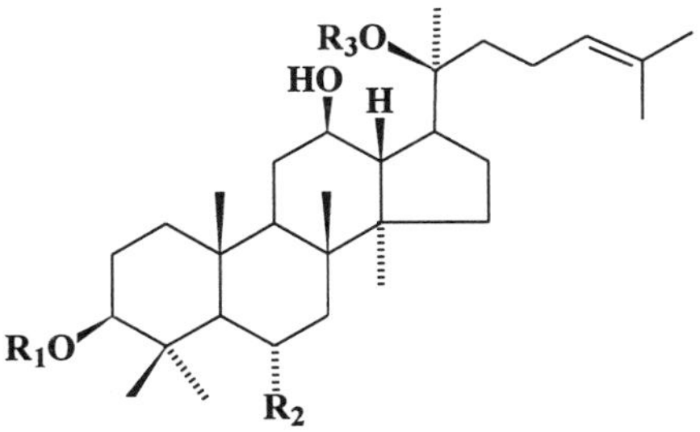

20(*S*)-Protopanaxadiol		
R_1 = –Glc2-Glc, R_2 = H		
		R_3
20	Rb_1	–Glc6-Glc
21	Rb_2	–Glc6-Ara(p)
22	Rc	–Glc6-Ara(f)

20(*S*)-Protopanaxatriol			
R_1 = -H, R_2 = $-OR_4$			
		R_3	R_4
23	Re	–Glc	–Glc2-Rha
24	Rg_1	–Glc	–Glc
25	Rh_1	–H	–Glc

Glc = β-D-glucopyranoside Ara (f) = α-L-arabinofuranoside
Ara (p) = α-L-arabinopyranoside Rha = α-L-rhamnopyranoside

produced almost entirely by organized cultivation (*33*). To cultivate ginseng of high quality, rather strict conditions must be adhered to. The annual average temperature should be in the range 0.9-13.8 °C, and in the summer, 20-25°, with a rainfall of 1,200 mm/year, and little or no snow. Dispersed sunlight is required (1/8 to 1/13 of direct sunlight), and the soil should be at pH 5.5-6.0, and free of insect infestation. Hilly exposures to the north or northeast are preferable, but flat locations can be used for ginseng cultivation if the drainage is good (*36*). The aerial parts of the plant die back each year, and the roots are harvested in September of the fourth through the sixth year of growth, depending upon the quality of ginseng desired (*33,36*). Two major ginseng products are produced in Korea, namely, red ginseng and white ginseng, which are prepared by different methods of drying (*33,37*). After six years of growth, the fleshy roots of *P. ginseng* are about 35 cm long, of weight usually in the range 70-100 g (up to 300 g has been recorded), and the diameter of the roots is about 3 cm (*36*). The roots are dried in the shade, washed, and cut lengthwise into two. Red ginseng is prepared by wrapping with a cloth, steaming for about three hours, and drying in the shade (one day) and with hot air (20 days). White ginseng is dried in the shade (ten days) and with hot air (20 days) (*36*). The processed material is sold commercially in the form of whole or powdered roots, and as tablets, capsules, tea bags, oils, and extractives (*33*). Other varieties of ginseng are Chinese, Japanese, and American, representing a total of three to six distinct species of *Panax* (*33*). American ginseng (*Panax quinquefolius*), for example, is an important specialty crop in Wisconsin, of value $50 million per year (*38*).

Of all the plants used as phytomedicines in Europe, probably the most important one is *Ginkgo biloba* (Ginkgo), for which the leaves are supplied as a standardized extract (EGb 761) used to treat cerebral blood flow insufficiency and other conditions (*30,39*). The gingko market in Europe is large, and in a recent year amounted to some $500 U.S. (*39*). Ginkgo leaf preparations are available in at least 30 countries (*40*). *G. biloba* is a gymnosperm native to the People's Republic of China and Japan that is one of the oldest known species on earth. The plants are extremely long-lived, and may grow as trees to more than 30 meters tall. While therapy with Ginkgo preparations can be traced back to Chinese traditional medicine, the more modern usage began in Germany in the mid-1960s (*40*). EGb 761 is an extractive of *G. biloba* leaves produced with acetone and water, followed by the removal of strongly lipophilic compounds and condensed polyphenols, and it has been suggested that it is the combined effects of the various leaf secondary metabolite constituents which are responsible for the observed clinical activities of Ginkgo. These constituents are complex and embrace various terpenoid lactones [e.g., ginkgolides A-C, J and M (**26-30**) and bilobalide (**31**)], biflavonoids (e.g., armentoflavone, **32**; bilobetin, **33**; ginkgetin, **34**; isoginkgetin, **35**; sciadopitysin, **36**; 5′-methoxybilobetin, **37**), as well as flavonol glycosides (e.g., mono-, di-, and triglycosides of kaempferol, isorhamnetin, and quercetin), coumaric acid esters of flavonols, and other phenols (*39-41*). The standard plant extract EGb 761 contains 24% w/w of flavonoids and 6% w/w of terpene lactones (*39*). Short reviews have appeared recently on the botany (*42*), constituents (*41*), pharmacology (*43*), and standardization (*39*) of *G. biloba*. The ginkgolides, which are unusual diterpenoids possessing six five-membered rings and a *tert*-butyl group that are found only in *G.*

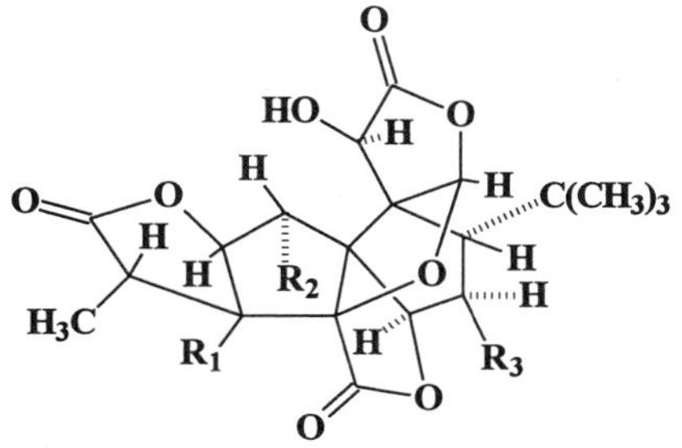

Ginkgolide		R_1	R_2	R_3
26	A	OH	H	H
27	B	OH	OH	H
28	C	OH	OH	OH
29	J	OH	H	OH
30	M	H	OH	OH

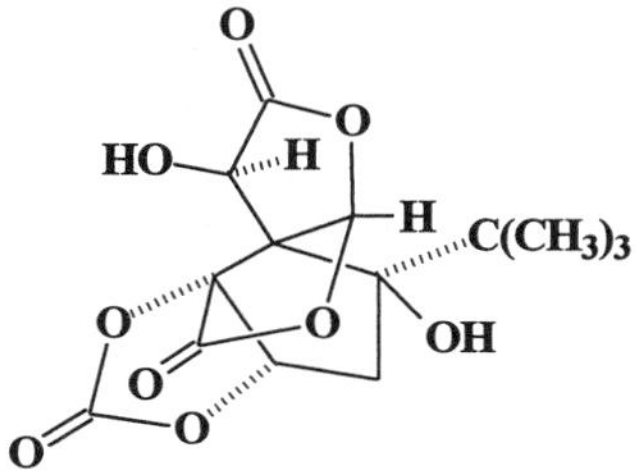

31 Bilobalide

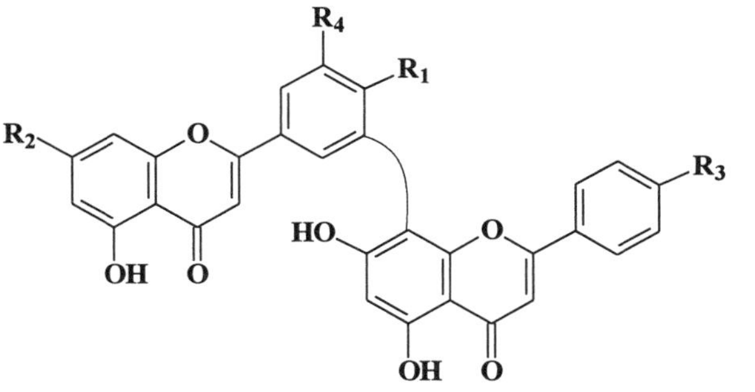

	R_1	R_2	R_3	R_4
32	H	H	H	H
33	OCH_3	OH	OH	H
34	OCH_3	OCH_3	OH	H
35	OCH_3	OH	OCH_3	H
36	OCH_3	OCH_3	OCH_3	H
37	OCH_3	OH	OH	OCH_3

biloba, are potent inhibitors of platelet-activating factor (PAF), a substance involved in the process of inflammation (*44*). The range of clinical applications for which extract EGb 761 is officially used in Europe includes headache, hearing loss, mood disturbance, short-term memory loss, tinnitus, and vertigo, when associated with degenerative angiopathy and cerebrovascular insufficiency (*39,40*). In addition to their prescription use in Europe, Ginkgo preparations with low terpene lactone and flavonoid glycoside contents are used in certain countries for self-medication to treat symptoms such as loss of concentration and memory, reduced mental and physical efficiency, and vertigo (*39*).

It is of interest to note that large amounts of the *Ginkgo biloba* leaves currently used in European phytomedicines are cultivated in the United States, at a 1,200-acre plantation in Sumter, South Carolina. Cultivation of *G. biloba* began at this site in 1982, and seedlings are planted in 40 inch rows with about 10,000 plants per acre. The seedlings are in turn grown from seed purchased in Korea, Japan, or the People's Republic of China, and cultivated in a nursery for two years. More mature plants, which are from six to 13 years old when harvested, are pruned annually, and shaped into a bush or shrub form. They are appropriately irrigated and fertilized, and are resistant to insect attack. The leaves of the plant are harvested mechanically in July through September of each year, and after being artificially dried and then baled, they are transported to Europe for further processing (C.H. Johnson, Garnay, Inc., Sumter, South Carolina, private communication).

Summary and Conclusions

Plants have long been important sources of medicinal agents, and their secondary constituents have served to provide chemical ideas for the molecular design of many synthetic drugs. The prospects for the discovery of further drugs from plants has recently attracted significant attention from major scientific publications outside this specialist field (*45,46*). This wider interest in plant drugs has been attributed to the undisputed efficacy of the plant-derived drugs already known, their frequent novel mechanisms of biochemical action, their value in drug semi-synthesis, and their use as extractives in phytomedicine (*46*). Because of the well-established chemical and biological diversity of plant constituents, coupled with the present availability of high-throughput bioassays and sensitive methods for compound structure elucidation, there has been an increased effort in plant drug discovery in industrial, academic, and industrial laboratories in many countries (e.g., *5,46,47*). However, despite the sophistication of modern organic synthesis, it is not always economically feasible to produce plant secondary metabolite drugs commercially by synthesis. Accordingly, most plant drugs are currently produced by cultivation, and may be used clinically either as semi-purified extractives or as pure crystalline substances. The cultivation of these plants as renewable resources tends to provide more stable supplies of plant drugs than collection in the wild. It has recently been suggested, however, that in future, plant sources of new drugs should be cultivated in plantations in the countries where they were originally collected, in order to return some of the revenues that their commercial exploitation will generate (*9*).

Acknowledgment

The authors are grateful to Dr. Cecil H. Johnson, Garnay, Inc., Industrial Farming, Sumter, South Carolina, for providing certain unpublished data on the cultivation of *Ginkgo biloba*.

Literature Cited

1. Farnsworth, N. R.; Akerele, O.; Bingel, A. S.; Soejarto, D. D.; Guo, Z. *Bull WHO* **1985**, *63*, 965-981.
2. Balandrin, M. F.; Klocke, J. A.; Wurtele, E. S.; Bollinger, W. H. *Science* **1985**, *228*, 1154-1160.
3. Tyler, V. E.; Brady, L. R.; Robbers, J. E. *Pharmacognosy*; 9th ed.; Lea and Febiger: Philadelphia, PA, 1988.
4. Farnsworth, N. R.; Soejarto, D. D. In *The Conservation of Medicinal Plants*; Akerele, O.; Heywood, V.; Synge, H., Eds.; Cambridge University Press: Cambridge, U.K., 1991; pp 25-51.
5. *Human Medicinal Agents from Plants*; Kinghorn, A. D.; Balandrin, M. F. Eds.; Sym. Ser. 534; American Chemical Society: Washington, DC, 1993.
6. Bruneton, J. *Pharmacognosy, Phytochemistry, Medicinal Plants*; Lavoisier Publishing, Inc.: Secaucus, NY, 1995.
7. Balandrin, M. F.; Kinghorn, A. D.; Farnsworth, N. R. In *Human Medicinal Agents from Plants*; Kinghorn, A. D.; Balandrin, M. F., Eds.; Symp. Ser. 534; American Chemical Society: Washington, DC, 1993; pp 2-12.
8. Towers, G. H. N.; Ellis, S. In *Human Medicinal Agents from Plants*; Kinghorn, A. D.; Balandrin, M. F., Eds.; Symp. Ser. 534; American Chemical Society: Washington, DC, 1993; pp 56-78.
9. Baker, J. T.; Borris, R. P.; Carté, B.; Cordell, G. A.; Soejarto, D. D.; Cragg, G. M.; Gupta, M. P.; Iwu, M. M.; Madulid, D. R.; Tyler, V. E. *J. Nat. Prod.* **1995**, *58*, 1325-1357.
10. Mendelsohn, R.; Balick, M. J. *Econ. Bot.* **1995**, *49*, 223-228.
11. Cragg, G. M.; Boyd, M. R.; Cardellina II, J. H.; Grever, M. R.; Schepartz, S. A.; Snader, K. M.; Suffness, M. In *Human Medicinal Agents from Plants*; Kinghorn, A. D.; Balandrin, M. F., Eds.; Symp. Ser. 534; American Chemical Society: Washington, DC, 1993; pp 80-95.
12. *Physician's Desk Reference*; 49th Edn.; Medical Economics Data Production Co.: Montvale, NJ, 1995.
13. Mancano, M. A. *Pharmacy Times* **1995**, March issue, pp 23-43.
14. *Taxane Anticancer Agents: Basic Science and Current Status*; Georg, G. I.; Chen, T. T.; Ojima, I.; Vyas, D. M., Eds.; Symp. Ser. 583; American Chemical Society: Washington, DC, 1995.
15. *TAXOL®: Science and Applications*; Suffness, M., Ed.; CRC Press: Boca Raton, FL, 1995.
16. Cragg, G. M.; Schpartz, S. A.; Suffness, M.; Grever, M. R. *J. Nat. Prod.* **1993**, *56*, 1657-1668.

17. Anonymous. *Chem. Eng. News* **1994**, December 19 issue, p 32.
18. Donovan, T. A. *Taxane J.*, **1995**, *1*, 28-41.
19. Holland, K. *Pharm. J.* **1994**, *253*, 645-649.
20. Wall, M. E.; Wani, M. C. In *Human Medicinal Agents from Plants*; Kinghorn, A. D.; Balandrin, M. F., Eds.; Symp. Ser. 534; American Chemical Society: Washington, DC, 1993; pp 149-169.
21. *Camptothecins: New Anticancer Agents*; Potmesil, M.; Pinedo, H., Eds.; CRC Press: Boca Raton, FL, 1995.
22. Lizzio, M. J.; Besterman, J. M.; Emerson, D. L.; Evans, M. G.; Lackey, K.; Leitner, P. L.; McIntyre, G.; Morton, B.; Myers, P. L.; Sisco, J. M.; Sternbach, D. D.; Tong, W.-Q.; Truesdale, A.; Uehling, D. E.; Vuong, A.; Yates, A. *J. Med. Chem.* **1995**, *38*, 395-401.
23. Sankawa, U. In *Trends in Traditional Medicine Research*; Chan, K. L.; Hussin, A. H.; Sadikun, A.; Yuen, K. H.; Asmawi, M. Z.; Ismail, Z., Eds.; The School of Pharmaceutical Sciences, University of Science Malaysia: Penang, Malaysia, 1995; pp 13-16.
24. Klayman, D. L. In *Human Medicinal Agents from Plants*; Kinghorn, A. D.; Balandrin, M. F., Eds.; Symp. Ser. 534; American Chemical Society: Washington, DC, 1993; pp 242-255.
25. White, N. J. *Trans. Roy. Soc. Trop. Med. Hyg.* **1994**, *88* (Suppl. 1), 3-4.
26. Roche, G.; Helenport, J.-P. *Trans. Roy. Soc. Trop. Med. Hyg.* **1994**, *88* (Suppl. 1), 57-58.
27. Laughlin, J. C. *Trans. Roy. Soc. Trop. Med. Hyg.* **1994**, *88* (Suppl. 1), 21-22.
28. Ammon, H. P. T.; Müller, A. B. *Planta Med.* **1985**, *51*, 473-477.
29. de Souza, N. J. In *Human Medicinal Agents from Plants*; Kinghorn, A. D.; Balandrin, M. F., Eds.; Symp. Ser. 534; American Chemical Society: Washington, DC, 1993; pp 331-340.
30. Tyler, V. E. In *Human Medicinal Agents from Plants*; Kinghorn, A. D.; Balandrin, M. F., Eds.; Symp. Ser. 534; American Chemical Society: Washington, DC, 1993; pp 25-37.
31. Anonymous. *Phytomedicine* **1994**, *1*, 173-176.
32. *Herbal Drugs and Phytopharmaceuticals. A Handbook for Practice on a Scientific Basis*; Bisset, N. G., Ed.; CRC Press: Boca Raton, FL, 1994.
33. Phillipson, J. D.; Anderson, L. A. *Pharm. J.* **1984**, *232*, 161-165.
34. Shibata, S.; Tanaka, O.; Shoji, J.; Saito, H. In *Economic and Medicinal Plant Research*; Wagner, H.; Hikino, H.; Farnsworth, N. R., Eds.; Academic Press: London, 1985, Vol. 1; pp 217-284.
35. Rahman, A.; Houghton, P. *Pharm. J.* **1995**, *254*, 150-152.
36. Anonymous. Korean Ginseng and Tobacco Research Institute: Tajeon, Korea, 1989; pp. 12, 14-15, 71-72.
37. Kitagawa, I.; Taniyama, T.; Shibuya, H.; Noda. T.; Yoshikawa, M. *Yakugaku Zasshi* **1987**, *107*, 495-505.
38. Walters, C. *Acres* **1995**, June issue, 1ff-2ff.
39. Sticher, O. *Planta Med.* **1993**, *59*, 2-11.

40. DeFeudis, F.V. Ginkgo biloba *Extract (EGb 761): Pharmacological Activities and Clinical Applications*; Elsevier: Paris, 1991.
41. Hölzl, J. *Pharm. Uns. Z.* **1992**, *5*, 215-223.
42. Melzheimer, V. *Pharm. Uns. Z.* **1992**, *5*, 206-214.
43. Oberpichler-Schwenk, H.; Kriegelstein, J. *Pharm. Uns. Z.* **1992**, *5*, 224-235.
44. Braquet, P; Hosford, D. *J. Ethnopharmacol.* **1991**, *32*, 135-139.
45. Abelson, P. H. *Science* **1990**, *247*, 513.
46. Anonymous. *Lancet* **1994**, *343*, 1513-1515.
47. McChesney, J. D.; Clark, A. M. In *Emerging Technologies for Materials and Chemicals for Biomass*; Rowell, R. M.; Schultz, T. P.; Narayan, R., Eds.; Symp. Ser. 476; American Chemical Society: Washington, DC, 1992; pp 437-451

Chapter 13

Application of Transgenic Plants as Production Systems for Pharmaceuticals

G. D. May[1], H. S. Mason[1], and P. C. Lyons[2]

[1]Plants and Human Health Program, Boyce Thompson Institute for Plant Research, Cornell University, Tower Road, Ithaca, NY 14853–1801

[2]Institute of Biosciences and Technology, Texas A&M University, Houston, TX 77030–3303

Transgenic plants that produce foreign proteins or secondary metabolites with intrinsic industrial or pharmaceutical value represent a cost-effective alternative to fermentation- or organic synthesis-based production systems. There have been a number of recent demonstrations that transgenic plants are capable of producing functional foreign proteins, or that secondary metabolic pathways can be altered for the production of proteins or compounds of known or potential pharmaceutical importance. These examples portend many new and exciting possibilities in the ancient science of plant medicinal chemistry. Here we discuss current progress in the production of phamacologically important proteins and secondary metabolites in transgenic plants.

Genetic transformation techniques are now applied routinely to a variety of plant species. Through this technology, numerous genes that confer agronomically important traits such as pest resistance or herbicide tolerance have been introduced into plants. The successes of modern plant biotechnology exemplified by these achievements in crop improvement have created considerable interest in further exploiting the remarkable biosynthetic capacity of plants by developing transgenic plants that will produce valuable new products. With a number of recent demonstrations that transgenic plants are capable of producing functional foreign proteins and peptides of known or potential pharmaceutical importance, new and exciting possibilities have been created in the ancient science of plant medicinal chemistry.

Biologically active peptides and proteins have many potential pharmaceutical applications, including use as vaccines, immunomodulators, growth factors, hormones, blood proteins, inhibitors, and enzymes. Efforts to produce these compounds economically and in adequate quantities have become increasingly reliant upon the use of various prokaryotic and eukaryotic cell culture expression systems. In this article, we will consider the use of transgenic plants as an alternative eukaryotic expression system for the production of compounds or recombinant protein pharmaceuticals, and the advantages that plants may have over other more traditional expression systems.

0097–6156/96/0647–0194$15.00/0

Production of Recombinant Proteins in Transgenic Plants

For many reasons transgenic plants are a feasible, and in some cases preferable, eukaryotic expression system for the production of valuable pharmaceuticals. The number of plant species amenable to genetic transformation is now quite large (1,2). These plants have the capacity to express foreign genes from a wide range of sources, including viral, bacterial, fungal, insect, animal, and other plant species. In addition, plants are also capable of high-levels of protein production; foreign protein production levels as high as 30% of the total soluble protein have been reported in transgenic plants (3) and expression levels in excess of 1% are often obtainable (4-8). With rapid advances being made in the manipulation of foreign protein expression through the development of novel vectors and through improved understanding of folding, assembly and protein processing in plants, one can expect a more predictable and higher level of production of a wide variety of functional proteins and peptides.

Plants have an advantage over other protein production systems in the low cost involved in growing large amounts of biomass. One of the traditional drawbacks of plants (difficulty in achieving adequately purified protein) has become less problematical with the advancements made in protein purification technology. Furthermore, should it prove feasible to utilize transgenic plants as an edible source for the oral delivery of recombinant pharmaceuticals, as some evidence already suggests (9,10), the need for costly purification and delivery procedures would, in such instances, be eliminated.

Plant cells perform complex post-translational modifications (including cleavage, glycolsylation, etc.). These capabilities are often needed for proteins of pharmaceutical value, and are not possible in prokaryotic expression systems. For example, attempts to produce mammalian proteins in prokaryotic cells may result in the production of a protein that does not retain its biological activity. In addition, plants may sometimes be more tolerant to the alteration of biosynthetic pathways or the addition of protein modifying enzymes, when the new gene products are sequestered in sub-cellular compartments.

Foreign proteins can be expressed either transiently or as stably inherited traits (1,2,10). Both approaches have been applied to the production of protein pharmaceuticals. Transient expression of foreign proteins in plants is most often achieved by infection of the host plant with a genetically modified plant virus (10). This infection allows the production of a foreign or fusion protein, typically consisting of the coat protein of the virus fused to a small foreign peptide sequence. Conversely, stable transformation has been achieved through incorporation of the foreign gene(s) encoding the protein(s) of interest into the plant genome *via* *Agrobacterium*-mediated transformation (1,2,10).

Recombinant Vaccines

A variety of recombinant vaccine antigens produced in plants through either stable or transient foreign gene expression have been reported. While viral diseases are the most frequent target for plant-recombinant vaccines, other non-viral antigenic proteins have also been produced.

Vaccines Against Viral Diseases

Hepatitis B Surface Antigen. The first viral vaccine protein produced in transgenic plants was the hepatitis B surface antigen (HBsAg) (11). Plant-derived recombinant HBsAg (rHBsAg) was expressed in transgenic tobacco plants (an important model plant species). Although the expression levels were low in these experiments (~0.01% of total soluble protein), it was shown that the plant-derived rHBsAg assembles into virus-like particles (VLPs). These particles were similar in size and other physical properties to yeast derived rHBsAg, the commercial source of hepatitis B vaccine (sold as Recombivax™ -- Merck, Sharpe & Dohme). Furthermore, studies performed with tobacco-derived rHBsAg provided the first evidence that a plant-derived recombinant vaccine protein is immunogenic (12). When a partially purified preparation of the tobacco-derived rHBsAg was used for parenteral immunization of mice, it resulted in an antibody response that mimicked the response obtained with the commercial Recombivax™ vaccine. In addition, lymph node T-cells isolated from mice which were primed with the tobacco-derived rHBsAg could be stimulated to proliferate *in vitro* by both tobacco- and yeast-derived rHBsAg. These experiments demonstrate the close antigenic relatedness of the plant-recombinant protein to a known, highly effective recombinant vaccine produced in another eukaryotic expression system.

Norwalk virus capsid protein. Another potential vaccine which has been expressed in transgenic plants is the Norwalk virus capsid protein (NVCP) (Mason, H. S.; et. al., *PNAS*, in press.). Norwalk virus causes epidemic acute gastroenteritis in humans. Recombinant NVCP (rNV) has been expressed stably in both tobacco leaves and potato tubers and self-assembles into VLPs. Expression and self-assembly of the rNV had first been achieved in recombinant baculovirus-infected insect cells (13) and this material has shown promise as an oral vaccine. In plant cells, the accumulation of rNV was nearly 0.3% of the total protein; for potatoes, this resulted in a yield of about 20 μg rNV per gram of tuber. Feeding mice with transgenic potato tubers which express rNV induced production of both humoral and mucosal anti-Norwalk virus antibodies. These results demonstrate that a plant-derived recombinant subunit antigen will cause oral immunization when consumed as a food source, thus lending viability to the concept of using plants as a source of edible vaccines.

Viral Peptide epitopes. Peptide epitopes of three different viruses have been expressed transiently in plants using the cowpea mosaic virus (CPMV) as an expression system: foot-and-mouth disease virus (14), human rhinovirus, and human immunodeficiency virus (15) epitopes have all been expressed in plants using this viral vector. Viral epitopes, up to 30 amino acids in length, were expressed as chimeras in the small coat protein (which contains a well-exposed surface loop) of the virus. CPMV has the advantages of producing high yields (2g of virus per kg of host tissue), and of being thermostable and easily purified. Antibodies raised against a CPMV-HIV chimera were able to neutralize three different strains of HIV (L. McClain, L.; et al., *AIDS Res. Human Retrovirus*, in press) indicating that epitopes expressed in this manner also can be immunogenic.

Vaccines Against Non-Viral Diseases

Non-Viral Peptide epitopes. Non-viral epitopes with the potential for use as vaccines have also been expressed in plants infected with recombinant virus. Epitopes derived from malarial sporozoites (16) and the zona pellucida ZP3 protein of mammalian oocytes (Fitchen, J.; et al., *Vaccine,* in press), a potential target for immuno-contraception, were expressed transiently as fusion proteins with the tobacco mosaic virus (TMV) capsid protein. Transiently expressed malaria epitope-fusion proteins were recognized by monoclonal anti-sporozoite antibodies, demonstrating their antigenicity. Antibodies to the zona pellucida ZP3 fusion protein produced in immunized mice were shown to be recruited to the zona pellucida of the these mice, although as of yet, no effect on the fertility of these mice has been observed.

Bacterial Antigenic Proteins. Potential vaccines for bacterial diseases have also been produced in transgenic plants. Indeed, the first report of a vaccine antigen produced in plants was presented in a patent application (17) that described a means to produce a surface protein (spaA) from the bacterium *Streptococcus mutans* by stable expression in tobacco. Unfortunately, no further data have been presented regarding this system to indicate whether the plant-derived protein is functional as a vaccine.

Bacteria that produce enterotoxins cause diarrheal diseases that are an important cause of mortality in developing countries. An oral vaccine composed of the cholera toxin B subunit (CT-B) and killed *Vibrio cholerae* cells has been shown to provide protection against cholera and related enterotoxigenic *Escherichia coli* (18). The heat-labile *E. coli* enterotoxin (LT) is structurally, functionally and antigenically very similar to cholera toxin (CT). The B subunit of LT (LT-B) has been expressed in transgenic tobacco and potato plants (19), where it appears to have at least partially assembled into the pentameric form that binds to the gangliosides on epithelial cells. When given orally to mice, the plant-derived antigen stimulated both humoral and mucosal immune responses with antibody titers comparable to those obtained with bacterial-derived LT-B. Antibodies produced against the tobacco-derived LT-B were able to neutralize LT activity, indicating the potential protective value of the observed immune response. Feeding mice with potato tubers expressing the recombinant LT-B also caused both mucosal and serum antibodies to be produced in mice, thus providing additional evidence that food sources containing a foreign antigen can cause oral immunization.

Bioactive Peptides

In addition to the classes of proteins detailed above, smaller bioactive peptides with a variety of other potential pharmaceutical applications have been expressed in transgenic plants. Stable expression of a synthetic gene coding for the human epidermal growth factor (hEGF), a small mitogenic peptide which stimulates *in vitro* proliferation of animal cells, was achieved in transgenic tobacco (20). Although native hEGF is processed proteolytically in human cells from a larger precursor, the synthetic gene encodes only the active peptide portion. Nevertheless, incorporation of this synthetic gene into the plant genome resulted in expression of a peptide that reacted with hEGF-specific antibody. Unfortunately, the highest measurable level of hEGF produced in transgenic plants was only 0.001% of total soluble proteins.

Recombinant enkephalin has been produced both in tobacco protoplasts (21) and in regenerated transgenic tobacco plants (22). In protoplasts, enkephalin was expressed as a fusion protein with the TMV coat protein. Yields of plant-recombinant leu-enkaphalin, a pentapeptide from brain that exhibits opiate activity, were considerably higher than hEGF (22). This peptide was produced in *Arabidopsis thaliana* and *Brassica napus* (canola or oilseed rape) as a fusion protein with a 2S albumin seed storage protein. Since 2S albumins represent up to 60% of total seed protein, in some plant species, expression of foreign proteins in this manner should greatly facilitate protein purification. Using a 2S albumin-rich fraction from either *Arabidopsis* or canola, it was possible to obtain the purified peptide by tryptic digestion followed separation on HPLC. It has been estimated that the yield obtained from *B. napus* would be equivalent to 15-75 g of peptide per hectare. Furthermore, the authors of this study suggested that by screening individual transformants for higher expression levels, by using lines of oilseed rape with higher protein:oil ratios, and by utilizing standard breeding techniques, yields of the pentapeptide could be increased substantially.

Human lysosomal enzyme glucocerebrosidase (acid β-glucosidase; hGC) has been produced in transgenic tobacco plants (Cramer, C. L., CropTech Development Corp., Virginia Tech Corporate Research Center, unpublished data.). These researchers generated transgenic tobacco plants expressing the human glucocerebrosidase cDNA under the regulatory control of an inducible promoter system. It was reported that hGC accumulated to levels greater than one milligram of protein per gram of leaf tissue, and that plant-derived hGC is glycosylated and co-migrates with the placental-derived protein during electorphoresis. Human GC is administered intravenously to patients in the treatment of Gaucher disease and is currently derived from human placentae or CHO cells. Declared the "world's most expensive drug" by the media, hGC is an ideal candidate for production in transgenic plants as a means to provide a safe, low-cost source of this enzyme.

A recombinant inhibitor for the 12-peptide angiotensin-I-converting enzyme was transiently expressed in tobacco and tomato (9). The native inhibitor can be found in the tryptic hydrolysate of milk and has anti-hypertensive effects when orally administered. In plants, the inhibitor was expressed transiently as a fusion protein with the TMV coat protein. Yield of the fusion protein in tomato fruit was approximately 10 μg fusion protein per g fresh weight of plant tissue. It has been suggested that the tomato fruit could be administered as a dietary antihypertensive since the peptide inhibitor could be released from the fusion protein in the intestine by trypsin digestion.

Recombinant Toxins

Many toxins are proteins, usually derived from bacteria or plants, which kill cells by interfering with metabolism, often by the inhibition of protein synthesis. Due to the extreme cytotoxicity of many of these proteins, they have been the subject of intense investigation as tumor controlling agents. Most toxins exhibit very little site specificity, and therefore, in order to be used therapeutically, they require modification of the binding domains so that they will be preferentially directed to the appropriate target site (23). This modification usually involves inactivation or removal of the binding domain and expression of the toxin as a fusion protein with a carrier that can deliver the toxin to the preferred site of action.

Two recombinant toxins have been expressed in transgenic plants (4,7), both of which are plant-derived eukaryotic ribosome-inactivating proteins (RIPs). The first of these, α-trichosanthin, a 27 kDa protein from *Trichosanthes kirilowii*, inhibits the replication of HIV in acutely infected $CD4^+$ lymphoid cells and in chronically infected macrophages. Although α-trichosanthin had been expressed previously in *E. coli*, the amount of protein recovered was low (less than 0.01% of total cellular protein). Transient expression of this protein in tobacco plants using recombinant TMV as the vector resulted in accumulation of α-trichosanthin in leaves to a level of at least 2% of total soluble protein (4).

Ricin is a RIP from castor bean with potential therapeutic use in the treatment of AIDS and cancer. Development of recombinant toxins utilizing ricin has been hampered by low expression levels and by difficulties in achieving proper folding and processing in several expression systems. However, in transgenic tobacco, stable expression of active, processed recombinant ricin was obtained (7). As with the transiently expressed α-trichosanthin, levels of the stably expressed recombinant ricin reached approximately 2% of the total soluble protein. The results obtained with these two toxins, together with the successful expression of fusion proteins detailed elsewhere in this review, suggest that transgenic plants may be useful for the production of site-selective toxins obtained by fusing the toxin to a targeting protein as suggested above.

Plantibodies (Recombinant Antibodies)

Antibodies, or fragments of antibodies, have important therapeutic applications and are produced as recombinant molecules in mammalian, bacterial, yeast, and plant expression systems. The first report of antibody production in transgenic plants (24) involved stable expression of a complete and assembled catalytic monoclonal IgG1 antibody which binds a low-molecular weight phosphonate ester and catalyzes the hydrolysis of certain carboxylic esters. To achieve assembly of the antibody, the γ- and κ-chain of this antibody were each expressed separately in transgenic tobacco plants. When transgenic plants expressing either the γ- or κ- IgG1 chains were genetically crossed, a portion of progeny from these crosses were obtained that expressed both chains. The two chains were assembled in these progeny to yield gamma-kappa complexes that exhibit an affinity to the phosphonate ester that equals that of the hybridoma-derived antibody. Interestingly however, although the recombinant protein levels were low in plants expressing only a single γ- or κ- chain, production levels of the assembled antibody in progeny expressing both chains were as great as 1.3% of the total soluble protein. Thus, it appears that assembly of antibody complexes may also have resulted in increased stability of the recombinant molecules.

A similar approach to whole antibody production in plants was utilized to express a bivalent IgG antibody (25). The monoclonal antibody (mAb) Guy's 13 is a mouse IgG1 class of antibody which recognizes a cell-surface protein of *Streptococcus mutans* -- the principal cause of dental caries in humans. Topical immunotherapy with mAb Guy's 13 prevents colonization of this bacterium. In this example, the antigen-binding region of mAb Guy's 13 was conserved in the expression vector, but the IgG constant regions were replaced with domains from an IgA heavy chain since passive immunotherapy with secretory IgA has proved more effective than with IgG. As with the example above, the light and heavy chains were first expressed separately in tobacco plants and progeny from genetic crosses made between plants expressing the different chains were screened to identify those producing both chains. Even though the heavy chain had been modified to contain the IgA domain, the two chains expressed in these progeny assemble into functional bivalent antigen-binding molecules that recognize the native antigen from *S. mutans*, and cause aggregation of cells of this bacterium.

This approach for combining multiple recombinant genes in the same plant was used (26) to simultaneously express as many as four distinct proteins in the same transgenic tobacco plant. Remarkably, all four proteins competently assembled into a functional secretory IgA-IgG antibody to a cell-surface protein of *S. mutans*. It has been suggested by the authors that it may be possible to formulate these 'plantibodies' into toothpaste that would aid in the prevention tooth decay.

To circumvent the difficulties involved in genetically crossing and selecting for progeny as entailed in the procedures described above, other approaches have been utilized to produce antibodies in plants. In one such approach (27), an expression vector was designed that contained chimeric genes of both the light and heavy chains of mAb B 1-8 fused with a plant signal-peptide. However, both genes for the light and heavy chains were each placed under the control of different regulatory elements. Tobacco plants transformed with this vector produced both the light and heavy chains which assemble into immunologically detectable antibodies that bind to the appropriate hapten. The level of expression, however, was not reported in this study.

Another approach first developed for the production of antibodies in *E. coli* (28) has been adapted to transgenic plants. This approach involves expression of single-chain Fv antibodies in which the light and heavy chain variable domains of an immunoglobulin are fused together by a flexible peptide linker. This flexible linker facilitates folding and assembly of the light and heavy chains resulting in a functional synthetic antibody. The first instance in which this approach was used in plants involved the expression of a functional anti-phytochrome single-chain Fv protein in transgenic tobacco (29). This approach was also successfully used in tobacco to express an antibody which recognizes a coat protein of the artichoke mottled crinkle virus (30). Production of this antibody resulted in a reduced incidence of infection and delayed symptom development in transgenic tobacco plants inoculated with this virus. In both of the above examples, expression levels of the single-chain Fv antibodies were approximately 0.1% of the total soluble leaf protein.

One further approach utilized for antibody production in plants involved the synthesis of a general purpose expression vector containing a multiple cloning site that facilitates the insertion of heavy-chain variable (VH) domains (31). VH domains expressed in *E. coli* have been shown to fold correctly and often retain antigen-binding activity. Using this vector, a 'single-domain antibody' consisting of the HV domain of an antibody that recognizes substance P (a neuropeptide) was expressed in transgenic tobacco and accumulated to levels of approximately 1% of the total soluble protein.

For a detailed review on the production of antibodies in plants and their uses, see (32).

Serum Proteins

A number of proteins (including coagulation and anti-coagulation factors) present in human serum are of vital importance to medicine. Due to the nature of donated plasma and the necessity for a sufficient supply of highly purified blood products, alternative production systems have been studied to meet these needs. In transgenic tobacco and potato plants, recombinant human serum albumin (rHSA) that is indistinguishable from the authentic human protein has been produced (33). The significance of this study rested not only in its demonstration of the expression of a valuable protein in transgenic plants, but also that it was possible to achieve proper processing by fusion of HSA to a plant pre-sequence that resulted in cleavage and secretion of the correct protein. This accomplishment was particularly relevant in view of the difficulties that have been encountered in other systems used to express recombinant HSA. The level of expression in transgenic potato plants was 0.02% of total soluble leaf protein.

Human protein C, a highly-modified and glycosylated serine protease zymogen which undergoes proteolytic processing, has also been produced in transgenic tobacco plants (Weissenborn, D. L., Virginia Polytechnic Institute and State University, unpublished data.). These researchers have generated transgenic tobacco plants that express a human protein C cDNA, with the resultant mRNAs being of the appropriate size. In addition, Western analysis of tobacco leaf protein extracts detect the presence of both single-chain and heavy-chain forms of the protein. Therefore, it appears that the product of the human protein C cDNA undergoes (with reduced efficiency) proper proteolytic cleavage in this plant expression system.

Plant Secondary Metabolites

Current resurgence of interest in traditional medicinal plants can benefit from cell culture and/or gene transfer technologies as well. Plants have the ability to produce a number of pharmacologically important compounds such as alkaloids, steroids, and analgesics as well as others that are too complex or expensive to produce synthetically.

However, difficulties in harvesting, or the insufficient availability of usable biomass, make some compound-producing plants species an unfeasible source of starting materials for the pharmaceutical industry. The application of cell-culture techniques to these important plant species, however, can help overcome the logistical problems associated with their use.

Through cell-suspension systems, compounds such as anthraquinones (34), benzophenanthridine alkaloids (35) and thiophenes (36) have been produced. Methods utilized to redirect transport of, or to enhance the elicitation of secondary metabolites have been reviewed (37).

Plants or plant cell cultures have the ability to tolerate new or dramatically altered biosynthetic pathways. For example, plants have been engineered to include a bacterial biosynthetic pathway that facilitates the synthesis and storage of a biodegradable thermal plastic (38). By utilizing cell-culture and gene transfer techniques, metabolic pathways can be altered to enhance production of secondary metabolites and sufficient amounts of biomass of these modified cell lines can then be made available through propagation of cell cultures. Alternatively, entirely new biosynthetic pathways can be introduced into a common crop species, such as tobacco, as a means to provide low cost production of pharmacologically important compounds.

Conclusions

It is clear from this brief review that transgenic plants are capable of synthesizing a variety of pharmacologically active recombinant proteins (see Table I). Levels of expression of these proteins have varied considerably, ranging over three orders of magnitude from about 0.001% to greater than 2% the of total soluble protein. Reports of foreign protein expression levels in plants as high 5% of the total soluble protein have been described in other instances. At these higher levels of production, plants become increasingly attractive as alternative expression systems for recombinant pharmaceuticals. Particularly encouraging is the high degree of success obtained thus far in producing active, processed forms of the expressed candidate protein pharmaceuticals in transgenic plants. Evidence of appropriate post-translational processing steps such as folding, assembly, secretion, and proper cleavage of precursor molecules indicate that even highly complex foreign proteins can be produced in plants and are likely to be functional. Although limitations exist (e.g. plants are unable to glycosylate with sialic acid) it may be possible to engineer plants with enhanced capacity for protein modification.

The prospect of delivering orally administered vaccines and other pharmaceuticals *via* edible plant tissues expressing these bioactive compounds is an appealing aspect of the use of transgenic plants. The economics and logistics of this approach are ideally suited to developing countries where transportation and an adequate cold-chain necessary for most current vaccines and many other drugs are lacking. The use of transgenic plants is, of course, not strictly limited to protein or peptide pharmaceuticals. As complex biosynthetic pathways for additional non-protein plant pharmaceuticals are characterized at the molecular level, and as knowledge of protein expression in plants continues to improve, the application of plant genetic transformation techniques should eventually lead to a pharmaceutical industry that will rely more heavily upon transgenic plants for the production of a large number of protein and non-protein drugs.

Table I. Pharmaceutical Proteins Produced in Transgenic Plants

Protein or peptide	*Application*
Hepatitis B surface antigen	Vaccine
Norwalk virus capsid protein	Vaccine
Foot-and-mouth disease virus	Vaccine
Human rhinovirus14	Vaccine
Human immunodeficiency virus	Vaccine
S. mutans surface protein	Vaccine
E. coli enterotoxin, B subunit	Vaccine
Malarial circumsporozoite epitopes	Vaccine
Mouse ZP3 protein epitope	Vaccine
Mouse catalytic antibody 6D4	Antibody
Mouse mAB Guy's 13	Antibody
mAB B 1-8	Antibody
anti-phytochrome Fv protein	Antibody
anti-substance P	Antibody
Human serum albumin	Serum protein
Human protein C	Serum protein
a-trichosanthin	Cytotoxin
Ricin	Cytotoxin
Human epidermal growth factor	Growth factor
Leu-enkephalin	Neuropeptide
Human acid β-glucosidase (hGC)	Enzyme

Literature Cited

1. Gasser, C. S.; Fraley, R. T. *Sci. Amer.* **1992**, *266*, 62.
2. Uchimaya, H.; Handa, T.; Brar, D.S. *J. Biotech.* **1989**, *12*, 1.
3. McBride, K. E.; Schaaf, D. J.; Daley, M.; Stalker, D. M. *Proc. Natl. Acad. Sci. USA*. **1994**, *91*, 7301.
4. Kumagai, M. H.; Turpen, T. H.; Weinzettl, N.; Della-Cioppa, G.; Turpen, A. M.; Donson, J.; Hilf, M. E.; Grantham, G. L.; Dawson, W. O.; Chow, T. P.; Piatak, Jr., M.; Grill, L.K. *Proc. Natl. Acad. Sci. USA* **1993**, *90*, 427.
5. McBride, K. E.; Svab, Z.; Schaaf, D. J.; Hogan, P. S.; Stalker, D. M.; Maliga, P. *BIO/TECH.* **1995**, *13*, 362.
6. van Rooijen, G. J. H.; Moloney, M. M. *BIO/TECH.* **1995**, *13,* 72.
7. Sehnke, P. C.; Pedrosa, L.; Paul, A.-L.; Frankel, A. E.; Ferl, R. J. *J. Biol. Chem.* **1994**, *269*, 22473.
8. Shade, R. E.; Schroeder, H. E.; Pueyo, J. J.; Tabe, L. M.; Murdock, L. L. Higgins, T. J. V.; Chrispeels, M. J. *BIO/TECH.* **1994**, *12,* 793.
9. Hamamoto, H.; Sugiyama, Y.; Nakagawa, N.; Hashida, E.; Matsunaga, Y.; Takemoto, S.; Wantanabe, Y.; and Okada, Y. *BIO/TECH.* **1993,** *11,* 930.
10. Mason, H. S.; Arntzen, C. J. *TIBTECH.* **1995**, *13*, 388.
11. Mason, H. S.; Lam, D. M.-K.; Arntzen, C. J. *Proc. Natl. Acad. Sci. USA* **1992**, *89*, 11745.
12. Thanavala,Y.; Yang, Y.-F.; Lyons, P.; Mason, H. S.; Arntzen, C. J. *Proc. Natl. Acad. Sci. USA* **1995**, *92*, 3358.
13. Jiang, X.; Wang, M.; Graham, D. Y.; Estes, M. K. *Science* **1990**, *250*, 1580.
14. Usha, R.; Rohll, J. B.; Spall, V. E.; Shanks, M.; Maule, A. J.; Johnson, J. E.; Lomonossoff, G. P. *Virology* **1993**, *197*, 366.
15. Porta, C.; Spall, V. E.; Loveland, J.; Johnson, J. E.; Barker, P. J.; Lomonossoff, G. P. *Virology* **1994,** *202,* 949.
16. Turpen, T. H.; Reinl, S. J.; Charoenvit, Y.; Hoffman, S. L.; Fallarme, V.; Grill, L. K. *BIO/TECH* **1995,** *13*, 53.
17. R. Curtiss and G. A. Cardineau, *World Intellectual Property Organization PCT* **1990**, *US89*, 03799.
18. Clemens, J. D.; Hartzog, N. M.; Lyon, F. L.; *J. Infect. Dis.* **1988**, *158*, 372.
19. Haq, T. A.; Mason, H. S.; Clements, J. D.; Arntzen, C. J. *Science* **1995**, *268*, 714.
20. Higo, K.-I.; Saito, Y.; Higo, H. *Biosci. Biotech. Biochem.* **1993**, *57*, 1477.
21. Takamatsu, N.; Watanabe, Y.; Yanagi, H.; Tetsuo, M.; Shiba, T.; Okada, Y. *FEBS* **1990**, *269*, 73.
22. Vandekerckhove, J.; Van Damme, J.; Van Lijsebettens, M.; Botterman, J.; De Block, M.; Vandewiele, M.; De Clercq, A.; Leemans, J. Van Montagu, M.; Krebbers, E. *BIO/TECH.* **1989**, *7*, 929.
23. Kreitman, R. J.; Pastan, I. *Adv. in Pharm.* **1994,** *28,* 193.
24. Hiatt, A.; Cafferkey, R.; Bowdy, K. *Nature*, **1989**, *342*, 76.
25. Ma, J. K.-C.; Lehner, T.; Stabila, P.; Fux, C. I.; Hiatt, A.; *Eur. J. Immunol.* **1994**, ***24,*** 131.
26. Ma, J. K.-C.; Hiatt, A.; Hein, M.; Vine, N. D.; Wang, F.; Stabila, P.; van Dolleweerd, C.; Mostov, K.; Lehner, T. Science, **1995**, *268*, 716.
27. During, K.; Hippe, S.; Kreuzaler, F.; Schell, J. *Plant Mol. Biol.* **1990**, *15*, 281.
28. Bird, R. E.; Hardman, K. D.; Jacobson, J. W.; Johnson, S.; Kaufman, B. M.; Lee, S.-M.; Lee, T.; Pope, S. H.; Riordan, G. S.; Whitlow, M. *Science* **1988**, *242,* 423.
29. Owen, M.; Gandecha, A.; Cockburn, B.; Whitelam, G. *BIO/TECH.* **1992**, *10*, 790.
30. Tavladoraki, P.; Benvenuto, E.; Trinca, S.; De Martinis, D.; Cattaneo, A.; Galeffi, P. *Nature* **1993**, *366*, 469.

31. Benvenuto, E.; Ordas, R. J.; Tavazza, R.; Ancora, Giorgio, A.; Biocca, S.; Cattaneo, A.; Galeffi, P. *Plant Mol. Biol.* **1991**, *17*, 865.
32. Ma, J. K.-C.; Hein, M. B. *TIBTECH* **1995**, *13*, 522.
33. Sijmons, P. C.; Dekker, B. M. M.; Schrammeijer, B.; Verwoerd, T. C.; van den Elzen, P. J. M.; Hoekema, A. *BIO/TECH.* **1990**, *8*, 217.
34. Robins, R. J.; Rhodes, M. J. C. *Appl. Microbiol. Biotechnol.* **1986**, *24*, 35.
35. Byun, S. Y.; Pederson, H.; Chin, C.-K. *Phytochem.* **1990**, *29*, 3135.
36. Buitelaar, R. M.; Langehoff, A. A. M.; Heidstra, R.; Tramper, J. Enzyme *Microb. Technol.* **1992**, *13*, 487.
37. Brodelius, P; Pederson, H. *TIBTECH* **1993**, *11*, 30.
38. Nawrath, C.; Poirier, Y.; Somerville, *C. Mol. Breed.* **1995**, *1*, 105.

Chapter 14

Human Plasma Proteins from Transgenic Animal Bioreactors

R. K. Paleyanda[1], W. H. Velander[2], T. K. Lee[1], R. Drews[1], F. C. Gwazdauskas[2], J. W. Knight[2], W. N. Drohan[1], and H. Lubon[1,3]

[1]Holland Laboratory, American Red Cross, 15601 Crabbs Branch Way, Rockville, MD 20855
[2]Department of Chemical Engineering, Virginia Polytechnic Institute and State University, 153 Randolph Hall, Blacksburg, VA 24061

We have evaluated the transgenic animal bioreactor (TAB) system for the production of human Protein C (HPC), an anticoagulant plasma protein. Mouse whey acidic protein (mWAP) gene regulatory sequences targeted expression of mWAP/HPC hybrid genes to the mammary gland of transgenic mice and pigs. The transgenes were stably transmitted through several generations and expression did not adversely affect the health of animals. We purified recombinant HPC from milk which was structurally and enzymatically similar to the human protein, demonstrating the potential of TABs for the production of plasma proteins. The active fraction was found to be only a part of the total protein secreted. At mg/mL levels of secretion, modifications like proteolytic maturation and γ-carboxylation became limiting factors in the production of functional protein. We have proposed and shown that these limitations may be overcome by engineering the posttranslational protein modification capacity of selected organs of the TAB.

In the 1980s, genes encoding human plasma proteins traditionally used in transfusion and replacement therapy, such as serum albumin (HSA), Factor VIII (FVIII) and Factor IX (FIX) were cloned. The coding sequences of several other therapeutic proteins, namely human tissue plasminogen activator (tPA), erythropoietin (EPO), α_1-antitrypsin (α_1AT), antithrombin III (AT-III), Protein C (HPC), Protein S (HPS), Factor VII (FVII), Factor X (FX), prothrombin, hemoglobin, fibrinogen, lactoferrin and collagen were also cloned. Advances in cell biology, recombinant DNA technology and cell culture reactor technology have resulted in improvements in the expression of complex proteins *in vitro*. However, the complexity of many therapeutic proteins, coupled with limitations in host cell

[3]Corresponding author

0097–6156/96/0647–0205$15.00/0

intracellular processing and secretory pathways have limited the utility of bacterial or yeast cell production systems, although nonglycosylated proteins like albumin are considered to be good candidates for these systems. Mammalian cell culture systems have been used successfully to produce recombinant FVIII and EPO, which are required in small doses for human administration. Even so, such production methods for FVIII are not as cost-effective as purification by traditional means from pooled human plasma. It is also unlikely that mammalian cell culture will be cost-effective for the production of proteins needed in large quantities namely HSA, α_1AT, hemoglobin, lactoferrin and collagen.

One approach is to direct the expression of human proteins to specific cells or tissues in transgenic animals which are easily accessible and collectible. An example of this is the expression of human hemoglobin in the erythrocytes of mice and pigs (*1, 2*). Another approach is to produce heterologous proteins in the bodily fluids of animals. This has been demonstrated by the secretion of several human proteins into milk (reviewed in *3*), of rFIX (*4*) and α_1AT (*5*) into blood, of a C-terminal peptide of FVIII into saliva (*6*) and of rHPC into urine (Lubon et al., unpublished observations) of transgenic mice and livestock. Some of these attempts have been successful in producing foreign proteins at g/L levels. Leading examples of the potential of this technology are the transgenic sheep secreting 60 g/L of α_1AT into milk (*7*) and pigs expressing 24% or 32 g/L of human hemoglobin in blood (*8*). Proteins like tPA (*9, 10*), α_1AT (*7*), lactoferrin (*11*) and HSA (*12*) are good candidates for synthesis in the mammary glands of such "transgenic animal bioreactors" (TAB). All these proteins have a common feature - simple posttranslational modifications. For instance, albumin requires proteolytic processing for maturation, α_1AT and lactoferrin require glycosylation, and tPA requires both. As there are other, more complex proteins involved in hemostasis, we and others undertook to challenge the TAB system to express some of these proteins, including HPC (*13-15*), FVIII (*16*, Lubon *et al.*, unpublished observations), FIX (*4, 17*) and hemoglobin (*2*).

Human Protein C

We have generated transgenic animals for HPC (*13-15*), FVIII, FIX and fibrinogen (Velander and Lubon, unpublished observations). This report will focus on the expression of HPC in TABs. HPC is an anticoagulant vitamin K-dependent protein that is synthesized in the liver and undergoes extensive co- and/or post-translational modification (*18*). This processing includes the proteolytic cleavage of a signal peptide and a propeptide, propeptide-directed vitamin K-dependent γ-carboxylation of glutamic acid residues (GLA), glycosylation at four N-linked sites, β-hydroxylation of Asp^{71} and disulfide bond formation. Endoproteolytic removal of the Lys^{156}-Arg^{157} dipeptide produces a disulfide-linked heterodimer composed of a 21 KDa light chain, and a 41 kDa heavy chain containing the N-terminal activation peptide and serine protease domain. After activation of the zymogen by a thrombin-thrombomodulin complex at the endothelial cell surface, activated HPC (APC) regulates the coagulation cascade by inactivating coagulation factors VIII_a and V_a, which are necessary for the efficient generation of factor X_a and thrombin, respectively. As described earlier, HPC has been used in replacement therapy for homozygous and heterozygous HPC deficiency and in the treatment of coumarin-induced skin necrosis (*15*). APC has also been shown to prevent the extension of venous thrombi in dogs and rhesus monkeys, to protect baboons from septic shock due to lethal *E. coli* infusions and to delay thrombotic occlusion in a baboon arterial shunt model (*15*). Inhibition by APC of disseminated intravascular coagulation and microarterial thrombosis in rabbits has also been reported (*19*).

Animals Transgenic for HPC

Most mammalian cell lines engineered to express rHPC secrete only partially active molecules or low levels of fully active molecules (*20-22*). We therefore decided to target the expression of HPC to the mammary gland of transgenic animals, as proteins secreted within this organ are largely separated from the bloodstream and considerable amounts of protein, up to 100 mg/mL are found in the milk of certain species. In our initial experiments we used the promoter and gene of a mouse whey acidic protein (mWAP) to direct mammary-specific and lactogenic hormone-inducible expression. The DNA construct contained a 2.5 kb mWAP promoter, the 1.5 kb HPC cDNA inserted at the KpnI site in the first exon of the 3.0 kb mWAP gene, and 1.6 kb mWAP 3' flanking sequences (*13*). Mice transgenic for the cDNA construct secreted 3-10 μg/mL rHPC into the milk. To increase the concentration of rHPC in the milk of transgenic mice, we cloned additional 5' flanking sequences of the mWAP promoter (*23*) and used a construct containing 4.1 kb mWAP promoter sequences, the 9.0 kb HPC gene and 0.4 kb of HPC 3' flanking sequences (*15*). mWAP-promoter directed expression of the HPC gene was mainly restricted to the lactating mammary gland as detected by northern blot analysis, with "leakage" of expression in the salivary gland and kidney, at less than 0.1% of mammary expression. In the mammary gland, we detected rHPC in the epithelial cells lining the alveoli and in the milk-filled lumina, Figure 1. The developmental pattern of transgene expression differed from that of the endogenous mWAP gene, in that rHPC transcripts appeared earlier in pregnancy than mWAP, with no major induction during lactation. This suggested that the 4.1 kb promoter fragment also did not contain all the regulatory elements responsible for developmental regulation of the transgene similar to the mWAP gene. The precocious expression of the transgene did not compromise the health or nursing abilities of the transgenic females, but this may not hold true when proteins with potent biological activity, like EPO, are expressed. Thus, the lack of strict tissue-specific and developmental regulation of transgene expression could be a limitation of the TAB system. In fact, the health of the animals will be a major guideline in the stable production of therapeutic proteins from TABs.

Characterization of rHPC

Structure. Using immunoaffinity chromatography, we purified rHPC from pooled milk of HPC cDNA mice which had about 70% of the anticoagulant activity of HPC, as assayed by activated partial thromboplastin time (APTT) assays (*13*). Low concentrations of protein in the milk of cDNA mice did not allow further characterization. rHPC was secreted at higher concentrations of 0.1 to 0.7 mg/mL,

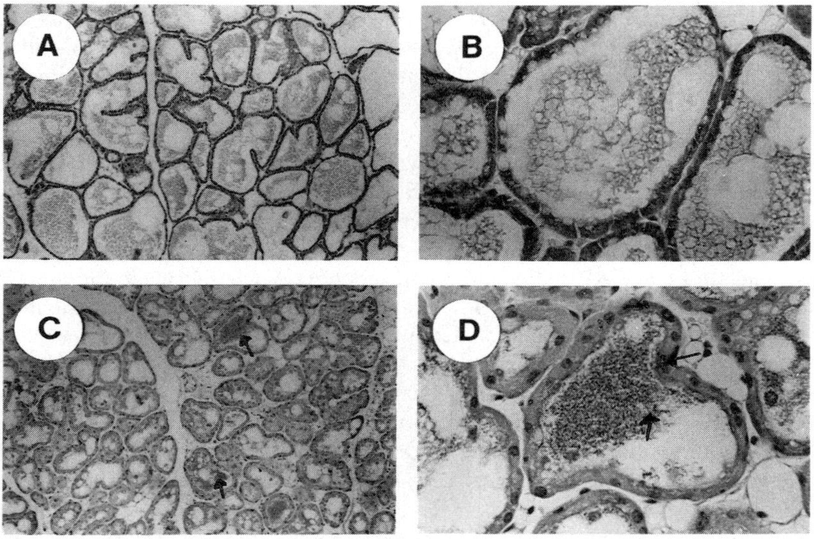

Figure 1. **Immunohistochemical localization of rHPC in mammary gland**

Paraffin-embedded sections of mammary gland from mid-lactation (day 10) mice were probed with a sheep anti-HPC polyclonal antibody and detected by the indirect immunoperoxidase technique (Vector Labs.). Sections were counterstained with hematoxylin. The arrows denote the darkly staining rHPC detected predominantly in the alveolar lumina and localized in secretion vacuoles of alveolar epithelial cells.

(A, B) Mammary gland of control mouse (C, D) Mammary gland of mWAP/HPC transgenic mouse. Initial magnification = 100 X (A, C), 400 X (B, D).

when the genomic sequences were expressed (*15*). This is the first report of the secretion of mg/mL amounts of a human vitamin K-dependent plasma protein into mouse milk. Western blot analysis of milk proteins separated by SDS-PAGE revealed rHPC single, heavy and light chain forms similar to plasma-derived HPC, although they migrated with slightly increased electrophoretic mobility, Figure 2, lanes 3, 7. rHPC consisted of 30 to 40% single chain form, which is two-to-three times more than that observed in plasma HPC. The two-dimensional gel electrophoresis of mouse whey proteins and western blot detection revealed that rHPC was more heterogenous than HPC (*24*). The predominant single and heavy chain polypeptides of mouse rHPC were more basic than that of HPC, Figure 3. The HPC heavy chain polypeptides resolved at an isoelectric point of 4.7-5.5, while mouse rHPC resolved at pH 5.3-6.2. The observed heterogeneity of rHPC was not attributable to alternate splicing of mRNAs, as confirmed by sequencing of the exon-exon junctions. However, this possibility cannot be ruled out entirely for all precursor RNA molecules. These differences could be due to species-specific posttranslational modifications of rHPC like proteolytic processing, glycosylation or γ-carboxylation in the mouse mammary gland. For example, a decrease in sialic acid or GLA content could affect the electrophoretic mobility of mouse rHPC. Enzymatic deglycosylation with N-glycosidase F, which cleaves between the internal N-acetylglucosamine and asparagine residues of most glycoproteins, showed that these molecular weight and charge disparities were in part due to glycosylation (*15*). The carbohydrate composition of rHPC produced in cell culture systems was also responsible for the altered electrophoretic mobility as compared to HPC (*21, 22*). Thus, recombinant proteins produced in transgenic animals have different patterns of glycosylation from their human counterparts.

Amino-terminal sequence analysis of rHPC purified by immunoaffinity chromatography showed that in 60-70% of the purified protein, the cleavage of the signal peptide and propeptide, and removal of the connecting KR dipeptide had occurred at the appropriate sites. The increased amount of single chain protein and the presence of propeptide on 30-40% rHPC suggested that at mg/mL levels of expression, mammary epithelial cells were incapable of the efficient proteolytic processing of the rHPC precursor. This result was not unexpected as studies with several different mammalian cell lines, including CHO, C127 and BHK-21 cells have shown inefficient proteolytic processing of rHPC. Hepatocytes of the human liver also do not completely process the KR dipeptide, as 5-15% of the single chain form has been detected in plasma (*20-22*).

Activity. Despite the observed differences between mouse rHPC and HPC, the enzymatic domain was functional, as the K_m of rAPC for the synthetic tripeptide substrate <Glu-Pro-Arg-p-nitroanilide, 0.27 mM, was similar to that of human APC (*15*). Both the α and β glycoforms of rHPC were active. The specific amidolytic activity of rAPC was lower than that of plasma APC, and is probably due to the incompletely processed propeptide. rHPC purified from the milk of mice expressing low levels of 3-10 μg/mL rHPC had about 74% of the activity of HPC in APTT assays (*13*), but only trace anticoagulant activity was detected in mice secreting 0.5 to 0.7 mg/mL rHPC (*15*). The anticoagulant activity of HPC is particularly

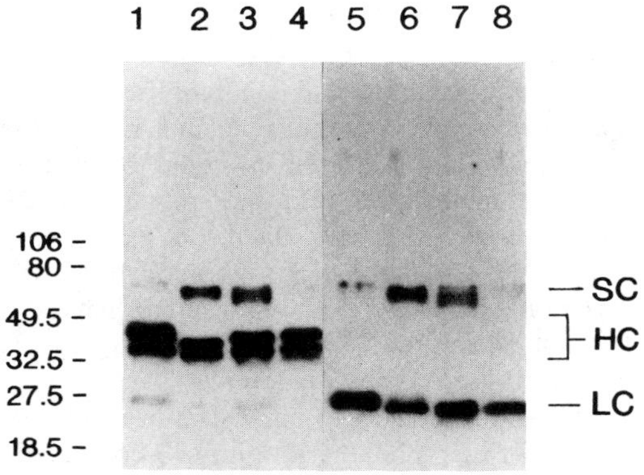

Figure 2. **Comparison of rHPC from milk of transgenic animals**
Purified HPC and rHPC were denatured, reduced and separated by 12% SDS-PAGE (Novex). Western blot analysis was carried out using the 8861 monoclonal antibody (lanes 1-4) directed against HPC heavy chain, and the 7D7 monoclonal antibody directed against HPC light chain (lanes 5-8) and the HRP/ECL system (Amersham).
Lane 1, 5: plasma-derived HPC; lane 2, 6: rHPC from transgenic pig; lane 3, 7: rHPC from HPC transgenic mice; lane 4, 8: rHPC from HPC/PACE bigenic mice. Molecular weight markers are described on the left. SC: single chain; HC: heavy chain; LC: light chain.

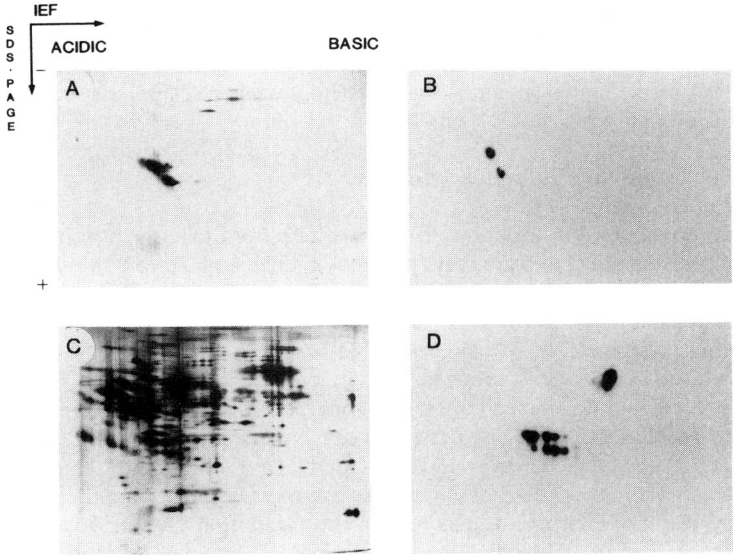

Figure 3. **Heterogeneity of rHPC**
Whey proteins were denatured, reduced and separated by isoelectric focussing, followed by SDS-PAGE. (A, C) Total proteins were detected by silver staining, (B, D) while rHPC was detected using the 8861 monoclonal antibody directed against HPC heavy chain and the HRP/ECL system (Amersham). A, B) Human protein C, 0.3 and 0.5 μg plasma-derived HPC, respectively. C, D) Mouse whey proteins, 10 and 30 μg protein, respectively.

dependent upon the degree of vitamin K-dependent γ-carboxylation. N-terminal sequence analysis of rHPC showed good recovery of glutamic acid residues at positions 6, 7 and 14, that are known to be γ-carboxylated in HPC. As GLA residues are not extracted during Edman degradation, this confirmed the insufficient γ-carboxylation of rHPC at high levels of expression in the mouse mammary gland. As yet, we have been unable to conclusively define the range of expression at which the enzymes carrying out γ-carboxylation become limiting in the efficient modification of rHPC in the mammary gland. We have generated transgenic mice secreting different amounts of rHPC in milk using combinations of either the 2.5 or 4.1 kb mWAP promoters and the HPC cDNA, minigene or gene (*25*). Mice derived from the cDNA constructs secreted consistently lower amounts of rHPC than mice containing the genomic construct. rHPC expressed at 0.03 to 0.3 mg/mL had 36% of the anticoagulant activity of HPC, while rHPC expressed at 0.5 to 0.7 mg/mL had only 1.4 to 21% activity.

Protein Production in Transgenic Pigs

At this point, we did not know whether results from laboratory animals could be directly extrapolated to transgenic livestock bioreactors. As we will discuss below, results obtained in mice are not always predictive of those in larger farm animals. Technology for the creation of transgenic rabbits, pigs, sheep, goat and cows is currently available (*26, 27*). However, the best species for the production of large amounts of heterologous proteins has not been defined and may be difficult to define. Selection will depend upon the properties of each individual protein and the animal species. In addition to productivity issues, species-specific limitations in posttranslational machinery may become the key criterion in the choice of a livestock host for a given protein, as stated above. We chose the porcine species for several reasons. First of all, pigs have a short gestation time of 4 months, a short generation time of 12 months and give birth to a large number of offspring, 10-12 piglets, per pregnancy. It is estimated that as much as 200 to 400 L of milk may be collected a year from a lactating sow. The annual yield of recombinant protein per animal has been estimated to be about 1 kg, at expression levels of 5 g/L milk. This implies that 80-100 pigs could produce the annual U.S. requirement for HPC, 2 pigs would suffice to produce the 2 kg annual demand for FIX and just one pig could produce the 120 g of FVIII required. Secondly, transgenic pigs are being intensively studied as potential donors for xenotransplantation, due to similarities in the physiology of pigs and humans (*28*). Thirdly, we believe that porcine embryonic stem cells will be the first livestock stem cells to be commercially developed.

rHPC from Porcine Milk. We used the mWAP/HPC cDNA (*14*) and the mWAP/HPC gene constructs to generate transgenic pigs. Although pigs do not possess an endogenous mWAP gene, the mouse WAP regulatory elements directed expression of rHPC to the pig mammary gland. To our surprise, pigs transgenic for the cDNA construct secreted from 0.2 to 2.0 mg/mL of rHPC (*14*). These data show species-specific differences in the recognition of the same regulatory sequences in mice and pigs, and are a good example of some difficulties in directly applying results from mice to pigs. On the other hand, 0.2 to 5 mg/mL rHPC was secreted as expected, into the milk of pigs transgenic for the HPC gene. These levels are equivalent to that of mouse WAP, which is normally present in mouse milk at 1 to 2 mg/mL (*3*). Another encouraging observation in pigs was the high anticoagulant activity of rHPC. In sharp contrast to mouse rHPC, 30% to 60% of rHPC produced in transgenic pig milk at 0.2 to 1.0 mg/mL had 70% to 150% of the anticoagulant activity of HPC (*14*). This again suggested species-specific differences in the posttranslational γ-carboxylation capacity of mammary epithelial cells. A possible explanation is that the level and/or substrate-specificity of the vitamin K-dependent γ-carboxylase differs in the mammary glands of mice and pigs. N-terminal sequencing of pig rHPC confirmed that it was efficiently γ-carboxylated. As in the mice, we found that 10% to 25% of rHPC still retained the propeptide and 30-40% was in the single chain form, Figure 2, lanes 2 and 6, demonstrating that the endoprotease(s) of the swine mammary epithelial cells also did not fully process the rHPC precursor. Here the mouse and pig mammary glands exhibit similar limitations in proprotein maturation capacity. A portion of pig rHPC had novel N-termini at positions -1, 152 and 157, in addition to the expected N-termini beginning at residues -24, +1 and 158. Since rHPC was found to be stable both in the milk and after purification, these additional cleavages on the amino terminal side of arginines at the dipeptide sequences Lys^{-2}-Arg^{-1}, Lys^{151}-Arg^{152} and Lys^{156}-Arg^{157}, probably occurred in the mammary gland (*29*). Thus, proteolytic processing of the rHPC precursor in the pig mammary gland may differ in part from the human liver and other expression systems.

These cleavages appeared not to adversely affect the amidolytic and anticoagulant activities or to alter the structure of rHPC. We compared the thermal stability and domain-domain interactions of rHPC from pig milk and plasma HPC, by scanning microcalorimetry and spectrofluorimetry (*30*). Both proteins exhibited similar heat absorption peaks corresponding to the melting of the catalytic module, when heated in the presence of 2 mM EDTA, at pH 8.5. The serine protease module was found to consist of two domains that unfolded independently. A second peak corresponding to the melting of the EGF-like domains was obtained at high temperatures and pH 3.0, indicating the presence of two thermostable and interacting domains. rHPC melted at a temperature 20^0C lower than HPC, suggesting that the EGF-like domains of the rHPC light chain are less stable. However, all the domains in rHPC underwent Ca^{2+}-induced conformational changes and folded into compact structures similar to HPC, indicating structural similarity.

After establishing that transgenic pigs can produce large quantities of biologically active protein in milk, we selected the pig as our model production animal. Pigs are not normally raised for milk, so we had to develop a procedure and a milking machine for its collection. Our next challenge was to develop a cost-effective large-scale purification process to isolate rHPC (*31*). Briefly, milk was defatted by centrifugation, and caseins and whey proteins were selectively precipitated with polyethylene glycol. A viral inactivation step using solvent/detergent was carried out, then rHPC was enriched by barium/citrate precipitation and fractionated by pseudoaffinity chromatography on an ion-exchange column using $CaCl_2$ step elution. This process gave an overall recovery of 24-35%. rHPC was 95% pure and had enzymatic and anticoagulant activities similar to those of plasma-derived HPC. Taken together, these results demonstrate that transgenic pigs are promising bioreactors for the production of rHPC, and the purification of recombinant proteins from milk by conventional methods can be scaled-up for economical manufacture.

Engineering the Mammary Gland

As mentioned earlier, the posttranslational processing of heterologous proteins is a major concern facing researchers involved in the production of recombinant human proteins (*18*). Recombinant α_1AT from sheep milk was shown to be fully N-glycosylated, but the sequence of oligosaccharides differed from human α_1AT (*7*), while recombinant hemoglobin from pig erythrocytes was well processed, had no unwanted modifications and was structurally and functionally equivalent to the human protein (*32*). In spite of the rHFIX secreted into sheep milk being biologically active and therefore probably γ-carboxylated, it had a lower molecular weight than the plasma protein (*17*). rtPA secreted into mouse milk was also active, but consisted of predominantly the proteolytically cleaved two-chain form, unlike that derived from Bowes melanoma cells (*33*). The longer-acting tPA variant (LAtPA) was secreted into goat milk mainly in the two-chain form, had 84% activity and significantly dissimilar oligosaccharides from C127 cell-derived LAtPA (*9, 10*). The fact that sometimes only a part of the recombinant protein is fully processed, and the differences in protein processing in the mammary gland, detract from some of the advantages of TABs. Hence, if TABs are to become the preferred expression systems of the future, efforts will have to be made to generate TABs with improved protein processing capabilities.

We proposed that the posttranslational protein processing capacity of specific animal organs could be enhanced by the coexpression of selected processing enzymes. Several vitamin K-dependent protein precursors, prohormones and complement proteins contain the consensus sequences for subtilisin-like serine proteases, and endoproteolysis at these paired basic residues yields the mature protein (*34*). In order to overcome the insufficient proteolytic processing of rHPC in the mammary gland, we coexpressed a cDNA encoding the proprotein processing enzyme, furin/paired basic amino acid cleaving enzyme (PACE), along with the HPC gene (*35*). We also generated double-transgenic mice that express HPC and an enzymatically inactive mutant of furin, PACEM. Coexpression of PACE with HPC resulted in the efficient conversion of the rHPC precursor to mature rHPC, in contrast to mice expressing HPC alone, or both HPC and PACEM. Analysis of rHPC from HPC/PACE bigenic mice demonstrated that the amount of single chain form in the milk of bigenic HPC/PACE mice was reduced on an average to less than 5%. This indicated that efficient processing of the precursor polypeptide to the mature two-chain form had occurred, Figure 2, lanes 4, 8. Amino acid sequence analysis of rHPC also revealed that cleavage of the rHPC precursor had occurred at the appropriate sites, suggesting the involvement of PACE in the *in vivo* processing of HPC. This is the first example of the engineering of the mammary gland into a new generation of bioreactor with enhanced recombinant protein processing capacity. If these observations can be repeated in transgenic livestock with these and other combinations of heterologous proteins and processing enzymes, this will imply much broader application of TABs in the pharmaceutical industry and in agriculture, than considered so far.

Summary

We have provided a critical overview of our experience in developing transgenic animal bioreactors for the production of human plasma proteins. Several technical and possibly ethical questions will need to be addressed in the future, as have been discussed by others (*26, 27, 36-38*). We believe that TABs will with time challenge the production of recombinant proteins by other expression systems, although progress in this field will not be as rapid and spectacular as in the case of bacterial or mammalian cell culture. Our vision for the next century is of pharmaceutical processing units adjacent to green farmlands covered by transgenic livestock producing unconventional proteins.

Acknowledgments

The authors would like to thank all members of the transgenic animal program of the American Red Cross, Rockville and Virginia Polytechnic Institute and State University, Blacksburg.

Literature Cited

1. Logan, J.S. *Curr. Opin. Biotech.* **1993**, *4*, 591-595.
2. Swanson, M.E.; Martin, M.J.; O'Donnell, K.; Hoover, K.; Lago, W.; Huntress, V.; Parsons, C.T.; Pinkert, C.A.; Pilder, S.; and Logan, J.S. *Bio/Technology*, **1992**, *10*, 557-559.
3. Hennighausen, L. *Protein Expr. and Purif.* **1990**, *1*, 3-8.
4. Choo, K.H.; Raphael, K.; McAdam, W.; and Peterson, M.G. *Nuc. Ac. Res.* **1987**, *15 (3)*, 871-883.
5. Kelsey, G.D.; Povey, S.; Bygrave, A.E.; and Lovell-Badge, R.H. *Genes Dev.* **1987**, *1*, 161-171.
6. Mikkelsen, T.R.; Brandt, J.; Larsen, H.J.; Larse, B.B.; Poulsen, K.; Ingerslev, J.; Din, N.; and Hjorth, J.P. *Nuc. Ac. Res.* **1992**, *20 (9)*, 2249-2255.
7. Wright, G.; Carver, A.; Cottom, D.; Rewes, D.; Scott, A.; Simons, P.; Wilmut, I.; Garner, I.; and Colman, A. *Bio/Technology* **1991**, *9*, 830-834.
8. Sharma, A.; Matin, M.J.; Okabe, J.F.; Truglio, R.A.; Dhanjal, N.K.; Logan, J.S.; and Kumar, R. *Bio/Technology* **1994**, *12*, 55-59.
9. Ebert, K.M.; Selgrath, J.P; DiTullio, P.; Denman, J.; Smith, T.E.; Memon, M.A.; Schindler, J.E; Monastersky, G.M.; Vitale, J.A.; and Gordon, K. *Bio/Technology*, **1991**, *9*, 835-838.
10. Denman, J.; Hayes, M.; O'Day, C.; Edmunds, T.; Bartlett, C.; Hirani, S.; Ebert, K.M.; Gordon, K.; and McPherson, J.M. *Bio/Technology* **1991**, *9*, 839-843.
11. Platenburg, G.J.; Kootwijk, E.P.A.; Kooiman, P.M.; Woloshuk, S.L.;

Nuijens, J.H.; Krimpenfort, P.J.A.; Pieper, F.R.; de Boer, H.A.; and Strijker, R. *Transgenic Res.* **1994**, *3,* 99-108.

12. Shani, M.; Barash, I.; Nathan, M.; Ricca, G.; Searfoss, G.H.; Dekel, I.; Faerman, A.; Givol, D.; and Hurwitz, D.R. *Transgenic Res.* **1992**, *1,* 195-208.
13. Velander, W.H.; Page, R.L.; Morcol, T.; Russell, C.G.; Canseco, R.; Young, J.M.; Drohan, W.N.; Gwazdauskas, F.C.; Wilkins, T.D.; and Johnson, J.L. *Ann. N.Y. Acad. Sci.* **1992**, *665*, 391-403.
14. Velander, W.H.; Johnson, J.L.; Page, R.L.; Russell, C.G.; Subramanian, A.; Wilkins, T.D.; Gwazdauskas, F.C.; Pittius, C.; and Drohan, W.N. *Proc. Natl. Acad. Sci. USA* **1992**, *89*, 12003-12007.
15. Drohan, W.N.; Zhang, D-W.; Paleyanda, R.K.; Chang, R.; Wroble, M.; Velander, W.; and Lubon, H. *Transgenic Res.* **1994**, *3*, 355-364.
16. Halter, R.; Carnwath, J.; Espanion, G.; Herrmann, D.; Lemme, E.; Niemann, H.; and Paul, D. *Theriogenology*, **1993**, *39*, 137-149.
17. Clark, A.J.; Cowper, A.; Wallace, R.; Wright, G.; and Simons, J.P. *Bio/Technology* **1992**, *10*, 1450-1454.
18. Yan, B.S.; Grinnell, B.W.; and Wold, F. *Trends in Biological Sci.* **1989**, *14*, 264-268.
19. Arnljots, B.; Berggvist, D.; Dahback, B. *Thromb. Haem.* **1994**, *72,* 415- 420
20. Suttie, J.W. *Thromb. Res.* **1986**, *44*, 129-134.
21. Grinnell, B.W.; Walls, J.D.; Berg, D.T.; Boston, J.; McClure, D.B. and Yan, S.B. In *Genetics and Molecular Biology of Industrial Microorganisms*; Hershberger, C.L.; Queener, S.W. and Hegeman, G., Eds.; Amer. Soc. Microbiol., Washington, D.C., 1989, 226-237.
22. Grinnell, B.W.; Walls, J.D.; Gerlitz, B.; Berg, D.T.; McClure, D.B.; Ehrlich, H.; Bang, N.U. and Yan, S.B. In *Protein C and Related Anticoagulants*; Bruley, D.F. and Drohan, W.N., Eds; Gulf Publishing Company, Houston, TX, 1990, 29-63.
23. Paleyanda, R.K.; Zhang, D-W.; Hennighausen, L.; McKnight, R.; Drohan, W. N.; and Lubon, H. *Transgenic Res.* **1994**, *3*, 335-343.
24. Paleyanda, R.K. *Ph.D. Thesis*, **1994**.
25. Paleyanda, R.K.; Russell, C.G.; Chang, R.R.; Johnson, J.; Velander, W.; Drohan, W.N. and Lubon, H. *Miami Bio/Technology Short Reports*, Advances in Gene Technology; IRL Press at Oxford University Press, 1994, Vol. 6, 108.
26. Wall, R.J. and Seidel, G.E., Jr. *Theriogenology* **1992**, *38*, 337-357.
27. Seidel, G.E., Jr. *J. Anim. Sci.* **1993**, *71 (Suppl. 3)*, 26-33.
28. Cooper, D.K.C. *Transplant. Proc.*, **1992**, *24 (6)*, 2393-2396.
29. Lee, T.; Drohan, W.N. and Lubon, H. *J. Biochem.*, **1995**, *118*, 81-87.
30. Medved, L.V.; Orthner, C.L.; Lubon, H.; Lee, T.K.; Drohan, W.N. and Ingham, K.C. *J. Biol. Chem.*, **1995**, 270, 13652-13659.
31. Drohan, W.N.; Wilkins, T.D.; Latimer, E.; Zhou, D.; Velander, W.; Lee, T.K.; and Lubon, H. In *Proceedings of the First International Symposium on Bioprocess Engineering*; Galindo, E. and Ramirez, O.T., Eds.; Kluwer Academic Publishers, Netherlands, 1994, 501-507.

32. Rao, M.J.; Schneider, K.; Chait, B.T.; Chao, T.L.; Keller, H.; Anderson, S.; Manjula, B.N.; Kumar, R. and Acharya, A.S. *Art. Cells, Blood Subs., and Immob. Biotech.* **1994**, *22 (3)*, 695-700.
33. Pittius, C.W.; Hennighausen, L.; Lee, E.; Westphal, H.; Nicols, E.; Vitale, J. and Gordon, K. *Proc. Natl. Acad. Sci. (USA)* **1988**, *85*, 5874-5878.
34. van de Ven, W.J.M.; Roebroek, A.J.M. and van Duijnhoven, H.L.P. *Crit. Rev. Oncog.* **1993**, *4 (2)*, 115-136.
35. Drews, R.; Paleyanda, R.K.; Lee, T.K.; Chang, R.R.; Rehemtulla, A.; Kaufman, R.J.; Drohan, W.N. and Lubon, H. *Proc. Natl. Acad. Sci. USA,* **1995**, *92,* 10462-10466.
36. Basu, P.; Masters, B.; Patel, B. and Urban, U. *J. Anim. Sci.* **1993**, *71 (Suppl. 3)*, 41-42.
37. Thompson, P.B. *J. Anim. Sci.* **1993**, *71 (Suppl. 3)*, 51-56.
38. O'Connor, K.W. *J. Anim. Sci.* **1993**, *71 (Suppl. 3)*, 34-40.

Specialty Applications

Chapter 15

Risk in Bioenergy Crops: Ameliorating Biological Risk by Using Biotechnology and Phytochemistry

B. H. McCown[1], K. F. Raffa[2], K. W. Kleiner[2], and D. D. Ellis[1]

[1]Department of Horticulture, University of Wisconsin, 486 Plant Sciences Building, 1575 Linden Drive, Madison, WI 53706–1590
[2]Department of Entomology, University of Wisconsin, Madison, WI 53706

In the Northcentral U.S., plans call for 50 MW electric powerplants fueled 50% by dedicated biomass crops, especially poplar plantations. Plantations totaling 32,000 acres would be required to generate 150,000 dry tons of biofuel yearly. A major deterrent is the very real risk of reduced production by natural stresses such as pests. One solution is to grow poplar selections that have a high degree of genetic resistance to pests. By engineering poplars with BT genes, pest-resistant selections have been created. However, the deployment of these trees in plantations without incurring the additional risk of the catastrophic emergence of pest populations resistant to the engineered controls must be solved. 'Stacking' engineered BT with native phytochemical controls already existing in poplar selections offers one attractive approach. The complexity of the biology requires the intelligent planning of the use of varied control mechanisms in plantations to provide sustainable biofuel sources.

Throughout the Northcentral Region of the U.S., interest is increasing in the use of tree plantations to meet a variety of needs, including energy, fiber, and industrial feedstocks. This interest is driven by a projected shortage of fiber and a corresponding increase in stumpage prices for trees, by increasing demands to relieve our native forest resources from non-sustainable uses, and by a need for renewable energy sources to lessen this region's dependence on environmentally-harmful and economically volatile imported petroleum and coal.

In addition to the traditional use of trees as a pulp source for paper production, a very significant demand for tree biomass can be created if bioenergy uses are realized. Currently, one focus in the Northcentral part of the U.S. is on the use of biological feedstocks for electricity production in direct combustion powerplants. The first powerplants are envisioned to be small and decentralized and generate about 50 MW of electrical power. To fuel such a plant using only biofuels, 300,000

0097–6156/96/0647–0220$15.00/0

dry tons of biomass would be needed per year. The powerplants would be fed a variety of biofuels with about 50% of the demands supplied by dedicated energy crops like tree plantations. Calculations indicate that if bioenergy plantations produce a minimum of 5 dry tons per acre per year, then 1300 acres would be needed to fuel each MW of electrical generating capacity on a sustainable basis. Thus for a 50 MW powerplant relying on dedicated bioenergy crops for 50% of its fuel, plantations totaling 32,000 acres (or about 50 square miles) would be needed. To minimize transportation costs, these plantations would be within a 50 mile radius of the powerplant. Plantations spanning 32,000 acres of densely planted trees would require approximately 230 million trees!

Poplar Plantations as a Biofuel

For more than 3 decades, the Department of Energy's Office of Renewable Energy based at Oak Ridge National Laboratories has been researching alternative crops for biomass production in the U.S. Various species and hybrids of poplar (*Populus*) have consistently emerged as a preferred genus to plant for such uses, particularly in the northern tier of states (1). In addition, the U.S. Forest Service, especially at its Rhinelander Forest Research Station, supported projects in the improvement and culture of poplars under plantation situations.

If the demands for biomass and the biological potential of poplar plantations are combined, the prospect for dramatically increasing poplar planting and production is real and significant. However, an important problem is that poplars suffer from some important pest infestations and these intensify in plantation settings. With a crop that requires 3 to 7 years before income can be realized, the threat of loss of the plantation by pests is an exceedingly forceful deterrent to becoming a large grower of poplars for biomass. Effective pest controls need to be found before massive poplar production can appear economically attractive. Without the development of appropriate pest management methods, the long term profitability of poplar plantations will appear dubious and unacceptably risky.

The traditional pest control method relies heavily on the use of chemical pesticides. Not only is the use of petroleum-based pesticides a contradiction to the goal of using alternative energy sources, but such methods tend to be less effective in non-intensive agricultural production systems like forests. Agrochemical use is typically less cost effective in more extensive systems such as tree plantations that generate marketable output only after years of cultivation. The regular use of pesticides also poses unacceptable hazards of groundwater contamination, the possible loss of beneficial insect predators, and toxic exposure to non-target organisms.

Thus a major question arises: *How can one minimize the risks of growing dedicated biomass tree plantations without a reliance on the use of agrochemicals?* A proven alternative to chemical pest control is through the development of genetic plant resistance to pests. Such resistance can be achieved by both selecting for existing resistance in tree clones or by introducing novel sources of resistance using biotechnology. Traditional breeding has the advantage of exploiting a broad diversity of chemical, morphological, and physiological traits that have co-evolved with trees. Biotechnology has the advantage of considerably speeding the genetic selection process as it allows for the introduction into trees of novel forms of pest resistance derived from other organisms. However, even genetically-based controls against biological hazards can be as misused as the agrochemical controls. A worse-case example of such misuse would be the planting of thousands of acres of poplars all containing the same gene conferring resistance to pests. Since the trees would be dominant in their environment for decades (unlike annual crops which can be rotated yearly), the pest population would be intensely and continuously impacted by these pest-resistant trees. Realistically, only those members of the pest population that harbor some form of tolerance to the genetically-based controls in the trees would be able to vigorously reproduce. The eventual, and inevitable, outcome is a pest population that is no longer controlled without the use of agrochemicals, a situation that puts us back where we started.

A multidisciplinary research team at UW-Madison has been exploring these issues for the last 5 years. The overall goal has been to explore alternative strategies for deploying poplars genetically-engineered for insect resistance without incurring a high risk of ecological hazards such as the emergence of insects tolerant to the engineered trait. This research is far enough along to begin to show the complexities involved and the critical need for intelligent and well-reasoned planning in the use of such biomass crops.

Pest Resistant Poplars: A Mixed Bag

Poplars can be attacked by a myriad of insect pests, the principal groups being foliar feeders and stem borers. Foliar damage can be inflicted by caterpillars (Lepidoptera) such as tent caterpillars and gypsy moth, and by beetles (Coleoptera) such as cottonwood leaf beetles. Both of these pest groups can be periodically highly destructive of foliage. However it is rare that the trees will be killed by such attacks. Heavy feeding on foliage will decrease the growth during the years of heavy attack and thus may interfere with achieving the goal of producing the anticipated productivity of 5 dry ton per acre per year. More importantly, recurring attack by insect pests can weaken trees so that they become more susceptible to infection by pathogens, a major cause of catastrophic and permanent loss of plantations. Thus control of major insect pest outbreaks is an important aspect in reducing the risk of growing poplar bioenergy crops in large plantations.

A common and useful approach using biotechnology for pest control is genetically-engineering plants with BT genes. The genes were originally isolated from the ubiquitous soil bacterium, *Bacillus thuringiensis* or *B.t.* This bacterium produces a protein that when ingested by a susceptible insect, causes severe disruption in the digestive tract which leads to reduced feeding and often eventual death. Sprays of spore suspensions of the *B.t.* organism have been used worldwide for more than 40 years to control insect pests in a very wide range of crops. The safety is so proven and trusted that *B.t.* is even approved for use on totally organic farms. This safety stems from the unique specificity of the *B.t.* proteins which is due to the requirement of very specific binding sites found only in the insect gut. Although a wide variety of different *B.t.* proteins have been identified, each has maximal insecticidal activity for a narrow range of insects species. The most commonly used *B.t.* biopesticides are effective on the caterpillar or Lepidopteran pests, although others effective against beetle pests are becoming more commonplace.

A number of genes have been isolated from *B.t.*; this diversity has been further expanded by molecular engineering of the genes. Thus a very wide range of BT genes are now available for use in biotechnology. After considerable re-engineering of a BT gene so that its expression in plants is enhanced, plants engineered with the gene have been shown to be amazingly resistant to damage by specific insect pests. More than 5 years ago, the UW-Madison team successfully employed this approach for poplar and showed that trees could be produced that were highly resistant to the feeding and injury by both forest tent caterpillars or gypsy moths (Figure 1; *2, 3, 4*). Field plots have now been planted with these genetically-engineered trees and the pest resistance has been found to be stably maintained in plantations over at least one season of growth (*4*).

Does the deployment of BT containing trees remove susceptible insect pests as a biological risk? Unfortunately, as discussed above, if these poplars were planted widely and indiscriminately in large plantations, the very real risk of the emergence of insect pests that had developed tolerance to the BT genes would be apparent (*5*). How can this problem be minimized?

One solution would be to present the insect pest population in a plantation with a myriad of controls on its population growth. For example, if a plant contained two or more independent modes of pest resistance, the chances of an insect population simultaneously developing tolerance to both modes is very much less than if only one control mode was utilized. Fortunately, other research conducted at Madison has pinpointed the existence of other pest control mechanisms. In a survey of a wide diversity of poplar selections, the occurrence of inherent or native pest resistances were identified; that is, some poplar selections had evolved mechanisms of resisting the feeding of certain insect pests (*6*). The phytochemical basis of this native pest resistance, at least against caterpillar pests, usually involves a class of secondary biochemical compounds called phenolic glycosides (*7*). Several such compounds naturally occurring in poplars and known to be important in native pest resistance are shown in Figure 2. These compounds appear to accumulate in the newest foliage on actively growing shoots and provide protection against the feeding of such insects as gypsy moths (Figure 3). What would be the effect of combining or 'stacking' both the native (phenolic glycosides) and the engineered (BT) controls in the same plant? This question has recently been addressed by the Madison team and the early results appear to indicate that the BT adds to the pest resistance provided by the native mechanisms (Figure 4).

A plantation of poplar trees containing both the BT and the phenolic glycoside mechanisms of pest control should theoretically be (a) highly resistant to the attack by such insects as gypsy moths and (b) be much less prone to the emergence of pest populations tolerant of these controls. Further 'stacking' of pest controls can be

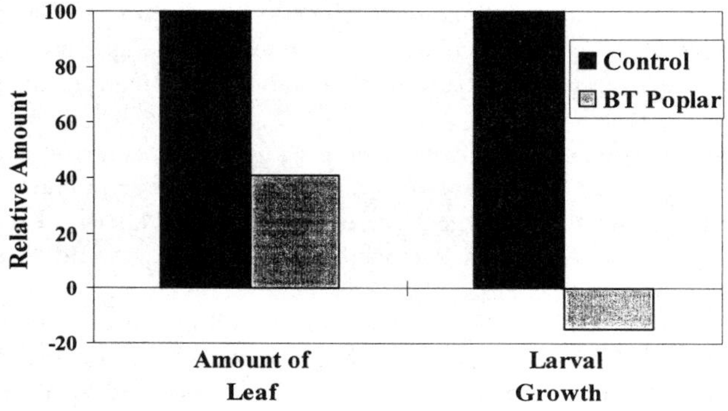

Figure 1. The relative amount of feeding of the Lepidopteran pest, gypsy moth, on leaves of non-transformed poplar (selection NC5339) and the same selection genetically-engineered with a BT pest resistance gene. Leaves were collected from one-year old field-grown plants and fed to insects in the laboratory where the growth of the larvae and the amount of leaf tissue consumed were measured for 3 days.

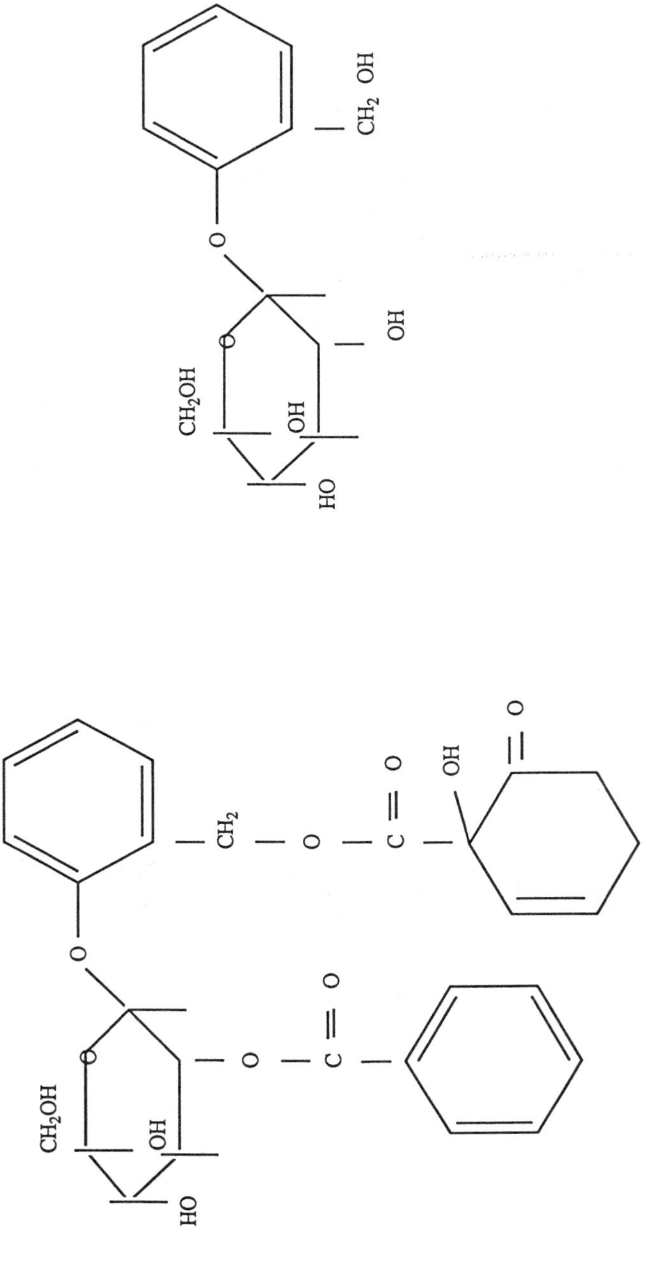

SALICIN

TREMULACIN

Figure 2. Several phenolic glycosides that are commonly found in some poplar selections and are known to be of importance in the native resistance of poplars to some insect pests.

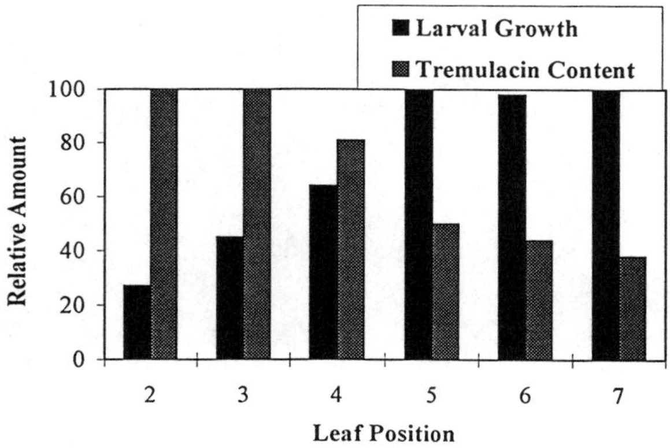

Figure 3. The relative amount of larval growth of gypsy moth and the relative content of the phenolic glycoside, tremulacin, as influenced by the leaf position on actively-growing shoots of poplar (selection NC5339). The leaves were collected from one-year old field grown plants and evaluated in the laboratory. Peak tremulacin levels reached 10% of the leaf dry weight and peak larval weight gain reached 8 mg.

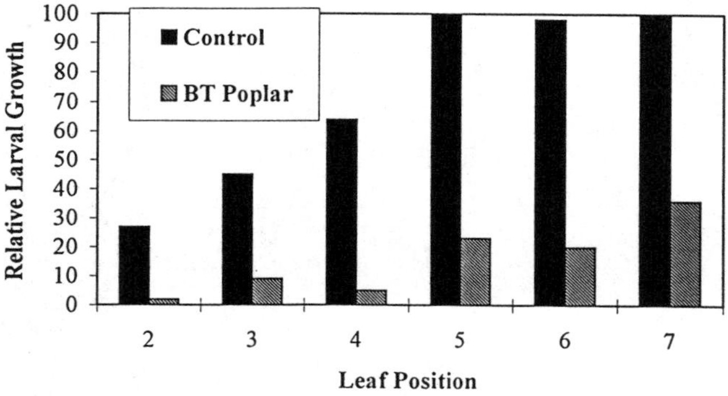

Figure 4. The growth of gypsy moth larvae on the leaves of non-transformed poplar (selection NC5339) and the same selection genetically engineered with BT as influenced by leaf position along the actively growing shoot. The leaves were collected from one-year old field-grown plants and bioassayed in the laboratory using 3 day feeding trials. Peak larval weight gain reached 23 mg for control and 8 mg for the BT poplar.

achieved by interplanting a plantation with 2 to 3 groups of trees, each group containing sets of different pest control mechanisms. Thus stacking can be integrated with a second approach, 'mosaics', which also advantage of different modes of traditionally bred and transgenic resistances. An example of this approach is illustrated in Table I. Thus the plantation as a whole could provide a wide array of controls on multiple groups of insect pests while ecologically giving a high degree of stability and predictability to the plantation viability.

Table I. An example of combining or 'stacking' various sources of resistances to insect pests in different poplar trees

Poplar clone	Source of resistance[a]	
	Caterpillar pests	Beetle pests
NM6	Native	Bt: CryIA(a)
NC5271	Bt: CryIIIA	Native

[a]Native refers to resistance inherent in the individual selection (often secondary phytochemcials) and Bt to specific proteins produced by genes genetically engineered into the trees.

Biological systems are inherently complex, and this is certainly the case for plant/pest dynamics. Such complexity makes the actual application of concepts as 'stacked resistances' not a trivial matter. The allelochemical composition of *Populus* consists primarily of condensed tannins and phenolic glycosides (*7*). While some of these compounds may complement the insecticidal activity of BT (*8, 9*) as appeared to occur above in the studies with genetically-engineered poplars, other compounds appear to interfere with BT efficacy (*9-13*). Further, some of the compounds that appear to offer pest resistance to caterpillar insects may actually attract other insect pests such as beetles (*14*). Although these interactions between the biochemical bases for pest controls may at times seem bewilderingly complex, it's this very diversity and complexity that offers the necessary tools to design a sustainable pest resistant biofuel system.

Conclusions

Although still early in its actual deployment, the production of electrical energy using the direct burning of biomass feedstocks is promising. Benefits include the reduction on a reliance on imported energy sources, a lessening of CO_2 and other emissions, and a stimulus for rural development by providing new and alternate sources of income. However, relying on biofuel resources presents yet another economic risk in the already volatile energy market. Plant biotechnology, coupled with conventional breeding and selection, can be used to generate trees that have a lower inherent biological risk of catastrophic loss or reduced productivity. However, biotechnology is no magic solution. The actual realization of a sustainable bioenergy electric power industry in the U.S. will require the best thinking and planning available, for the issues are both complex and highly multidisciplinary. This challenge is daunting to some; to others, it is incredibly exciting!

Literature Cited

1. Ranney, J.; Wright, L.; Layton, P. *J. Forestry* **1987**, *85*, 18-28.
2. McCown, B. H.; McCabe, D. E.; Russell, D. R.; Robison, D. J.; Barton, K. A.; Raffa, K. F. *Plant Cell Reports* **1991**, *9*, 590-594.
3. Robison, D. J.; McCown, B. H.; Raffa, K. F. *Environ. Entomol.* **1994,** *23*, 1030-1041.
4. Kleiner, K. W; Ellis, D. D.; McCown, B. H.; Raffa, K. F. *Environ. Entomol.* **1995**, *24*, 129-135.
5. Raffa, K. F. *BioScience* **1989**,*39*, 524-534.
6. Robison, D. J.; Raffa, K. F. *For. Sci.* **1994**, *40*, 686-714.
7. Lindroth, R. L.; Bloomer, M. S. *Oecologia.* **1991**, *86*, 408-413.
8. Arteel, G. E.; Lindroth, R. L. *Great Lakes Entomol.* **1992**, *25*, 239-244.
9. Hwang, S. Y.; Lindroth, R. L; Montgomery, M. E; Shields, K. S. *Environ. Entomol.* **1995**, *88*, 278-282.
10. Appel, H. M. 1993. *J. Chem. Ecol.* **1993**, *19*, 1521-1552.
11. Lord, J. C.; Undeen, A. H. *Environ. Entomol.* **1990**, *19*, 1547-1551.
12. Lüthy, P.; Hofmann,C; Jaquet,F. *FEMS Microbiology Letters* **1985**, *28*, 31-33.
13. Navon, A.;,Hare, J. D; Federici, B. A. *J. Chem. Ecol.* **1993**, *19*, 2485-2499.
14. Bingaman, B. R.; Hart, E. R. *Environ. Entomol.* **1993**, *22*, 397-403.

Chapter 16

Sugar Beet and Sugarcane as Renewable Resources

Margaret A. Clarke and Leslie A. Edye

Sugar Processing Research Institute, Inc., 1100 Robert E. Lee Boulevard, New Orleans, LA 70124

Sugarcane (Saccharum officinarum) is grown, generally as a perennial crop, in tropical and subtropical areas; some 750 million tonnes are produced each year. Food, feed and energy are the major products of the sugarcane plant; sugarcane fiber, bagasse, fuels the cane processing plants and provides electricity to local grids through cogeneration. Sugarbeet (Beta vulgaris) is produced annually on the order of 400 million tonnes, in temperate climates. The primary product is sugar (sucrose); other products include feeds (molasses and beet pulp), and raffinose, pectin and arabinan. Recently, production of paper from sugarbeet pulp has begun.

A range of chemicals is available from chemical and microbial reactions on process streams and sugars. Chemical transformations reviewed herein include production of sucrose mono-, di- and poly-esters, polyurethanes, carboxylic acid derivatives, and thermally stable polymers. Products of microbial processes include polymers to use as biodegradable plastics and others for food and non food use (levan, dextran). Basic chemicals, including citric acid and lactic acid, and amino acids, notably lysine, are produced from sugar sources. The production of ethanol, as fuel or as beverage, is well known. Products and processes are outlined, and recent developments are emphasized.

Among essential requirements of a crop that is to serve as a useful renewable resource are availability, accessability, amenability to cultivation and collection, and profitability - a high ratio of value of final products to cost of crop production, harvest and processing. Both sugarcane (Saccharum officinarum) and sugarbeet (Beta vulgaris) crops meet these criteria and each crop has, indeed, served as a renewable resource, and renewable source of food,

0097–6156/96/0647–0229$15.00/0

feed and chemicals, throughout the tropical and temperate zones, respectively, for at least two hundred years. Uses for these renewable resource crops are the subject of this chapter.

Annual production of sugarbeet and sugarcane and yields of sucrose and co-products will be presented as requisite preliminary information. Composition of the crops as harvested, and of their co-products will be outlined. The processes by which the crops are converted are also outlined. Uses of the fibrous co-products, bagasse from sugarcane and beet pulp from sugarbeet, will be described.

In the major section of the chapter, utilization of sucrose as a raw material for chemical or microbial products is the subject. Sucrose either isolated or in juices or molasses will be considered. Production of ethanol, either as fuel alcohol or beverage alcohol is not discussed; voluminous literature is available on both subjects.

The chapter emphasizes processes and products currently on the market, with information about production quantities and locations where available.

Production of Sugarbeet, Sugarcane and Sugar

Sugarcane is grown in the tropical and semitropical areas of the world. It requires some 60" in rainfall (or irrigation equivalent) per year, and absence of a killing freeze (below 28° F) to grow to maturity. Sugarbeet is grown in temperate zones, particularly in North America and Europe (including Russia), with some production in southern South America, North Africa, Pakistan and North China. World and USA estimates of crop production and sugar yield are shown in Tables I and II (*1, 2*).

Viewed as a renewable resource, fueled by solar energy and water (plus fertilizer and cultivation), sugarcane is highest among cultivated crops at capturing solar energy in reusable form: over 2% of solar energy falling on cane is converted to potential energy in chemical bonds in the plant's components (*4*). The degree of fixation of solar energy in sugarbeet is somewhat lower, but comparable to that found in corn (maize) or potato.

Composition of the crops, as millable sugarcane (without roots, leaves and trash) and beet (without leaves and soil) is shown in Table III. Under good agricultural conditions for each crop, sucrose content of beet is higher, but average cane tonnage per acre is higher.

Processing of cane or beet entails extraction of sucrose (sugar) with two major by-products or co-products: residual fiber, and molasses. Molasses is traditionally defined as the final concentrated cane or beet juice, after several sucrose crystallizations, from which no more crystalline sugar may be economically extracted. The process called "molasses desugarization," an ion exclusion separation of sucrose from beet molasses developed and installed by most U.S. beet sugar companies since the 1970's, has changed this definition for beet molasses in the U.S. For sugarbeet outside North America and for cane, the definition remains the same.

Outlines of the sugarcane and sugarbeet factory processes, to explain preparation of co-products, are shown in Figures 1 and 2.

Production quantities of molasses, cane and beet, are listed in Table IV, and general sugars composition of cane and beet molasses in Table V. Molasses, as defined by the American Association of Feed Control Officials, is at a minimum of 80 Brix (% solids by weight) and 46% total sugars (sucrose, glucose and fructose). Other oligosaccharides and

Table I. Current Annual and U. S. Production of Sugarbeet and Sugarcane

Annual Production		1993-94, Tonnes (metric)
World	Sugarcane	735×10^6
	Sugarbeet	410×10^6
U. S.	Sugarcane	27×10^6
	Sugarbeet	27×10^6

Table II. Production of Sugar From Sugarbeet and Sugarcane

	Annual Production (Tonnes x 10^6)	1993-94	1992-93
Cane Sugar	World	71.5	73.1
(96° pol basis)	USA		
	Florida	1.61	1.64
	Louisiana	0.80	0.80
	Hawaii	0.64	0.60
	Total	3.23	3.14
Beet Sugar	World	39.3	44.0
	USA	3.3	3.6

Table III. General Composition of Sugarcane and Sugarbeet

	Cane, %	Beet, %
Sucrose	9-18	11-20
Water	70-80	70-80
Fiber	8-18	4-8
"Non-Sugars"	2-4	2-4

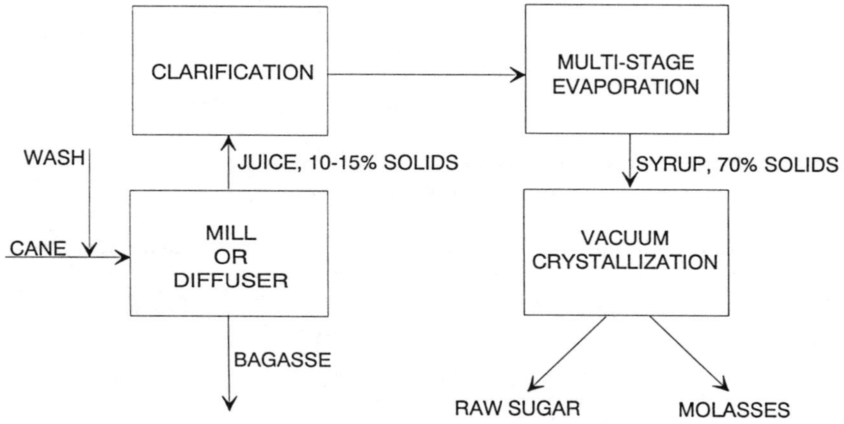

Figure 1. Raw cane sugar manufacturing process in sugar mill or factory

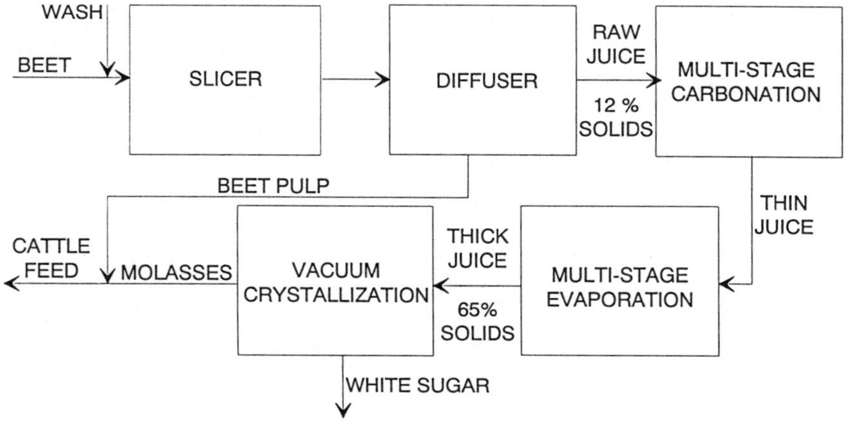

Figure 2. Beet sugar manufacturing process

Table IV. Annual Production of Cane and Beet Molasses

Annual Production (Tonnes)		1993-94	1992-93
World, Total		38.8×10^6	39.8×10^6
	Cane	24.3×10^6	24.3×10^6
	Beet	14.6×10^6	15.5×10^6
U.S			
	Cane	1.0×10^6	0.9×10^6
	Beet	0.63×10^6	0.8×10^6

polysaccharides present may add to the analysis of "total fermentables," which is often higher than the total sugars figure (*8*).

Fiber composition varies in the two crops, as cane is a giant grass, and beet a root. General composition of fibers, dry weight, is shown in Table VI.

Sugarcane Fiber

The fibrous residue of sugarcane, after sugar-containing juice is extracted by milling or diffusion, is known as bagasse. Regular mill-run bagasse contains about 50% moisture. Detailed composition is shown in Table VII (*3, 6*)

The major use for bagasse is as fuel for the cane factory that produced it. The fuel value of bagasse is: 1 ton dry bagasse (2 tons mill-run) approximately equals 2 barrels of fuel oil. Bagasse has a minimum value of some $20 per ton at 50% moisture going up to $50/ton depending on area and fuel price. Sugarcane factories are energy self-sufficient. Depending on steam utilization and boiler efficiency, a factory burning only bagasse for its own fuel may have 20%-30% bagasse in excess of its requirement. Excess bagasse is often burned as fuel for an adjacent refinery or, outside the U.S., distillery, but the increasingly common use for excess bagasse is in the production of electricity (in excess of factory needs) for sale to the local grid. Cogeneration at an efficient factory can produce 100 kilowatt hours per ton of cane; technological developments in conversion of this renewable resource can improve this yield (*5*). Cogeneration currently is in effect in Hawaii, Florida, Brazil, Colombia, Cuba, Egypt, India, Mauritius (cane is a major source of electricity on this island), Pakistan, The Philippines, and other cane-growing countries. It is particularly cost-effective in tropical areas that must import fuel, and always ecologically positive.

Bagasse is a major source of fiber for papers and boards in cane-growing countries. The higher lignin content of sugarcane fiber makes it better for paper and board use and poorer for feed use (without pretreatment) than sugarbeet fiber. Storage, drying, depithing (separation of long fibers from very short) and bleaching of bagasse are all prerequisite for paper and/or board manufacture; many processes and variations have been developed for each step and are well described in the literature (*6*). World production of bagasse pulp was some 2,200,000 tonnes in 1990 at 96 mills (*6*). Bagasse pulp production utilized about 7×10^6 tonnes bagasse, or 15-20% of world bagasse production. About 20% of this was made in the People's Republic of China (*7*), where wood is in short supply and most bagasse is used for paper; Chinese sugar factories burn low grade coal to save bagasse for paper and board use. Mexico, Peru and Venezuela are other major consumers of bagasse pulp. All forms of paper from coarse brown papers through disposable cups and plates (Taiwan) to high grade newsprint (India) and glossy high quality paper (South Africa and Louisiana) have been made from bagasse fiber. Board production includes high density hardboard, medium density-particle boards and fiberboards, with total installed capacity of 600,000 tonnes worldwide (*6*). South Africa, Cuba, China and India are major manufacturers. The Celotex Corporation plant, in Marrero, Louisiana USA, is one of the longest running examples of renewable resource utilization. It has been manufacturing bagasse-containing fiberboard for over 60 years.

Table V. Composition of Molasses

	Sucrose	Invert (Glc + Fru)	Total Sugars
Cane Molasses	25-33	9-18	35-48
Beet Molasses	45-50	1-3	6-53

Table VI. Average Composition of Fiber From Sugarbeet and Sugarcane Plants

	Cane, %	Beet, %
Cellulose	37-43	23-27
Hemicellulose	30-45	26-29
Lignin	20-25	3-5
Pectin	0	24-29

Table VII. Average Composition of Bagasse from 12-15 Month Old Cane (%)

	Cellulose	Hemicellulose	Ligin	Ash
Whole bagasse	36	26	20	2.2
Fiber	41	27	21	1.2
Pith	34	29	21	3.2

Fiber Content: 40 - 60 %
Pith Content: 20 - 30 %

Bagasse can be hydrolyzed to its components: 85-95% xylose and a few percent of arabinose and glucose. Furfural is made by steam hydrolysis of bagasse and subsequent dehydration to furfural (furan aldehyde). Production of furfural is described in Table VIII. Furfural is used as a selective solvent in refining of specialty oils, or as raw material for production of furfuryl alcohol (world production about 50,000 tonnes) and furan resins. Diacetyl (2,3-butanedione), which in very dilute solution is an artificial butter flavour, it is a co-product of furfural production in South Africa. Xylitol, a non-cariogenic sweetener, is made from reduction of bagasse xylose in Brazil and in China. Other products that can be made from bagasse include thermosetting and thermoplastic resins, although those made to date are low quality, and carboxymethyl cellulose (*6*).

Bagasse is used in a range of animal feed, most successfully for dairy cattle. Digestibility is increased either by alkaline (NaOH) hydrolysis or steam hydrolysis combined with alkali treatment: the latter is practiced at some fifteen plants in Brazil, and in China. Hydrolyzed bagasse is mixed with molasses and a nitrogen source (usually urea) to make a complete feed (*6*)

Hydrolyzed bagasse has long been a growth medium for production of single cell protein (*3*); this may become a commercial reality in South Africa to produce chicken feed (*6*).

Sugarbeet Fiber

Sugarbeet pulp, the material remaining after extraction of sucrose containing juice, is pressed at the factory and emerges at about 30% dry solids. Composition of pressed pulp is shown in Table IX (*9,10*). Variations arise from variety, soil and fertilizer differences. A major difference from cane fiber is the protein (8% on average) in beet pulp. This protein makes pulp a valuable animal feed. In some areas, especially in Europe, pulp is used directly as feed, either fresh or stored in containers or small silos on the farm, sometimes with addition of molasses before ensilage. Pulp drying, to 90% dry matter (6%-15% sugars) is general in the U.S. and increasingly popular elsewhere, although restrictions on energy requirements are bringing growing emphasis to efficient dryers. Dried pulp, mixed with molasses and often pelleted, is a popular feed for ruminants: over 75% of pulp in Europe and the U.S. is sold dried.

Sugarbeet pulp can be processed into a high fiber food ingredient; several companies in the U.S. and Europe have developed processes for this (*10*). The major commercial product is Fibrex, developed by Sockerbolaget (*11*), and marketed in Europe and the U.S. as an ingredient for breads, vegetarian foods and prepared mixes. The composition of Fibrex is shown in Table X. The high proportion of soluble dietary fibers is held responsible for the beneficial effect of sugarbeet fiber in lowering blood cholesterol levels in humans (*9,11*).

Fermentation processes have been used to produce enzymes on beet pulp, to be sold as product or to enhance the protein and feed value of beet pulp; in the latter case, fungi are the organisms of choice (*9,10*). Many processes have been developed, but few have been cost-effective. Ensilage processes of course entail fermentation; this is the only widespread use of pulp fermentation for animal feed use.

Table VIII. Furfural From Sugarcane Fiber

Where:	Dominican Republic	38,000 t/year
	USA	10,000 t/year
	South Africa	10,000 t/year
	Brazil	6,000 t/year
	India	5,000 t/year
	Venezuela	4,000 t/year
	China	?
	Cuba	?
	Taiwan	3,000 t/year
	The Philippines	3,000 t/year

Table IX. Composition of Sugarbeet Pulp (Pressed) Results are on Dry Matter. Pulp Contains an Average 30% Dry Matter.

Component	Content, % on dry matter
Cellulose	18-36
Hemicellulose	18-32
Arabinan	~20
Galactan	~7
Lignin	3-6
Pectin	15-32
Protein	5-10
Ash (insoluble - soluble)	3-12
Residual sugar	2-6

Table X. Composition of Fibrex (Sugarbeet Fiber)

Component	Content, % total weight
Dietary fiber	73
Cellulose	18
Hemicellulose	29
Pectin	22
Lignin	4
Water	10
Protein	10
Minerals	3.5
Sugar	3.5
Fat	0.3

Arabinan can be made from sugarbeet pulp hemicellulose, generally by alkaline extraction (*10*). Arabinan can be used as a flavor encapsulant (replacement for gum arabic), a confectionery base and a binding agent. When enzymically debranched, to a linear, less soluble form, its rheology allows its use as a non-digestible fat replacement compound. Hydrolysis yields about 80% L-arabinose (*10*).

Pectin in sugarbeet pulp has a low methoxyl and high acetyl content, which give low gelling properties. Processes have been developed to increase gelling properties (*12*) and are now in commercial use in Eastern Europe. Much work has been done on the structure and function of sugarbeet pectin, particularly by Thibault in France (*13, 14*). Recently, Thibault (*15*) has developed a process to convert ferulic acid residues on sugarbeet pulp into vanillin, creating a naturally sourced vanillin.

A recent development is the use of sugarbeet pulp as an ingredient for paper. Dried pulp is ground by a hammer mill or a micronizing mill and added to a pulping mixture at up to 20% level (*17*). Whole beet pulp is used; cellulose is not extracted. High quality papers can be produced.

Among other products from sugarbeet fiber are odor-removing adsorbent materials, and biogas; the latter has been the subject of research, but is not believed to be currently in production.

Products from Sucrose

This section will describe products that can be made from sucrose per se (as sugar, or in molasses, syrups, juices or other forms) by chemical or microbiological processes.

The best known and longest known chemical produced from sucrose is ethanol, either as rum, or as fuel or industrial alcohol: that subject has voluminous literature, and will not be discussed here. Because ethanol can be converted to ethylene, almost all petrochemical products can be made from sucrose. Sugar crops are therefore a renewable resource that can supply almost all the products of the oil industry, including fuel. The route to chemicals production does not, however, begin to become economically favorable until the comparative price situation of ethylene exceeds $450 per tonne and molasses is available at under $35 per tonne, over a predictably secure long term. It is possible that the next two decades will see a reversal of the current situation as sugar crop plantings increase and if oil supplies become less reliable.

A major route, often referred to in the literature on synthesis of chemicals from sucrose, begins with the hydrolysis, by acid or enzymes, of sucrose to glucose and fructose. The hydrolysis of sucrose is called inversion, by sugar chemists, (who refer to the glucose-fructose mixture as invert sugar) with subsequent reaction of the monosaccharide. Because glucose can, in most countries, be produced from the hydrolysis of starch more cheaply than from the hydrolysis of sugar, and because fructose can now be made by the enzymic isomerization of glucose more cheaply than from the hydrolysis of sugar, syntheses with inversion of sucrose as the initial reaction, as, for example, of sorbitol and mannitol or of hydroxymethyl furfural, will not be discussed here. Such products, generally known as sucrose degradation reaction products, which can be economically produced from sugars in countries that have a lot of sugar and no starch or foreign exchange funds, are covered extensively in the literature on sucrochemistry (*6, 7, 9, 10, 18-29*).

The cost of sucrose varies from 9¢/lb in the lowest priced molasses to 25¢/lb as refined sugars; purities of these raw materials range from 30% sucrose in molasses to 99.9% sucrose in refined sugar.

Sucrose is a non-reducing disaccharide with multiple functionality because of its eight hydroxyl groups. The reactivity of sucrose has been both a bane, because of its ready hydrolysis and decomposition, its limited solubility in organic solvents and its resistance to protecting group strategies, and a blessing, because of the great number of reactions it can undergo. The theoretical multiplicity of derivatives from the three primary and five hydroxyl groups of sucrose is limited by steric hindrance. Reaction conditions are limited by the molecular susceptibility to inversion and heat decomposition. Many compounds classified as chemicals from sucrose are degradation products, e.g., lactic acid, hydroxymethyl furfural, sorbitol and mannitol.

Among the classes of derivatives that provide marketable chemicals are (*20*):

1. Ethers (alkyl, benzyl, silyl, allyl (discussed below), alkyl, and the internal ethers or anhydro derivatives, which last have generated a new sweetener.
2. Esters of fatty acids, including surfactants, emulsifiers, coatings and a new fat substitute.
3. Other esters: sulfuric acid esters that polymerize well; sulfate esters, including an antiulcerative, discussed below, and other mixed esters.
4. Acetals, thioacetals and ketals that act as intermediates and may have biocide activity.
5. Oxidation products, most of which are products of the hydrolyzed monosaccharide products and reduction products, including mannitol and sorbitol, and reductive aminolysis products including methyl piperazine.
6. Halogen and sulfur derivatives and metal complexes, some with applications in water soluble agricultural chemicals.
7. Polymers and resins: polycarbonates, phenolic resins, carbonate-, urea- and melamine-formaldehyde resins, acrylics, and polyurethanes which are discussed below.

There are hundreds of sucrose derivatives, many with commercial applications, and they are amply described in the literature (*17-25*). The rest of this section is devoted to discussion of a few compounds, selected because they are in commercial production, or appear likely to be in the near future.

Sucrose Esters. A wide variety of sucrose esters, from mono- to octa-substituted, most usually with fatty acids, have been synthesized for a variety of uses. The major esters in commercial production are the monoesters of long-chain fatty acids, which are approved as food additives in the U.S. The non-ionic, non-toxic, 100% metabolizable and biodegradable esters find application as surfactants, and emulsifiers in coffee whiteners, whipped toppings and ice creams (*29, 30*), and for margarines. They have a wide range of emulsifying and stabilizing capacity and can inhibit freeze denaturation of proteins.

Other applications are as fruit and vegetable preservatives (*33*), crystallization inhibitors, antibacterial agents, and wheat flour improvers (*25*). The commercial products, Superfresh and Pro-long, fruit coatings to extend storage life, contain sucrose esters (*32, 33*).

There are a number of processes described in the literature for the production of sucrose monofatty acid esters (*18, 25*). The first process, "The Nebraska-Snell Process," reacts sucrose with a triglyceride in N, N-dimethylformamide in the presence of a basic catalyst at 90°C (*31*). The reaction product is a mixture of mono- and di-glycerides and sucrose esters. With some modification the process is used by Mitsubishi Kasei in their production of Ryoto sugar esters. In order to avoid the use of toxic and expensive N,N-dimethylformamide, Dai-Ichi Kogyo Seiyaku Company Limited has developed a water-based process (*19*), but it involves several process steps and requires reduced pressure. Sucrose esters have been approved for food additive use in the USA since 1983.

Tate and Lyle plc developed a solventless process which simply involves the treatment of sucrose with a triglyceride in the presence of a basic transesterification catalyst, potassium carbonate or/and potassium soap, at 130°C at atmospheric pressure. The product mass contains less monoester (~50%), which increases the cost of purification. The Tate & Lyle process is, however, very rapid and suitable for continuous operation (*6, 18-21*).

The current price of sucrose mono-and di- stearates range from \$22 - \$29/kg on dry basis. It is estimated that about 4000 tonnes of sucrose esters are produced per year. Usage in Japan is greater than in the U.S. However, the safety and biodegradability of the food grade esters are a selling point in the U.S.

The crude sucrose glyceride mixture produced by a solventless process can be included in detergent formulations (*25*). Although the surfactant properties of the sucrose esters may not be as good as those of the synthetic detergents, the raw materials costs are lower than the price of synthetics. Hence, wherever the raw materials are available and foreign exchange is limited, the production of sucrose-based surfactants should be considered. The biodegradable detergent is currently produced in Brazil. The cost per tonne of raw materials for sucrose ester tallow based detergent runs from \$400 to \$800 whereas the sulfonate ester alone, the usual ester for commercial detergents, costs over \$1000/tonne.

Detergent compositions usually contain, in addition to a detergent active material (~15%), a detergent builder (~30%) whose role is to remove hardness from the wash liquor and to prevent re-deposition of dirt. Water soluble phosphates are extensively used as detergent builders. However, for reasons of eutrophication allegedly caused by sodium tripolyphosphate, and cost, an alternative adjuvant has been sought. Unilever plc has developed a phosphate-free detergent adjuvant containing 10 parts by weight (p.b.w.) of sodium carbonate, 20 p.b.w. of calcite, 4 p.b.w. of sucrose, and 4 p.b.w. of an anionic surfactant (*34*). We may speculate that such a detergent builder may also be compatible with the crude sucrose ester glyceride mixture.

A detergent bleaching booster, sucrose polyacetate (SUPA) has been developed by Eridania-Beghin Say (*35*), along with other sucrose based detergent additives.

Sucrose esters are also used in cosmetic creams and lipsticks, and in toothpaste. Sucrose acetate isobutyrate (SAIB) has been produced commercially for use in lacquers and decorative coatings and printing ink by Tennessee Eastman Co. from sucrose esterfied with acetic acid and isobutyric anhydride.

Sucrose Polyester. The market demand for low calorie fats and oils has long been recognized. Procter & Gamble Co. has identified a group of products to suit that market: fatty acid esters of sucrose, termed as sucrose polyester (SPE) and given the brand name "Olestra" (*26*). The ester is produced by a solventless process involving three steps: interesterification of sucrose and long-chain fatty acid methyl esters followed by refining and extraction (*36*). Sucrose polyester contains six to eight fatty acid ester groups per molecule as shown in Figure 2. Structure of the fatty acids determines the physical properties of the resulting fat which varies from a solid to a liquid. The viscosity of sucrose polyester is slightly higher than the triglyceride with the same fatty acids. Other properties such as taste, appearance, interfacial tension, aroma, lubricity, and immiscibility in water are claimed to be indistinguishable from triglyceride. The ester is not hydrolyzed by pancreatic lipases and, therefore, passes unmetabolized. The absence of hydrolysis prevents absorption and thus makes sucrose polyester a low calorie fat. It acts as a lipophilic solvent in the intestine to enhance the excretion of fat soluble compounds like cholesterol. Usage of SPE has recently been made more acceptable to the consumer (*37, 38*) and has recently received regulatory approval.

Sucralfate. A sucrose ester of chlorosulfonic acid, sucrose octa-sulfate, when treated with aluminum chloride in basic solution yields the SO_3 [$A1_2(OH)_3$) salt of octa-substituted sucrose, known as sucralfate. Sucralfate inhibits peptic hydrolysis and is sold in many countries as an antiulcerative medication (*39, 40*).

Polymers Containing Sucrose or Made From Sucrose.

Sucrose in urethane polymer manufacture. High molecular weight, low functionality polyols are used for flexible and elastomeric polyurethanes. Low molecular weight high functionality polyols, such as sucrose, are used for rigid and solid polyurethanes. Sucrose cannot directly be used in polyurethane manufacture because it leads to brittle products; thus the polyhydroxypropyl ether of sucrose, which confers miscibility with the blowing agent and imparts strength and flexibility to the finished foam, is used.

The use of sucrose in polyurethanes provides one of its most important industrial outlets. Although the total world usage of sucrose in polyurethane foams is not known, Bayer's estimate of the total world market for rigid foams of 946 million metric tonnes in 1986 may be taken into account. Potentially up to 20% of this material could be sourced by sucrose, provided it can compete in price and performance with those of petroleum chemicals, starches, sorbitol and others. In Europe the market for sucrose in polyurethanes has been estimated as approximately 10,000 tonnes, indicating that only a small proportion of rigid foams employs sucrose. Changes in the balance of raw materials prices could alter the demand for sucrose significantly (*25*). Sucrose based polyurethane was formerly produced in the U.S. by Union Carbide.

Allyl sucroses (from ether formation with allyl halides with varying substitution) are film forming materials used as paints and coatings. They polymerize under thermal and oxidative conditions (*25*).

Polyhydroxyalkanoates. The family of polyhydroxyalkanoates (PHA's) first came to public attention with polyhydroxybutyrate (PHB), a polyester produced by numerous microorganisms especially when provided with a nitrogen-deficient feed. Sucrose is converted to PHB in 70% yield (by weight) by the organism, Alcaligenes eutrophus (*9, 10, 26*). The product is being produced by Zeneca Corp., and is sold in the U.S. as Biopol. Its applications are both unique and limited: it has many similarities with polypropylene but is biodegradable. However, the production cost, using the existing technology, is considerably higher than that of polypropylene. It takes 3.5 tonnes of sugar to produce 1 tonne of PHB, so that the sugar price must be around U.S. $150 per tonne, with significant economics of scale, before this product can compete with the existing, large tonnage plastics. PHB has found low-volume high-value medical applications such as in surgical pins and sutures. Variations in raw materials and microorganisms are now multiple. Page (*41, 42*) has done extensive studies on the production of PHB's from sugarbeet molasses.

Dextrans. Dextran is produced from sucrose only, not from monosaccharides, by the organism Leuconostoc mesenteroides. Strain NRRL-512B, which is used industrially, contains an extracellular dextransucrase enzyme that produces a very soluble polymer of glucose with α-(*1,6*) linkages (over 90% in the industrial product) and α-(*1,3*) linkages at branch points, and molecular weight of 5 to 500 million (*26, 45*).

Dextran is mainly used as a blood plasma extender. For clinical purposes the native polymer is depolymerized to give dextrans of molecular weight 40,000 and 70,000. Dextran can be used as a stabilizer of ice cream, sugar syrup and other confectionery but at $40 to $150 per kg is generally too expensive. The total world production is approximately 100-500 tonnes per year with the purification stage accounting for a high proportion of costs (*43-45*). Dextran, produced mostly by the Pharmacia Co., Sweden and Pfeifer and Langen, Germany, is also used as a chromatographic support, when cross-linked. Both the blood plasma and chromatographic usage are increasing rapidly.

Polysucrose (Dormacoll). Sucrose partially cross-linked with epichlorohydrin produces a neutral, branched molecule that is very soluble, giving a low viscosity aqueous solution with high specific density. The solutions provide excellent media for cell separation density gradients (*45*).

Xanthan Gum. Xanthan gum, a polymer of glucose, mannose and glucuronic acid can be used as a viscosity and gelling agent in foods. Its major use is a lubricant for oil well drilling muds. Xanthan is a well developed commercial product, selling from $6 to $9 per kilo. It is made by Xanthomonas campestris grown on sucrose or glucose in the U.K., and glucose (starch) in the U.S. (*6, 17*), by Merck Corp. It is an example of a product that can use the cheapest available starting material.

Polyfructose. This levan, a β→(2→6) linked D-fructose polymer with 12% branching, is made from sucrose by B. polymyxa. It is proposed as an encapsulating or thickening agent for various food processing uses (*54*).

Thermally stable polymers. A series of thermally stable polymers has been prepared by Sachinvala (*56*) from methymethacrylate, and from styrene, using sucrose-based monomers and linking agents. The sucrose-containing polymers show better aging properties for weather resistant coatings and finishes, and for components e.g. doors, signs, exposed to the elements, than do current acrylic and styrenic polymers (*55*).

Specialty Sweeteners from Sucrose. Several alternate sweeteners are produced from sucrose: many are sold in Japan and/or Europe but not yet in the United States. Some are low-calorie, others are nutritive and equal in calories to sucrose, but are non-cariogenic. Xylitol is mentioned above, under "Sugarcane Fiber".

Isomaltulose. Isomaltulose, 6-0-(α-D-glucopyranosyl)-D-fructose, a stable reducing disaccharide with half the sweetness of sucrose, is known in the literature as "Palatinose" (Süddeutsche Zucker, Germany, and Mitsui Sugar Company, Japan) and as "Lylose" (Tate and Lyle plc, England). Isomaltulose is produced commercially from sucrose by the immobilized enzyme from Protaminobacter rubrum (*25, 46*). It costs roughly two and half times as much as sucrose per kilogram in the Japanese market; some 5000 tonnes/yr are used in production in non-cariogenic candies. It is also used in specialty chocolate, chewing gum and cookies.

Palatinit. Süddeutsche Zucker-AG of Germany has developed a process for the production of Palatinit from sucrose, whereby sucrose is microbially transglucosylated to palatinose which is then hydrogenated to give a mixture of 6-0-(α-D-glucopyranosyl)-D-sorbitol (*25, 27, 46-48*).

The sweetness of Palatinit is 0.45 times that of sucrose and the taste profile is "pure sweet" without the cooling effect of xylitol or sorbitol. Palatinit is claimed to be less cariogenic in rats than sucrose and lactose; and has an energy value of 2 Kcal/g.

Neosugar. Neosugar is a mixture of fructo-oligosaccharides with one, two or three fructose residues linked by way of β→ (1→2) bonds to the fructosyl moiety of sucrose (*25, 26*). The sweetness intensity of neosugar depends on the composition of the mixture: for example, when 95% of the mixture is nystose and kestose, its sweetness intensity is ~20% of that of sucrose. It is claimed that neosugar is a non-cariogenic and non-nutritive sweetener, suitable for diabetics, though there is some question about its caloric yield (*26*). Neosugar is produced commercially by Meiji Seika Company of Japan. The process involves the microbial transformation of sucrose using a fungal fructosyltransferase enzyme (*49*).

Leucrose. Leucrose, the disaccharide 5-D-(α-D-glucopyranosyl)-α-D-fructose, is produced from sucrose by Leuconostoc mesenteroides as an alternative product during dextran production. The yield of leucrose (rather than dextran) is increased by addition of fructose to the feedstock (*45, 51*).

Sucralose. Sucralose is a new high intensity sweetener that is a derivative of sucrose. It is chemically 4-chloro-4-deoxy-α-D-galactopyranosyl 1,6-dichloro-1,6-dideoxy-β-D-fructofuranoside (*25, 26, 50*). It has a pure sweet taste which is indistinguishable from that of sucrose, and is approximately 650 times sweeter than sucrose. It is resistant to enzymic hydrolysis and is sixty times more stable than sucrose to acid hydrolysis. Regulatory approval has been obtained in Canada, Mexico and Japan (Splenda), and applied for in the U.S. Sucralose is being developed by Tate & Lyle plc, in the United Kingdom, in collaboration with McNeil Laboratories of Johnson and Johnson of the U.S. (*25, 26, 50*).

Chemicals From Molasses. Several compounds are isolated commercially from sugarbeet molasses, but are not usually present in sugarcane molasses. Beet molasses containing an average of 48 to 55% sucrose, is higher in sucrose than cane molasses. Cane molasses contains 30 to 40% sucrose and 9 to 20% (usually 15%) invert sugar (glucose and fructose). For fermentation processes that require sucrose, beet molasses is therefore preferred. Yeast manufacture is a major industrial consumer of molasses and is well described in the literature of industrial fermentation (*6*). A potential product, similar to yeast, is single-cell protein for food and feed. This has been attempted on an experimental basis in many countries, but is operating, to the writer's knowledge, on an industrial scale only in the People's Republic of China. In general, more work has been done on utilization of sugarcane byproducts, in the countries where cane is grown because of the excess quantities available. More work on sucrochemistry has been done in the more industrialized countries where sugarbeet is grown and cane sugar is refined.

Raffinose and glutamic acid. Raffinose (O-α-D-galactopyranosyl (1→6) sucrose)is the major sugar (0.5 to 5.0%) other than sucrose in beet molasses. Raffinose can be extracted and recovered, although it is more readily available from other sources, such as cottonseed hulls. Glutamic acid is a major component (3 to 5%) of beet molasses. It was isolated, along with glutamine, from beet Steffen process molasses, by ion exchange resin systems in commercial processes in the U.S. and France for production of monosodium glutamate, a major flavor enhancer, until the late '70's, when Japanese microbial synthesis of MSG provided a cheaper source (*9*).

Betaine. Betaine (N-trimethyl glycine), present at 3 to 5% in molasses, is isolated by a resin chromatography process at the Finnish Sugar Company, Helsinki. This company has developed many resin based processes to isolate sugar fractions from molasses to prepare enriched syrups. Betaine is used in metal plating, and as a vitamin factor in chicken feeds and as an enhancer of fish flavors (*9, 10*). It is also used in special feed for minks, and has recently found application on crops as a stimulant to phytoalexin production, and in cosmetics. Several of the molasses desugarization plants, mentioned in the "Production" section now also produce betaine.

Lactic Acid. Lactic acid (2-hydroxypropionic acid) is used as a food acidulant and curing agent and in the tanning of leather, and as a dye additive. Lactic acid esters are becoming important in plastics manufacture, and for polylactide production. Over 12,000 tonnes per year are produced in the U.S. and Europe, with beet molasses and dextrose as major feedstocks, in a fermentation process. Lactobacillus delbruckii is the most commonly used microorganism. Several product formulations, at various concentrations and degrees of purity, are commercially available. Lactic acid is extremely corrosive; stainless steel equipment is required. Lactic acid in The Netherlands is produced from white sugar (sold at a low price in the E.U. when used as a chemical) to minimize product separation (*4*).

Citric Acid. Over 500,000 tonnes of citric acid (2-hydroxypropanetricarboxylic acid) are produced annually by fermentation of Aspergillus niger species on either beet or cane products. Some 150,000 t were produced in the U.S. in 1990. The submerged fermentation, generally used for cane molasses, is replacing the flotation fermentation, generally used for beet molasses. Much citric acid production now uses sugar, not molasses, as substrate, to simplify separation. Although there is considerable literature on microbial and chemical aspects of the process, most citric acid is manufactured by Haarman and Reimer Corp. or their licensees, in many countries (*6*, *20*).

Aconitic Acid. Aconitic acid (1,2,3-propene tricarboxylic acid), unlike citric and lactic, is isolated from sugarcane molasses rather than synthesized by fermentation. Calcium aconitate becomes concentrated in molasses from the second crystallization, and can be separated by precipitation (*6*) and recovered in about 90% yield.

The process economics depend upon the level of aconitic acid, which varies widely from less than 1% in some tropical areas to over 4% (on solids in molasses) in Louisiana, where manufacture was carried on from the late 40's to the late 60's, when the acid was replaced in the preparation of plasticizers.

Oxalic Acid. Oxalic acid is said to be manufactured in India from the oxidation of molasses with vanadium pentoxide catalyst.

Lysine. L-lysine, which is a restrictive amino acid in plant protein (other than soybean) used for animal feed, is produced, in amounts of some 140,000 tonnes per year year, > 40,000 t in E.U., by fermentation of Brevibacterium flavum or Brevibacterium lactofermentum, on molasses. Modern mutation and protoplast fusion techniques have been applied to the microorganism to increase yield to over 40% on sugar, and decrease sensitivity to other molasses components and inhibitors (*52*). Most of the work has been conducted, as has the L-lysine production, in Taiwan and Japan. The world's largest lysine plant (11,000 t) opened in South Africa in 1995.

Other fermentation products. Other chemicals produced by fermentation of sucrose (pure, raw or in molasses) include glutamic acid, which can also be extracted from beet molasses, glucurono-deltalactone, itaconic acid and propionic acid (*6, 7, 11*).

Trehalose. This $\alpha \rightarrow (1 \rightarrow 1)$ diglucose is used to protect protein from freeze damage and freeze-drying damage. The gene to convert glucose to trehalose may be expressed in sugarbeet (*53*).

Sugarcane Wax Products. Sugarcane stalks are coated with a mixture of waxy lipids, which are concentrated during process in the clarifier muds that are filtered out. The wax, when isolated and refined by an acetone process, is similar to carnauba or candelilla wax. It was manufactured in Louisiana until the year-round supply of crude wax from filter muds in Cuban factories became unavailable in the early 60's. Sugarcane wax is currently manufactured in China, India and Cuba. The yield is only about 380 kg wax from 1000 tonnes of cane (*6*).

Inside the cane stalk, throughout the plant, are fatty lipids, which must be removed from the wax fraction to have a hard crisp wax. This fatty fraction contains sterols (approximately 1.6×10^{-3} % on cane), notably sitosterol, which can be extracted and converted through fermentation with Arthrobacter globiformis to 17-ketosteroids (androstenedienediones) which are precursors for many steroid drugs (*6*).

Summary

This review has attempted to outline the production of chemicals from sugar crops. The basic production processes for sucrose from sugarbeet and sugarcane, and the byproducts, are briefly described. Chemicals, including sweeteners, that are in current or potential production from sucrose are reviewed. The isolation or synthesis of chemicals from the major byproducts of cane and beet sugar factories are described.

Literature Cited

1. Licht, F. O. *Sugar Yearbook*, 1994 (and) 1995. F. O. Licht. Ratzeburg, Germany, 1994 (and 1995).

2. U. S. Department of Agriculture. *Sugar and Sweetener Reports*. U. S. Dept. of Agriculture, Washington, D. C., 1995.

3. Clarke, M. A. Bagasse. In *Encyclopedia of Chemical Technology*; 3rd Ed. Kirk-Othmer, eds., John Wiley & Sons, Inc, New York, N. Y. 1978, Vol. 3; pp 434-438.

4. Clarke, M. A. *Sugar y Azucar*. **1989** (March), 24-34.

5. St. John, G. In *Proc. Workshop on Products of Sugarbeet and Sugarcane*; Clarke, M. A., Ed.; Sugar Proc. Res. Inst. Inc.: New Orleans, LA, 1994; 23-33.

6. Paturau, J. M. *By-Products of the Cane Sugar Industry*; Elsevier; Amsterdam, 1989.

7. Purchase, B. S. In *Proc. Workshop on Products of Sugarbeet and Sugarcane*; Clarke, M. A., Ed.; Sugar Proc. Res. Inst. Inc.: New Orleans, LA, 1994; 1-22.

8. *Composition, Properties and Use of Molasses*; Baker, B. P.; United Molasses Co.: London, UK, 1978.
9. Hallanoro, H.; Berghäll, S. M.; In *Processing Workshop on Sugarbeet and Sugarcane Products*; Clarke, M. A., Ed.; Sugar Proc. Res. Inst. Inc.: New Orleans, LA, 1994; 75-132.
10. Broughton, N. W.; Dalton, C. C.; Jones, C. C.; Williams, E. L.; In *Proc. Workshop on Products of Sugarbeet and Sugarcane*; Clarke, M. A. Ed.; Sugar Proc. Res. Inst Inc.: New Orleans, LA, 1994; 150-167.
11. Tjebbes, J. In *Chemistry and Processing of Sugarbeet and Sugarcane*; Clarke, M. A.; Godshall, M. A., Eds; Elsevier, Amsterdam, The Netherlands, 1988; 139-145.
12. LeGrand, F. *Sugar y Azucar Yearbook*; **1991**; 24-28.
13. Thibault, J. F.; Guillon, F.; *Proc. Conf. Sugar Process Research.* 1990, pp. 23-29.
14. Thibault, J. F.; Rovau, X. *Carbohydr. Polym.* **1990**, 13, 1-16.
15. Asther, M.; Lesage, L.; Thibault, J. F.; Lapierre, C.; Falconnier, B.; Brunerie, P.; Collona-Ceccaldi. In *Food Ingredients Europe - Conference Proceedings 1993, Porte de Versailles, Paris, 4-6 October 1993.* Maarssen, Netherlands; Expoconsult Publishers (1993) 227-228.
16. Vaccari, G.; Nicolucci, C.; Mantovani, G.; Monegato, A. *Proc. Workshop on Products of Sugarbeet and Sugarcane*; Clarke, M. A., Ed.; Sugar Proc. Res. Inst. Inc.: New Orleans, LA, 1994; 133-149.
17. *Sucrochemistry*; Hickson, J. L., Ed.; A. C. S. Symposium Series No. 4; American Chem. Soc.: Washington, D. C. 1977.
18. *Expansion of Sugar Uses through Research*; Stuart, S. S., Ed.; Int'l. Sugar Research Foundation: Bethesda, MD, 1972; 55-56.
19. Kollonitsch, V. *Sucrose Chemicals.* Int'l. Sugar Research Foundation: Washington, D. C., 1970.
20. Haas, H. B. Proc. Int'l. Soc. Sugar Cane Technol. **1959**, 10, 31-37.
21. Wiggins, L. F. *Sugar and its Industrial Applications*; Lectures, Monographs and Reports, No. 5; Royal Inst. Chem.: London, UK, 1960; pp 44-45.
22. Hough, L. *Chem. Soc. Rev.*; **1985**, 14, 357-374.
23. Khan, R. *Adv. Carb. Chem. Biochem.* **1976**, 33, 235-278.
24. Khan, R. *Pure and Applied Chem.* **1984**, 56, 833-844.
25. Khan, R.; Jones, H. F.; In *Chemistry and Processing of Sugarbeet and Sugarcane*; Clarke, M. A.; Godshall, M. A., Eds; Elsevier: Amsterdam, The Netherlands, 1988, pp. 367-388.
26. Khan R. In *Sucrose*, Mathlouthi, M.; Reiser, P., Eds; Blackie: London, U. K., 1994, 264-278.
27. Vogel, M. *Zuckerindustrie.* **1991**, 116, 265-270.
28. Vlitos, A. J.; *Proc. Int'l Soc. Sugar Cane Technol.* **1986**, 19(3), 1182-1183.
29. Walker, C. E. *Cereal Foods World.* **1984**, 29, 286-289.
30. McCormick, R. D. *Prepared Foods.* **1984**, 153, 127-128.
31. Bass, H. B.; Snell, F. D.; York, W. L.; Osipow, L. I. U. S. Patent 2,893,990, 1959.
32. Park, H. J.; Bunn, J. M.; Vergano, P. J.; Testin, R. F. *J. Food Process. And Preserv.* **1994**, 18(5), 349-358.

33. Banks, N. H. *J. Exp. Botany.* **1984**, 35, 127-137.
34. Davies, J. F.; Lee, F. W.; Travill, A. W.; Williams, R. J. P. U. K. Patent GB2 174 712A, 1986.
35. Mentech, J.; Beck, R.; Burzio, F. In *Carbohydrates as Organic Raw Materials* 2; Descotes, G., Ed.; VCH: Weinheim, Germany, 1993, pp. 185-202.
36. Boggs, R. W. *Fette Seifen Anstr.* **1986**, 88, 154-168.
37. Guffey, T. B.; Boatenan, D. N.; Abe, S. S.; Talkington, S. R.; Mijac, M. D. U. S. Patent 5,021,256, 1991.
38. Aggarwal, A. M.; Camilleri, M.; Phillip, S. F.; Schlayback, T. G.; Brown, M. C.; Thomforde, G. M. *Digestive Diseases and Science.* **1993**, 38(3), 1009-1014.
39. Sucralfate; Caspary, W. F., Ed.; Verlag Urban and Schwarzenberg: Munich, Germany, 1980.
40. *Merck Index,* 11th ed., Budavari, S. Ed., Merck & Co., Rahway, N. J., 1989.
41. Page, W. J. *Biotechnol. Lett.* **1992**, 14(5), 385-390.
42. Chen, G. O.; Page, W. J. *Biotechnol. Lett.* **1994**, 16(2), 155-160.
43. Hacking, A. J. *Economic Aspects of Biotechnology*; Cambridge University Press: Cambridge, U. K. 1986.
44. Alsop, R. M. *Prog. Ind. Microbiol.* **1983**, 18, 1-44.
45. Giehring, H. In *Proc. Workshop on Products of Sugarbeet and Sugarcane*; Clarke, M. A.; Ed.; Sugar Proc. Res. Inst. Inc.; New Orleans, LA, 1994; 168-178.
46. Suzuki, K. *New Food Industry, Jpn.* **1984**, 26, 14.
47. Grenby, T. H. In *Developments in Sweeteners,* Grenby, T. H.; Parker, K. J.; Lindley, M. G.; Eds.; Vol 2. Applied Science Publishers: London, U. K., 1983, 51-52.
48. Schiweck, H.; Munir, M. In *Carbohydrates in Industrial Synthesis,* Clarke, M. A., Ed.; Verlag Bartens: Berlin, Germany, 1992, pp. 37-55.
49. Speights, R.; Perna, P.; Downing, S. In *Carbohydrates in Industrial Synthesis,* Clarke, M. A., Ed.; Verlag Bartens, Berlin, Germany, pp. 7-17.
50. Hough, L.; Phadnis, S. P.; Khan, R.; Jenner, M. R. Brit. Pat. 1,543,167, 1979. U. S. Pat. 4,549,013, 1985.
51. Schwengers, D. In *Carbohydrates as Organic Raw Materials,* Vol. 1; Lichtenthaler, F. W., Ed.; VCH: Weinheim Germany, 1991, pp. 183-196.
52. Liu, Y. T. *Proc. Int. Soc. Sugar Cane Technol.* **1986**, 19, 1160-1181.
53. U. S. Patent 9,506,126. 1995.
54. Clarke, M. A.; Bailey, A. V.; Roberts, E. J.; Tsang, W. S. C. In *Carbohydrates as Organic Raw Materials,* Vol 1.; Lichtenthaler, F W., Ed.; VCH: Weinheim, Germany, 1991, pp. 169-182.
55. Sachinvala, N. D.; Ju, R. F.; Litt, M. H. In *Proc. Workshop on Products of Sugarbeet and Sugarcane*; Clarke, M. A., Ed.; Sugar Proc. Res. Inst. Inc.: New Orleans, LA, 1994, 204-234.

Chapter 17

Products from Vegetable Oils: Two Examples

Marvin O. Bagby[1]

Oil Chemical Research, National Center for Agricultural Utilization Research, Agricultural Research Service, U.S. Department of Agriculture, 1815 North University Street, Peoria, IL 61604

Vegetable oils serve various industrial applications such as plasticizers, emulsifiers, surfactants, plastics and resins. Research and development approaches may take advantage of natural properties of the oils. More often it is advantageous to modify those properties for specific applications. One example is the preparation of ink vehicles using vegetable oils in the absence of petroleum. They are cost competitive with petroleum-based inks with similar quality factors. Vegetable oils have potential as renewable sources of fuels for the diesel engine. However, several characteristics can restrict their use. These include poor cold-engine startup, misfire and for selected fuels, high pour point and cloud point temperatures. Other characteristics include incomplete combustion causing carbon buildup, lube oil dilution and degradation, and elevated NOx emissions. Precombustion and fuel quality data are presented as a tool for understanding and solving these operational and durability problems.

United States agriculture produces over 16 billion pounds of vegetable oils annually. These domestic oils are extracted from the seeds of soybean, corn, cotton, sunflower, flax and rapeseed. Although more than 12 billion pounds of these oils are used for food products such as shortenings, salad and cooking oils and margarines, large quantities serve feed and industrial applications. The latter applications include chemicals such as plasticizers, which add pliability to other substances; stabilizers, which help other substances resist chemical change; emulsifiers, which enable the mixing of normally unmixable liquids; surfactants, which reduce the surface tension of liquids and are commonly used in detergents; and esters, nylons and resins, which are basic ingredients in many products. Besides detergents and plastics, products that contain chemicals derived from vegetable oils include lubricants, coatings, corrosion inhibitors, adhesives, cleaners, cosmetics, water repellents and fuels. Other vegetable oils widely used industrially include palm, palm kernel, coconut, castor and tung. However, these are not of

[1]Current address: 209 South Louisiana Avenue, Morton, IL 61550

domestic origin. The three domestic oils most widely used industrially are soybean, linseed from flax and rapeseed. Soybean oil provides nearly 80% of the seed oils produced annually in the United States. Consequently, the relatively low cost of soybean oil and the dependable supply make it one of the more important sources of industrial oil.

Non-food uses of vegetable oils have grown little during the past 40 years. Although some markets have expanded or new ones added, other markets have been lost to competitive petroleum products. Development of new industrial products or commercial processes are objectives of continued research in both public and private interests. Through these efforts vegetable oils should maintain or even add to their market share as non-renewable petroleum becomes more expensive. One example of a recent new market is the expanded use of soybean oil (40 to 60,000,000 pounds annually, Private communication, American Soybean Association) in printing inks, with lithographic news leading the way.

Research and development approaches frequently take advantage of natural physical or chemical properties of the oils or their major constituents, consisting of fatty acids and glycerol. More often it is advantageous to modify those properties for specific applications (1). For applications such as cosmetics, lubricants and certain chemical additives, vegetable oils are too viscous and too reactive toward atmospheric oxygen to establish significant markets. Physical properties of vegetable oils such as viscosity, pour point and freezing point can be decreased by introducing branching groups or side chains on the straight chain fatty acids. An example is isostearic acid, which is a coproduct of commercial dimer acid manufacture and in much demand for a variety of uses to be discussed later. Isostearic acid, actually a complex mixture of saturated and branched-chain fatty acids in which isostearic predominates, has liquid properties of oleic acid but thermal and oxidative stability resembling stearic acid. Important product applications for derivatives of isostearic acid include textile lubricants, softeners and antistatics; coupling agents and emulsifiers; and greases and synthetic lubricants. Conversely, for certain markets vegetable oils should be made more reactive by introducing additional functionality or cleaving the fatty acid molecules. For example, products resembling imported tung oil can be prepared by selected isomerization reactions. Important markets for these reactive chemicals include coatings, resins, ink vehicles and plastics. Markets for these types of products are expected to grow with increasing worldwide consumer sophistication and with changing and more stringent product performance requirements. The following selected examples illustrate progress in identifying and developing new technologies based mainly on soybean oil.

Soybean Oil Inks

Soy inks, alternatives to conventional petrochemical based inks, were developed by the Newspaper Association of America (NAA, formerly the American Newspaper Publishers Association, ANPA) and introduced to the marketplace in 1987 when the Cedar Rapids Gazette (Iowa) printed the first press run. This hybrid soybean oil technology, consisting of about 35 to 50% degummed soybean oil, 20 to 25% petroleum resin and pigments (2), has enjoyed rapid acceptance by the newspaper publishers industry, especially so for the colored inks. However black inks formulated by the NAA technology were not cost competitive with typical petroleum-based offset news inks. Because the technology consists of a direct substitution of soybean oil for the mineral oil portion of the vehicle (entraining and dispersing agent for the pigments and other solid substances), other oils of similar fatty acid composition should be directly interchangeable. In fact, some formulators have prepared inks containing mixtures consisting of soybean and corn oils. Economic considerations and marketing strategies govern the final selection for applying that technology. Since then at the request of NAA and the American Soybean Association, USDA developed a technology in which the vehicle is totally derived from vegetable oils (3-7). Although soybean oil was emphasized because of dependable supply and economic factors, this new technology was demonstrated with several commodity oils. Those oils selected provided a broad range of unsaturation with representative fatty acids consisting of saturated, mono-unsaturated, di-unsaturated and tri-unsaturated acids. Besides elimination of petroleum, this technology permits formulation of inks over a desirable, broader range of viscosity and is cost competitive with conventional petroleum-based offset news inks. Further, inks formulated with this technology have low rub-off characteristics equal to those formulated and marketed as low rub inks (4). The following discusses the development and the characteristics of that technology.

We selected alkali refined canola, cottonseed, soybean, sunflower and safflower oils to demonstrate the technology (3,6-7). Alkali refining removes the gums, waxes and free fatty acids. The presence of excessive amounts of any one of these materials will interfere with the desirable hydrophobic characteristics of the vehicle and the ultimate ink formulation.

This hydrophobic characteristic deserves further comment. The off-set printer plate or cylinder consists of two distinct areas. One area has been rendered hydrophobic (image area) while the non-image area is hydrophilic. Thus, the typical off-set printing process involves a two phase system consisting of an oil phase (the ink) and aqueous phase (the fountain solution). During the printing process, these phases must not form stable emulsions, or they will not separate properly on the printer plates. Poorly separated phases lead to smudged or ill-defined print. Understanding of this characteristic directed our attention toward techniques for modifying vegetable oils that would provide relatively non-polar products, i.e., low oxygen content polymers.

The vehicles were prepared from vegetable oils by two methods (3). In the first method, vegetable oils were heat polymerized at a constant temperature in nitrogen atmosphere to a desired viscosity. In the second method, the heat polymerization reaction was permitted to proceed to a gel point, and then the gel was mixed with vegetable oils to obtain a desired viscosity. The apparent weight average molecular weights (Mw) of both the heat-bodied vehicles and the gels were determined by Gel Permeation Chromatography (5).

Ink Methods

Vehicles were prepared in a four-necked reaction flask equipped with a mechanical stirrer. Two major methods were used in preparation of polymers. (A) Alkali-refined vegetable oil was polymerized with stirring at 330±3 °C under nitrogen atmosphere to the desired viscosity. Some polymers prepared by this method were used directly as vehicles, others having Gardner-Holdt viscosities as high as Z_8-Z_9, were admixed with low viscosity polymers and/or unmodified, alkali-refined vegetable oil at 65-75 °C in a reaction flask equipped with a mechanical stirrer. (B) Heat bodying was continued until the oil gelled. The reaction was discontinued at the transition point when clumps of gel began to climb up the shaft of the mechanical stirrer. The gel was blended in various ratios with unmodified alkali-refined vegetable oil at 330±3 °C. The heating softened the gel and promoted blending. Agitation was continued until a smooth vehicle was obtained. The proportions of the gel and unmodified oil determined the resultant vehicle viscosities.

The viscosities and color of the vehicles were established by ASTM D-1545-63 (Gardner-Holdt Bubble) and ASTM D-1544-63 (Gardner Color Scale). The apparent Mw was determined by Gel Permeation Chromatography as tetrahydrofuran solutions (5). For more detailed description of experimental approaches see References 3-7.

Ink Results and Discussion

The vehicles that we prepared typically had viscosity values in the range of G-Y on the Gardner-Holdt Viscometer Scale or about 1.6-18 poises (8). These viscosities correspond to apparent weight average molecular weights (Mw) of about 2600-8900 (5). As the oils are heated, they undergo polymerization and isomerization reactions. Thus, molecular weights and viscosities increase. The more highly unsaturated oils containing greater amounts of linoleic and linolenic and having the higher I.V., of course, react more rapidly (7).

Triglyceride, consisting of three fatty acids at which addition may occur, introduces the possibility of forming very complex structures and very large molecules. Thus, the reaction time necessary to reach a desired viscosity depends on mass, structure of the reactants, rate of heat transfer and agitation (7). Gelling times for safflower (I.V. = 143.1), soybean (I.V. = 127.7), sunflower (I.V. = 133.4), cottonseed (I.V. = 112.9) and canola (I.V. = 110.2) oils were 110, 255, 265, 390 and 540 min, respectively. Their uncatalyzed polymerization rate constants ($K \times 10^{-3}$ at 330 ± 3 °C) based on viscosity changes were, respectively, 17.23, 11.02, 7.07, 3.21 and 2.14 (7). Although iodine values of cottonseed and canola oil are similar, canola oil with its high oleic acid content reacts more slowly.

For comparison purposes, corn oil, with an I.V. of about 127 and an abundance of linoleic (62%), reactivity should closely parallel those responses shown for soybean and sunflower oils. Heat-bodied oils of different viscosities can be blended to produce a desired vehicle viscosity. Gels can, also, be blended with unmodified oils to give appropriate characteristics.

Vehicles, prepared by these technologies, are compatible with pigments for producing the four colors commonly used in the newspaper printing industry; namely, black, cyan, magenta and yellow. These vehicles are characterized by an exceedingly light coloration. Except for canola, they have values on the Gardner Color Scale of about 6 or less and typically are in the range of about 2-4. This property permits some reduction in the amount of pigment required for colored inks as compared to the pigment levels required by the much darker soy-hybrid or petroleum-based commercial vehicles (2,3) having Gardner Color Scale values of about 14 and greater.

Properties of ink formulated with soybean, cottonseed, canola, safflower and sunflower oils are characterized by viscosities in the range of 5 to 46 poises and by tack values of about 2 to 7 gram-meter (g-m) determined by Electronic Ink-o-meter, Model 101 (4). The typical viscosity for a black lithographic newsink is in the range of about 13 to 24 poises and about 5 to 12 poises for a black letterpress newsink. Tack values for lithographic inks are about 3.5 to 4.8 g-m and about 2.6 to 3.4 g-m for lettepress. The thickening effect of the pigment on the base vehicle should be considered when preselecting a vehicle viscosity. Because of the vehicle system we use (3), it is fairly easy to tailor the viscosity and tack values of the formulated inks. These inks, with a large range of viscosities and tack values, are suitable for both letterpress and lithographic applications. A variety of additives may be formulated into the inks including driers, lubricants, antioxidants and the like.

The water take up for lithographic ink performances was also tested, and the range of 20-30% water take up was well within the acceptable range of 20-36% (9). Inks having these properties were also characterized by acceptable or superior ruboff values. The majority showed blackness of less than 6% after 2 hr, thus they demonstrate good rub resistance (4). With one exception, all formulations tested had ruboff values lower than that formulated by NAA's soy-hybrid technology.

Properties of yellow, blue and red inks also meet the industrial standards. The addition of up to 5% thickening agent and/or optical brightener is an option but not a necessity for our color ink formulations. Elimination of the hydrocarbon resin from the vehicle results in significant savings for both black and colored inks. In addition, some reduction of pigment for colored inks due to the light coloring of vehicle lowers the price of the ink further.

Although the NAA ink formulations and many presumably similar formulations have been advertised as biodegradable, until recently data regarding degradation were unavailable. Erhan and Bagby (10) reported biodegradation data obtained by evaluating the USDA 100% soybean oil vehicle and commercial vehicles consisting of petroleum resins and either soybean oil (NAA soy-hybrid formulation) or mineral oil solvents. Both monocultures and mixed cultures of common soilborne organisms Aspergillus fumigatus, Penicillium citrinum and Mucor racemosus during a 25-day fermentation degraded the vehicles by about 86, 63 and 21%, respectively. An ANOVA model showed that the main effect of variation was due to duration of the fermentation and type of ink vehicle (11). Subsequent biodegradation studies of the same vehicles in the presence of pigments as formulated inks gave similar results. However, the degradation was somewhat slower and less complete showing about 80, 30 and 16% degradation, respectively, in 25 days (12). Similar, confirming results were obtained from using activated sewage sludge inoculum following the modified Sturm test procedure (13). Thus, it is clearly demonstrated that, in terms of biodegradation, the USDA soy vehicles are far superior to the NAA partial soy-hybrid which, in turn, is superior to the 100% petroleum vehicle.

In summary, the USDA lithographic newsink technology readily satisfies the initial requests of NAA and the American Soybean Association by being (a) cost competitive with petroleum based inks of comparable quality characteristics, (b) completely free of petroleum solvent or resin in the vehicle and (c) usable over a broad range of viscosity and tack by exceeding the industry needs at both the high and low extremes. Further the ink is formulated in the absence of volatile organic compounds, and the technology provides inks with the characteristics of low rub-off. During commercial printing tests, minimal accumulation appeared on machine rollers, and these residues are soft and easy to clean.

Diesel Fuel

Vegetable oils have potential as reliable and renewable sources of fuel for compression ignition engines (diesel)--a concept as old as the diesel engine itself. In fact, early engines were demonstrated running on peanut oil. Once cheap petroleum became readily available, the modern engine was designed to use petroleum fuel. Periodically, the alternative vegetable fuel concept has been reestablished, usually during times of petroleum shortages--and as petroleum shortages and prices eased, interest in alternatives again waned. Consequently, scientists do not yet understand how best to change the chemical and physical properties of vegetable oils to allow their trouble-free use as a fuel source. To fill this knowledge gap, NCAUR scientists are currently focusing on problems of high viscosity, low volatility, high pour and cloud points, incomplete combustion and exhaust emissions.

While evaluating vegetable oil fuels, NCAUR researchers have observed several characteristics that can restrict their use as motor fuels. These were grouped in two general categories, operation and durability. The former included ignition quality characteristics such as poor cold-engine startup, misfire and ignition delay, and the latter included characteristics of incomplete combustion such as carbon buildup in the combustion cylinder and on the injectors, ring sticking and lube oil dilution and degradation. In addition, the high viscosity of vegetable oil (more than 10 times that of number 2 diesel fuel) causes poor fuel atomization and inefficient mixing with air, further contributing to incomplete combustion (14,15). The influence of viscosity was effectively demonstrated with a high-pressure, high-temperature reactor designed to resemble, thermodynamically, conditions present in a diesel engine (14,15). High-speed photography of the injection event, with fuels at various viscosities, clearly demonstrated the effect of viscosity on fuel penetration, atomization and spray cone dispersion. This reactor also allowed the study of precombustion chemistry for fuel components at various temperatures and pressures and in atmospheres of both nitrogen and air. Samples were removed from the sprays of these fuels during the injection event before combustion would occur in an atmosphere of air (16,17). The samples collected were analyzed by gas chromatography/mass spectrometry. Rapid (less than 400 microseconds), significant chemical changes occur prior to combustion. Although most work was done with pure glyceryl esters of palmitic, stearic, oleic, linoleic and linolenic acids, the degradation products were similar. At the lower temperature (400C), fatty acids and aliphatic hydrocarbons were the more abundant products with lesser amount of aldehydes, alcohols, benzoic and succinic acids and glycerol. Higher temperature (450C) resulted in extensive decarboxylation and nearly complete absence of oxygenated compounds. Those compounds identified included aliphatic hydrocarbons, mainly unsaturated, aromatic and polyaromatic species, and small amounts of aldehydes and alcohols such as benzaldehyde and phenol. Presence of these products from both nitrogen and air atmospheres suggest that the predominant mechanism of precombustion is thermal and not oxidative/thermal.

NCAUR chemists in collaboration with Southwest Research Institute engineers reported fuel quality characteristics (cetane) of numerous fatty materials measured in a constant volume combustion apparatus (18). Selected data are summarized in Table I. Because these quality characteristics are the resultant of the starting substances plus their many degradation products formed during the period of ignition delay, the differences found to occur among the several monounsaturated esters together with the observed precombustion degradation products are significant mechanism indicators. Decarboxylation is an obvious event which likely provides a radical species as an early intermediate. A comparison between the saturated esters and alcohols suggest that the decarboxylation is more rapid than degradation of the corresponding alcohol, or at least the products of degradation for the latter are not ignited as quickly. Subsequently, the unsaturated radicals resulting from the acids might further degrade α,β to the double bond and rearrange or combine with other radicals to form the observed products. One possible product from the Δ-9 unsaturated acid is cyclohexane. Through a disproportionation reaction cyclohexane could be the source of the observed benzene. Additional work is ongoing to gain further insight into the mechanism of these reactions and subsequently, to develop additives for improved combustion and exhaust emissions. Because of their unsaturation, unmodified vegetable oils are more reactive than diesel fuel and are therefore much more susceptible to oxidative and thermal polymerization reactions that can interfere with combustion. These

Table I. Fuel Quality: Cetane or Lipid Combustion Quality Numbers*

Carbon:double bonds** (Position)	Compound Class	
	Methyl Ester	Alcohol
12:0	54	--
14:0	72	51
16:0	91	68
18:0	159	81
20:0	196	--
16:1 (9)	49.5	46
18:1 (6)	55.4	--
18:1 (9)	47.2	51
18:1 (11)	49.5	--
20:1 (11)	61.5	--
22:1 (13)	76.0	--
18:2 (9,12)	28.5	44
18:3 (9,12,15)	20.6	41

* $\sigma = 2.3$.
** cis-unsaturation.

properties are of major concern when the oils reach the hot surfaces of the engine and the crankcase.

NCAUR researchers have been more successful in reducing viscosity than they have been in increasing volatility. Four approaches have been tried with varying degrees of success (19): (a) transesterification (reaction of vegetable oil with alcohols to give smaller molecules consisting of methyl, ethyl or butyl esters), (b) dilution of vegetable oils with petroleum distillates, including diesel fuel, (c) pyrolysis (using heat to break chemical bonds and form new compounds of greater volatility) and (d) microemulsification or cosolvency (making heterogeneous mixtures like oil and water become a stable, homogeneous solution) (20).

Fuels produced by each technique have been tested in engines. However, those prepared by approaches (a), (b) and (d) have received the most attention and the more rigorous engine tests. All techniques have provided encouraging results, and the transesterification and dilution technologies have achieved some commercial success. In Brazil, fuels developed by dilution may be used under some very specific conditions without loss of engine warranty.

Alkyl fatty esters were studied extensively during the early 1980s in South Africa, Brazil and by USDA and their collaborators (19,21). Although ethyl and butyl esters were evaluated, methyl esters were emphasized. Warranties on some vehicles operated on esters are also honored in South Africa and parts of Europe. The oils used are those most readily available in the individual areas--soybean oil in Brazil, sunflower oil in South Africa and rapeseed oil in Europe. In the United States, fatty acid methyl ester fuels are being evaluated and demonstrated by numerous groups (22,23). Many of these activities are sponsored, at least in part, by the United Soybean Board with soybean checkoff funds. Although soybean oil is emphasized, other feedstocks, such as beef tallow, animal fats and waste cooking oils, are included. Unfortunately, the methyl esters present a further challenge of needing improved low-temperature properties (24). NCAUR efforts are continuing to further develop the technologies and improve the combustion, emissions, cold-temperature tolerance and cost-effectiveness of vegetable oils.

Literature Cited

1. Bagby, M. O.; Carlson, K. D. Chemical and Biological Conversion of Soybean Oil for Industrial Products, in Fats for the Future, edited by R.C. Cambie, published for International Union of Pure and Applied Chemistry by Ellis Horwood Limited, Chichester, 1989, Chapter 20, pp. 301-317.
2. Anon., American Newspaper Publishers Association, Manufacturing directions for soybean oil-based ANPA-ink, 1988, p. 1.
3. Erhan, S. Z.; Bagby, M. O. *J. Am. Oil Chem. Soc.* **1991** *68*(9), 635-638.
4. Erhan, S. Z.; Bagby, M. O.; Cunningham, H. W. *Ibid.* **1992** *69*(5), 251-256.
5. Erhan, S. Z.; Bagby, M. O. *J. Appl. Polym. Sci.* **1992** 46, 1859-1862.

6. Erhan, Sevim Z.; Bagby, Marvin O., Vegetable Oil-Based Printing Ink, U.S. Patent 5,122,188 (1992).
7. Erhan, S. Z.; Bagby, M. O. *J. Am. Oil Chem. Soc.* **1994** *71*(11), 1223-1226.
8. Mattil, K. F.; Norris, F. A.; Stirton, A. J.; Swern, D., Bailey's Ind. Oil Fat Prod., 1964, pp. 513.
9. Surland, A., A Laboratory Test Method for Prediction of Lithographic Ink Performance, Sun Chemical Corporation, 1980, pp. 5.
10. Erhan, S. Z.; Bagby, M. O., *TAGA Proceedings*, 1993, pp. 314-326.
11. Erhan, S. Z.; Bagby, M. O.; Nelson, T. C. Submitted to *J. Am. Oil Chem. Soc.*, 1996.
12. Erhan, S. Z.; Bagby, M. O. Taga Proceedings, 1994, Technical Association of the Graphic Arts, Rochester, NY, pp. 313-323.
13. Erhan, S. Z.; Bagby, M.O., Biodegradation of News Inks, in Environmental Challenges: The Next 20 Years, NAEP 20th Annual Conference Proceedings, 1995, National Association of Environmental Professionals, Washington, DC, pp. 847-855.
14. Ryan, T.W., III; Callahan, T.J.; Dodge; Moses, C. A. Final Report, USDA Grant No. 59-2489-1-6-060-0, Southwest Research Institute, San Antonio, TX, 1983.
15. Ryan, T.W., III; Dodge, L. G.; Callahan, T. J. *J. Am. Oil Chem. Soc.* **1984** *61*(10), 1610-1619.
16. Knothe, G.; Bagby, M. O.; Ryan, T.W., III; Callahan, T.J.; Wheeler, H.G., SAE Tech Pap Ser, Paper No. 920194, Warrendale, PA, 1992.
17. Ryan, T.W., III; Bagby, M.O. *Ibid.* **1993** 930933, Warrendale, PA.
18. Freedman, B.; Bagby, M. O.; Callahan, T. J. Callahan; Ryan, T. W. III, SAE Tech Pap Ser, Paper No. 90343, Warrendale, PA, 1990.
19. Schwab, A.W.; Bagby, M. O; Freedman, B. *Fuel* **1987** *66*, 1372-1378.
20. Dunn, R. O.; Bagby, M.O. *J. Am. Oil Chem. Soc.* 1994 *71*(1), 101-108.
21. Bagby, Marvin O.; Freedman, Bernard, Diesel Engine Fuels from Vegetable Oils in New Technologies for Value-Added Products from Protein and Co-Products, edited by L. A. Johnson, Symposium Proceedings of the Protein and Co-Products Divison of the American Oil Chemists' Society, Champaign, IL, 1989, pp. 35-42 and references therein.
22. Gallagher, Matthew; *Chemical Marketing Reporter* **1994** *246*(3), 7 and 10.
23. National Biodiesel Board, Biodiesel Report, December 1994, Published 10 times per year, United Soybean Board, Jefferson City, MO.
24. Dunn, R O.; Bagby, M. O. *J. Am. Oil Chem. Soc.* **1995** *72*(8), 895-904.

INDEXES

Author Index

Affiliation Index

Subject Index